Springer Tracts in Nature-Inspired Computing

The book series is aimed at providing an exchange platform for researchers to summarize the latest research and developments related to nature-inspired computing in the most general sense. It includes analysis of nature-inspired algorithms and techniques, inspiration from natural and biological systems, computational mechanisms and models that imitate them in various fields, and the applications to solve real-world problems in different disciplines. The book series addresses the most recent innovations and developments in nature-inspired computation, algorithms, models and methods, implementation, tools, architectures, frameworks, structures, applications associated with bio-inspired methodologies and other relevant areas.

The book series covers the topics and fields of Nature-Inspired Computing, Bio-inspired Methods, Swarm Intelligence, Computational Intelligence, Evolutionary Computation, Nature-Inspired Algorithms, Neural Computing, Data Mining, Artificial Intelligence, Machine Learning, Theoretical Foundations and Analysis, and Multi-Agent Systems. In addition, case studies, implementation of methods and algorithms as well as applications in a diverse range of areas such as Bioinformatics, Big Data, Computer Science, Signal and Image Processing, Computer Vision, Biomedical and Health Science, Business Planning, Vehicle Routing and others are also an important part of this book series.

The series publishes monographs, edited volumes and selected proceedings.

Mahdi Khosravy · Neeraj Gupta · Olaf Witkowski
Editors

Frontiers in Genetics Algorithm Theory and Applications

Springer

Editors
Mahdi Khosravy
Cross Labs, Cross Compass Ltd.,
Tokyo, Japan

Neeraj Gupta
Kevadiya Inc.,
Troy, MI, USA

Olaf Witkowski
Cross Labs, Cross Compass Ltd.,
Tokyo, Japan

ISSN 2524-552X ISSN 2524-5538 (electronic)
Springer Tracts in Nature-Inspired Computing
ISBN 978-981-99-8106-9 ISBN 978-981-99-8107-6 (eBook)
https://doi.org/10.1007/978-981-99-8107-6

This Springer imprint is published by the registered company Springer Nature Singapore Pte Ltd.
The registered company address is: 152 Beach Road, #21-01/04 Gateway East, Singapore 189721, Singapore

Preface

In concert with the ever-evolving landscape of science and technology, real-world problems are assuming increasingly intricate dimensions. This complexity is intrinsically tied to the proliferation of variables at play. This phenomenon is readily discernible across multifarious domains, encompassing politics, management, engineering, economics, and more. In response to these mounting challenges, the prevailing problem-solving paradigm has shifted decisively toward heuristic algorithms. These heuristic algorithms, profoundly informed by the elegance of natural systems, have emerged as potent tools for addressing the intricacies of non-linear and multifaceted problems. Notably, the Genetic Algorithm (GA), bearing the hallmark of inspiration from biological evolution and Darwinian principles, stands out as one of the most venerable and widely recognized heuristic techniques. It was introduced by John Holland in 1992, and since its inception, it has undergone continuous refinement and diversification in terms of GA operators. The Genetic Algorithm remains an enigmatic and ever-evolving domain within computational science. Particularly fascinating are its applications, where a single GA technique can be flexibly adapted in manifold ways to tackle a given problem, ultimately yielding varied levels of efficiency and performance. The aim of this scholarly endeavor is to offer a comprehensive and in-depth examination of the latest theoretical advancements in Genetic Algorithm research, as well as to explore its multifaceted applications in various scientific and engineering domains. This endeavor seeks to chart the frontiers of contemporary research in this domain and unravel the mysteries that persist, shedding light on how this powerful computational paradigm continues to shape our understanding and problem-solving capabilities in an ever-changing world.

This book comprises five distinct sections, each delineating critical facets of Genetic Algorithms (GA) and their wide-ranging applications. Part I is dedicated to the fundamental underpinnings of GA, elucidating the theoretical foundations that underlie these evolutionary computational paradigms. It also delves into contemporary advances in evolutionary computation and extends a gaze into the future, exploring the impending trajectories of this domain. Part II shifts the focus toward the real-world implications and applications of GA, as they manifest in the realms of logistics and transportation, where optimization plays an instrumental role in orchestrating the seamless flow of goods. It extends further into the broader societal and economic landscape, scrutinizing the far-reaching impact of GAs on various facets of human existence. Part III takes a more specialized perspective, centering on the pivotal role of GA in the optimization of power systems, a domain of paramount importance in ensuring the efficient and sustainable supply of electrical energy. Part IV steers the discourse into the burgeoning realm of electrical vehicle industries, elucidating the manifold ways in which GAs are leveraged to address the challenges and complexities inherent in the electric vehicle ecosystem. Finally, Part V advances into the domains of signal and image processing, wireless technology, and robotics, unraveling the intricate tapestry of applications where GAs have left indelible marks, revolutionizing these fields. Through these meticulously structured sections, this comprehensive exposition endeavors to shed light on the profound impact and diverse utility of Genetic Algorithms in the realms of theory, practice, and their enduring influence on science and technology.

Hanna Yoshida, in conjunction with her accomplished team, provides a comprehensive exposition on Evolutionary Computation in Chap. 1. Their work encapsulates a panoramic view of this field, commencing with a historical retrospective, meticulously tracing the trajectory of Genetic Algorithms (GA), and culminating in a visionary portrayal of the future, particularly concerning artificial life. In Chap. 2, Duggal et al. undertake the task of delivering a meticulous and all-encompassing examination of the technical infrastructure that underpins Genetic Algorithms (GA). Their work transcends mere scrutiny, offering readers a profound understanding of the intricate framework that forms the backbone of this evolutionary computational paradigm. In Chap. 3, Fahmi Abd Samad delves into the intriguing realm of 'mating' within Genetic Algorithms (GA), with a specific application in the domain of system identification. While frequently overshadowed by the more prominently discussed operators of selection, crossover, and mutation, 'mating' constitutes an indispensable, yet often underexplored, stage in the GA process. This process involves the pairing of chromosomes within a GA population prior to the onset of the pivotal crossover operation. Samad's insightful exploration brings to the fore this lesser discussed facet, shedding light on its profound significance in the context of GA applications.

Within Chap. 4, Cesario Times and Da Silva Firmino introduce an innovative approach—the Multi-Parent Biased Random-Key Genetic Algorithm (BRKGA)—engineered to tackle the intricacies of the Container Retrieval Problem (CRP). Their meticulously crafted variant, denoted as BRKGA-MP-IPR, represents an evolution within the BRKGA framework, distinguished by its utilization of multiple parents, each imbued with a tailored bias, to engender offspring. This novel strategy is underpinned by an implicit path-relinking technique, employed for local search, further enhancing its efficacy in addressing the CRP's multifaceted challenges. Chapter 5, authored by Yektamoghadam et al., embarks on an insightful exploration of the expansive landscape of genetic algorithm applications within the domain of power systems. Within this realm, genetic algorithms emerge as versatile problem solvers, adept at addressing a spectrum of intricate challenges. These encompass but are not confined to the intricacies of generation expansion planning (GEP), unit commitment, power flow optimization, and economic load dispatch. The particular allure of genetic algorithm-based solutions lies in their inherent capacity to grapple with issues characterized by elevated processing costs and an abundance of constraints. Their adeptness in navigating the complex, multifaceted terrain of power systems makes them invaluable tools for optimizing the efficiency and efficacy of these critical infrastructural networks.

Chapter 6, co-authored by Nepomuceno et al., presents a pioneering exploration into the fusion of the Political Optimizer and Genetic Algorithm frameworks. This innovative amalgamation is harnessed to tackle the intricate dynamics of the profit-based Unit Commitment Problem within the competitive electricity market. In a landscape characterized by cutthroat competition, dynamic pricing, and an ever-evolving energy marketplace, the utilization of such a hybrid approach offers a promising avenue for optimizing unit commitment strategies and decision-making processes. By leveraging the prowess of the Political Optimizer and Genetic Algorithm, this research contributes to the arsenal of tools available to practitioners and researchers in the quest for sustainable and profitable solutions within the complex domain of competitive electricity markets.

Ebrahimi and his colleagues offer a comprehensive exploration of Optimal Power Flow (OPF) in Chap. 7, introducing a well-defined perspective and delving into the application of Genetic Algorithms (GAs) for simulating OPF on a representative power system. OPF, as redefined in this context, becomes a critical tool for optimizing an objective function while maintaining compliance with technical and security constraints. This multifaceted approach places a strong emphasis on cost reduction, encompassing fuel costs and network losses among other factors. The significance of this chapter lies in its departure from the traditional, highly intricate mathematical approaches to OPF, which often grapple with issues of linearity and local minima. Instead, it champions the utility of meta-heuristic algorithms, prominently featuring the Genetic Algorithm, as a powerful means to address the challenges presented by OPF.

Chapter 8 by Nemouchi et al. is dedicated to a crucial endeavor in the realm of power systems: the multi-objective optimization of electrical networks incorporating both FACTS systems and renewable energy sources, notably offshore wind farms. Its significance lies in its focus on advancing methodologies designed to identify a collection of optimal solutions forming a Pareto front. This front represents a delicate balance between the minimization of investment costs associated with wind turbines and FACTS devices, the reduction of active losses within transmission lines, and the maximization of the stability and safety of node voltages. This chapter acknowledges the complex array of equality and inequality constraints that govern electrical network operation, ensuring strict compliance with system requirements and operational limitations. To tackle this multifaceted optimization challenge, two powerful meta-heuristic methods are deployed: the Ellitist Non-Dominated Sorting Genetic Algorithm II (NSGA-II) and NSGA-III. The results of this research, presented within the context of the Algerian network (GRTE) comprising 114 nodes, offer invaluable insights into the intricate interplay of various objectives in power system optimization. The pursuit of more efficient, sustainable, and reliable electrical networks, especially when integrating renewable energy sources, underscores the paramount importance of this chapter within the evolving landscape of electrical engineering and renewable energy integration.

Chapter 9 by Rudbari et al. serves as an exploration of immense significance in the realm of energy management, a domain grappling with the challenges posed by the rapid growth in energy demand driven by population expansion, industrialization, and technological progress. In this context, effective energy management is portrayed as a linchpin, maintaining the delicate equilibrium between the surging energy demand and its sustainable supply. Failure to manage this surge can lead to profound economic and environmental repercussions. The pivotal role of this chapter is underscored by its comprehensive examination of Machine Learning (ML)-based algorithms, particularly Genetic Algorithms (GAs), in the context of optimizing power grids. GAs emerge as potent tools for enhancing grid efficiency, bolstering reliability, curbing system costs, and mitigating adverse environmental impacts. These applications exemplify the profound importance of harnessing ML-based optimization algorithms to navigate the intricate energy landscape, an imperative in the face of rising energy demands and the quest for a sustainable and resilient energy future.

Chapter 10 by Duggal et al. provides a historical perspective on the evolution of autonomous vehicles (AVs) and their integration with electric vehicle (EV) technology. It traces the journey from early ideas in the 1950s, when General Motors proposed the concept of autonomous EVs guided by radio-controlled electromagnetic fields, to the breakthroughs in the 1980s that introduced completely self-sufficient AVs by notable entities like Mercedes Benz, Carnegie Mellon University, and the Eureka Prometheus Project. The significance of this chapter lies in the profound transformation that the automotive industry has witnessed with the advent of autonomous electric vehicles. It highlights the fusion of advanced technology, specifically Genetic Algorithms (GAs), with self-driving cars, leading to the creation of highly autonomous environments that can potentially operate independently. This represents a notable revolution within the automotive sector. Furthermore, the chapter underscores the wealth of data accumulated from autonomous electric vehicles, particularly those produced by Tesla, Inc. This data is a treasure trove for enhancing the electrical aspects of vehicles, including range, power delivery, efficiency, and battery health. Genetic algorithms emerge as versatile tools for optimizing and fine-tuning these electric maneuvers, offering promising prospects for the future of autonomous electric vehicles. This chapter is, therefore, a crucial contribution to the understanding of the intersection between autonomous technology and electric vehicles and its potential to reshape the automotive landscape.

Chapter 11 by Tousi et al. offers a comprehensive overview of the recent and noteworthy developments in the application of Genetic Algorithms (GAs) to the realms of autonomous vehicles and traffic control. It delves into the growing interest of research communities, industries, and urban organizations in optimizing traffic control and autonomous driving, which has given rise to numerous complex challenges. These challenges have sparked the use of various optimization algorithms to tackle them. The paramount importance of this chapter lies in its exploration of how GAs, as one of the most potent and versatile optimization methods, have been harnessed to enhance and stabilize the performance of autonomous vehicles and optimize traffic control in urban settings. This dual focus on urban traffic control problems and advancements in autonomous driving tasks underlines the critical role GAs play in addressing the complexity, reliability, and safety challenges inherent to these fields. The chapter serves as a vital resource for researchers, practitioners, and policymakers, shedding light on the impactful synergy between GAs and the evolving landscape of autonomous vehicles and traffic control. It holds the potential to propel innovations in these domains, improving efficiency, safety, and the overall quality of urban transportation systems.

Farmani et al, in Chap. 12 delve into the domain of wireless sensor networks (WSN), an emerging technology with wide-ranging applications in low-cost environmental monitoring. It emphasizes the fundamental components of WSNs, including sensor nodes, processing units, communication capabilities, and power sources, highlighting the essential role of these networks in data collection and transmission. WSNs present a myriad of challenges, ranging from deployment logistics and energy constraints to limited bandwidth, cost considerations, and security concerns. The significance of this chapter lies in its exploration of innovative approaches inspired by nature to address these multifaceted challenges within wireless sensor networks.

Chapter 13 by Pratama et al. underscores the critical importance of data security and ownership in an era of easy data exchange, where ensuring the confidentiality and integrity of digital information is of paramount concern. To address this challenge, the chapter introduces a sophisticated digital watermarking technique that leverages various mathematical transforms, including Discrete Cosine Transform (DCT), Discrete Wavelet Transform (DWT), Hessenberg Decomposition (HD), and Singular Value Decomposition (SVD). What makes their work particularly significant is its focus on optimizing the watermarking parameters using Genetic Algorithms, enhancing the method's overall performance in terms of both fidelity and robustness. This approach offers a resilient defense against a spectrum of potential threats, including filtering, noise, geometric distortions, and signal processing attacks. Notably, the chapter highlights the role of DCT in improving resistance, especially in the context of geometric attacks.

Finally, Davarpanah and her colleagues in Chap. 14 underscore the growing influence of Genetic Algorithms (GAs), as a prominent branch of Artificial Intelligence, in diverse scientific domains, including engineering and medical sciences. They place a particular spotlight on their application in optimizing Concentric Tube Continuum Robots (CTRs), a specialized type of robot characterized by its multiple curved tubes. The CTR's unique design poses intricate challenges in terms of inverse kinematics and control due to its redundancy, necessitating optimized solutions. The significance of this chapter lies in its demonstration of how GAs are harnessed to optimize both the construction parameters and the inverse kinematic solutions for CTRs. The primary focus is on achieving accuracy, a critical consideration for robots like CTRs, especially in precision-demanding applications such as medical surgery.

Tokyo, Japan — Mahdi Khosravy
Michigan, USA — Neeraj Gupta
Tokyo, Japan — Olaf Witkowski

Contents

About the Editors

Mahdi Khosravy was born in Birjand city, Iran, 1979. He received BSc. in Electrical Engineering (bio-electric) from Sahand University of Technology, Tabriz, Iran, and MSc. in Biomedical Engineering (bio-electric) from Beheshti University of Medical Studies, Tehran, Iran. Mahdi received his Ph.D. in the field of Information Technology from University of the Ryukyus, Okinawa, Japan. He was awarded by the head of University for his excellence in research activities. To grow his international experience in education and research, in September 2010, he joined University of Information Science and Technology (UIST), Ohrid, Macedonia, in the capacity of assistant professor. In 2016, he established a journal in information technology (ejist.uist.edu.mk) in UIST as currently hold its executive editorship. UIST professorship helped him a lot to extend his international collaborations. In July 2017, he became an associate professor. August 2018 he joined the Energy lab in University of the Ryukyus, Japan as a Visiting Researcher, and from April 2018 to September 2019, he was jointly a visiting associate professor in Electrical Engineering Department, Federal Universit of Juiz de For a in Brazil. From October 2019 to 2021 he was Specially Appointed Researcher in Osaka University. Since 2021, he is a senior research scientist in Cross Labs, Cross Compass Ltd., Tokyo, Japan. Dr. Khosravy is a member of IEEE.

Neeraj Gupta is Sr. Scientist at Kevadia Inc. Michigan, USA in the department of Kevadia Intelligence. where he is playing an instrumental role and lead in design and development of automated Deep learning platform to build edge enabled AI automatically for given data, without involving any expert (end-to-end digital expert). Prior to his industrial joining he served Oakland University, Rochester, Michigan in the Department of Computer Science and Engineering from 2017 to 2022, as Assistant professor, and Teaching & research staff. In addition to his academic service, Dr. Gupta was an Assistant. Professor at University for Information Science and Technology, "St. Paul the Apostle", Ohrid, Macedonia from 2014 to 2017. He received his Ph.D. from Indian Institute of Technology Kanpur, India in 2013 and geared up his carrier in Indian Institute of Technology Jodhpur, India as Postdoctoral Fellow and Guest Faculty until 2014. His research interests include Edge Enabled Deep Learning, Deep Learning, Computational Intelligence, Visual Computing, Evolutionary Computing, and Big Data Analytics.

Olaf Witkowski is a leading expert in AI and Artificial Life. He is the Founding Director of Cross Labs, a research institute in Japan, focusing on the study of intelligence in biological and synthetic systems. He is the President of the Board of Directors of the International Society for Artificial Life, Executive Officer at leading AI company Cross Compass, and External Lecturer at the University of Tokyo. He has cofounded multiple ventures in science and technology on three continents, including YHouse Inc. (nonprofit institute on the emergence of consciousness) and the Center for the Study of Apparent Selves (focusing on the ethics of agents). His research focuses on empathic, open-ended, collective, and hybrid AI built from a mathematical understanding of intelligence in any substrate.

Part I
Genetic Algorithms Foundations and Future Direction

Chapter 1
Evolutionary Computation: Perspectives on Past and Future

Hanna Yoshida, Alyssa M. Adams, and Olaf Witkowski

1 Summary

Evolutionary computation (EC) constitutes a sub-field of artificial intelligence that uses algorithms inspired by characteristics of Darwinian evolution to repeatedly refine a set of solutions over successive generations [1]. When weaker solutions (entities or individuals) are removed, and small, random changes are added to the system's dynamics or the solutions themselves, the quality of the remaining solutions continuously improves until an optimal solution is reached or the process is stopped. This computational process of iteratively generating and selecting solutions uses top-down and bottom-up approaches, making it flexible enough to apply to various fields, including engineering, medicine, and economics.

Genetic algorithms (GAs) are a well-known class of optimization algorithms that fall under the umbrella of evolutionary computation. GAs directly apply the principles of natural selection and genetic variation to machine learning. Their creation and implementation have enabled EC to expand to numerous forms that simulate different aspects of biological evolution. Still, it remains a challenge for researchers to derive a method for simulating "open-ended" evolution similar to what we see in natural speciation. Evolutionary computation bridges artificial intelligence and artificial life, both inspired by outcomes and aspects of natural evolution. Although challenges like simulating/re-creating open-ended evolution (OEE) remain, evolutionary computation remains a valuable approach in various research areas (Table 1).

H. Yoshida · O. Witkowski (✉)
College of Arts and Sciences, University of Tokyo, Tokyo, Japan
e-mail: olaf@crosslabs.org

A. M. Adams · O. Witkowski
Cross Labs, Cross Compass Ltd., Tokyo, Japan

M. Khosravy et al. (eds.), *Frontiers in Genetics Algorithm Theory and Applications*, Springer Tracts in Nature-Inspired Computing,
https://doi.org/10.1007/978-981-99-8107-6_1

Table 1 List of abbreviations

EC	Evolutionary computation
GAs	Genetic algorithms
OEE	Open-ended evolution
ES	Evolutionary strategies
EP	Evolutionary programming
POET	Paired Open-Ended Trailblazer

2 History and Development

The earliest research for evolutionary computation began in the late 1950s to the 1960s, when the first computational neural networks and artificial intelligence-related fields emerged [1]. From there, evolutionary computation branched into three main types: evolutionary strategies (ES), evolutionary programming (EP), and genetic algorithms (GAs).

2.1 Evolutionary Strategies (ES)

When evolutionary strategies (ES) were applied to computer science and engineering, more efficient solutions to multimodal optimization problems were easier to find. These solutions were previously challenging to find with trial-and-error methods using physical hardware. During the mid-1960s, three students from the Technical University of Berlin, Peter Bienert, Ingo Rechenberg, and Hans-Paul Schwefel, envisioned an automated system where the parameters for an experiment could be altered and optimized over time without researcher intervention [1]. Their procedure involved automating the cyclic process of running the experiment, measuring the outcome, and selecting the optimal parameters based on the outcome these parameters produced. As a result, the simplest type of ES, now called (1+1)ES, was created. Here, one parent produces one descendant selected among the rest of the population [1]. Mutations in the descendant are binomially distributed around the parent "ancestor." Today, variations of ES are more commonly used than the initial (1+1)ES version, including the (μ+λ) ES, which is an improved, more complex version of the initial (1+1)ES [2]. Despite all the variations, all ES generally follow three main principles:

1. In a given population, one or several parents are selected to produce new offspring via duplication and recombination.
2. The new offspring undergoes mutation and becomes new members of the population.
3. Environmental selection reduces the population to its initial size [2].

2.2 Evolutionary Programming

Another type of evolutionary computation is evolutionary programming (EP). Although Fogel designed EP separately from ES in 1960 [3], its aim is similar since it involves algorithms that optimize a solution to a problem in a fixed or variable environment. Fogel demonstrated a pioneering example of EP, where machine solutions were selected for their ability to predict a particular string of symbols. Parent machines generate offspring through a process of mutation. The selection of the offspring was probabilistic; the lesser-performing half of the machine population had a lower chance of surviving the next stage than being eliminated from the population based on performance [4]. This probabilistic selection process directly contrasts EP, where the offspring was directly selected from the population. According to Kantrowitz, EP also differs from genetic algorithms (see Sect. 2.3: Genetic Algorithms) because it does not use genetic operators—such as genetic crossover—to determine the extent of mutation. Instead, EP implements several types of mutation stages (from more minor to more significant mutations) that alter the function of offspring machines.

2.3 Genetic Algorithms

Genetic algorithms (GAs) were first introduced in John Holland's research in 1975 [1]. He introduced a GA to implement the selection methods of natural evolution in silico to automate optimization processes through trial and error. Unlike the previous two components of evolutionary computation (ES and EP), the aim of a GA is not necessarily to find an existing solution to a given problem. Instead, a GA simulates evolution and the "survival of the fittest" environment to derive the best possible outcome from a random parent population, which is then considered the optimal solution to the problem [5]. Hence, it may be more similar to biological evolution, where evolution is not directed toward a specific goal. In both GAs and nature, the outcome depends on how fitter individuals are more likely to propagate their genetic material to the next generation.

To simulate biological evolution, GAs involve several sources of "genetic" variation that cause mutations to the original "genome" (i.e., binary code). Sources of variation include natural selection, random mutation, and genetic crossover [6]—inspired by sources of genetic variation observed in nature. At a molecular level, chromosomes crossover in sexual reproduction, and random mutations in the DNA sequence occur relatively frequently. Today, genetic algorithms are used in artificial life simulators, industry, and engineering. GAs compute solutions to optimization problems by trial and error far more efficiently than manual methods, such as designing, building, and testing various prototypes (Fig. 1).

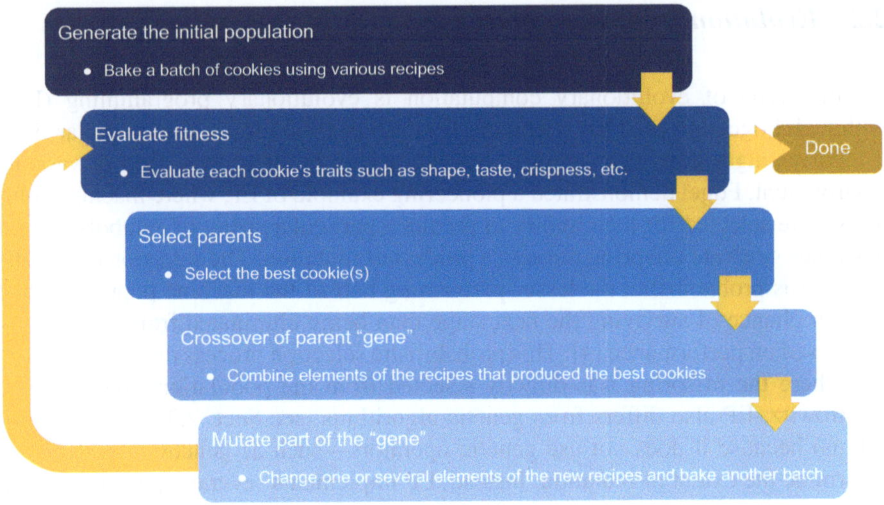

Fig. 1 Summary of a genetic algorithm (using cookies as an example), adapted from [6]

3 Existing Technologies and Applications

The three types of evolutionary computation algorithms mentioned in the previous section (ES, EP, and GA) are applied in many different ways to several fields, including artificial life, biomedical sciences, and industrial uses. This section discusses a few of these examples.

Digital Organism Simulators One topic of interest within artificial life is digitally simulating organisms and evolving their populations. In nature, organisms evolve over several generations because they have heritable and adaptable traits encoded in genetic material. Karl Sims' 1994 digital organism simulator uses a genetic algorithm to emulate evolution among virtual creatures in three-dimensional spaces for a specific physical goal, such as walking, swimming, or jumping [7].

Specifically, he employed genetic algorithms to allow the creatures to modify their physical traits based on the inheritance and modification of their corresponding code, or "genome." Sims applied the biological terms *genotype* and *phenotype* to describe how the creatures evolved. The *genotype* refers to the generated code that affects the physical traits or the *phenotype*. In artificial evolution, the genotype is where variations occur, such as mutation or crossover. On the other hand, the phenotype is where the selection occurs, where they are given a higher or lower chance of survival based on their fitness toward the goal [7].

Since Sims's research, many more genetic algorithms have been developed. One of these, called POET (Paired Open-Ended Trailblazer), can simulate co-evolution between organisms and their environment by allowing the virtual organisms to modify their environment as they evolve and adapt [5]. Virtual creatures in POET can learn to overcome increasingly complex environmental obstacles much faster than other GAs. Using a GA that does not modify the environment is likely to end with organisms that cannot move past unknown obstacles, hence preventing evolution from advancing to more complex directions [5]. Overall, research in digital organism simulators has led to intriguing resulting virtual creatures and potential for new ways to approach optimization problems in fields such as robotics.

Protein Structure Design and Prediction Genetic algorithms (GA) can also predict the structure of proteins from their amino acid sequence [8]. Finding an accurate and efficient way to predict structures for molecules as complex as a protein is challenging. The development of GAs (along with other methods such as deep learning and other machine learning techniques) has enabled biologists to access accurate models of folded proteins. While deep learning models dominate protein-folding research, GAs were prevalent earlier. GAs were employed to find the angles and orientation of each bond along the protein such that the overall structural energy was at a minimum [6]. Because proteins with lower structural energy are more likely to form naturally, GAs provided a quicker method to find the most probable solution. Although GAs can generate an accurate structure within the known optimal range for small proteins, more advanced algorithms are needed to predict larger structures accurately.

Artificial intelligence technologies such as deep learning algorithms dominate structural predictions of proteins. A well-known example of this is *AlphaFold 2* [9], a deep learning method that exploded in use among biologists after it was officially made public in July 2021 [10]. This algorithm is known to generate accurate predictions of protein structure and folding patterns and is potentially useful for drug design and medicine. However, because it is a deep learning algorithm, it requires a large set of foundational data for a confident prediction. Consequently, one issue in the current program is that it cannot predict the structure of proteins that mutate over time. Evolutionary computation can potentially contribute to constructing a more accurate structural prediction program by bringing a more "bottom-up" approach to simulate mutant proteins without needing massive data.

Optimization of Transportation Routes Evolutionary computation is used beyond the lab, such as in-car navigation systems where users may want to find the shortest path to their destination [11]. Typically, navigation systems use algorithms such as Dijkstra's algorithm, but it may not be able to compute solutions to common scenarios in some cases. For example, if the driver wants to stop at a gas station or a restaurant on their way home, the algorithm may need help finding the optimal route. The algorithm must consider too many possibilities within the given parameters, making it difficult to find a solution within a short time frame [11]. To solve this problem, methods such as genetic algorithms may be helpful. In one study by Kanoh, roads were considered "viruses" which would reproduce according to evolution and spread by infection under GA [11]. As a result, the navigation system using GAs determined the optimal route to the destination faster than the typical algorithm, even with the desired pit stops. These results suggest that evolution may be a faster way to solve real-life problems.

Other Applications The applications of evolutionary computation algorithms are certainly not limited to the ones mentioned above. For example, this approach has been used to evolve a neural network that controls a cart-like agent that balances a pole [12]. Since neural networks and deep-learning-based algorithms have demonstrated a lot of success for a wide variety of applications, evolutionary algorithms have been used as a method to evolve and fine-tune them. In addition, it has been used for particle swarm optimization, swarm intelligence, self-organization, art, finance, agriculture, and almost any other problem that requires an algorithm to be optimized or evolved over time.

4 *Top-Down* and *Bottom-Up* Approaches

Artificial intelligence overlaps with artificial life in many ways. Both aim to simulate aspects of life generated through natural processes. In particular, artificial intelligence focuses on simulating human intelligence and thought processes. Among their many similarities, there are some differences in their approach to problem-solving. Generally, artificial intelligence methods take a top-down approach, while those using artificial life tend to take a bottom-up approach [13]. Both approaches are based on the understanding that complex concepts consist of smaller, simpler building blocks [5]. Evolutionary computation involves both top-down and bottom-up approaches to some extent. Top-down approaches include selection processes and methods of evaluating the fitness of solutions or generated individuals. Bottom-up approaches generally involve genotype-to-phenotype mapping and an explicit relationship between the phenotypic expression of a gene and its contribution to an individual's/solution's fitness. Since both approaches are used, and EC approaches remain flexible to experimental design, it could bridge research between artificial intelligence and artificial life (Fig. 2).

4.1 Top-Down Approach

The top-down approach refers to a problem-solving process where an existing problem is understood as a collective quality of its parts. While it may be broken down into smaller elements, the problem is seen as a collective phenomenon rather than as the sum of the individual components. Traditionally, methods related to artificial intelligence are top-down. For example, a scientist may want to simulate the human brain, which is immensely complex. To achieve this using a top-down approach, they may use machine learning to obtain clues for how humans learn and output meaningful actions. This approach is also seen in some studies within artificial life, including evolutionary computation. During the development of evolutionary algorithms, there needed to be a solid understanding of how organisms evolve at a molecular level, such as genetic crossover and mutations. While the molecular processes weren't

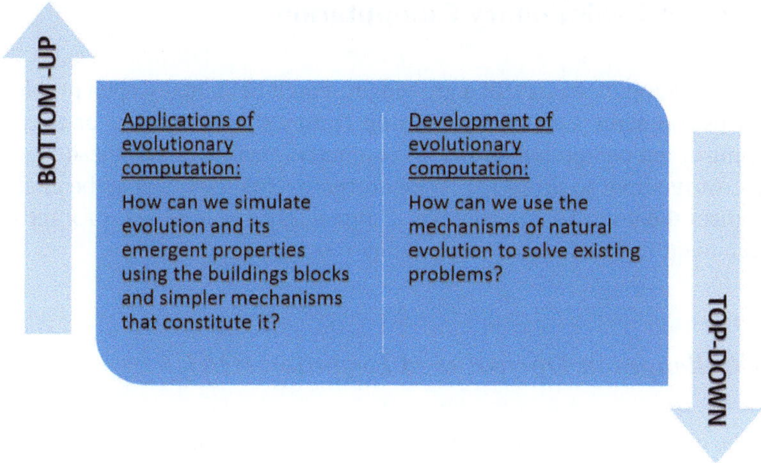

Fig. 2 Top-down versus bottom-up approach in evolutionary computation

implemented, they were generalized and abstracted. Consequently, scientists can observe how such processes lead to more complex systems, such as the evolution of a whole population and its properties.

4.2 Bottom-Up Approach

The bottom-up approach refers to a problem-solving process where we understand complex phenomena (or problems) by reconstructing them from their constituent building blocks. It focuses on the process of creating the complex phenomenon as much as the outcome itself. Artificial life fields apply this approach more, an example being the Boid program created by Craig Reynolds in 1986 [5]. This program simulates the flocking behavior of birds on a computer. In this case, the examined phenomenon was the emergent behavior of birds flocking together. Rather than explaining the problem as due to the behavior of individual birds, the Boid program reconstructed the flocking behavior as a whole. It confirmed that several rules between interacting individuals control flock behavior [5]. Evolutionary computation employs a similar bottom-up approach, where scientists can generate more complex phenomena, even entire artificial worlds, based on multiple components that interact. Another clear example of the bottom-up approach in evolutionary computation is digital organism simulators (see Sect. 3.0.1).

5 Future of Evolutionary Computation

Evolutionary computation is still a growing research field; many open problems and potential uses remain. One of the limiting factors that span most of these issues is the limit in the computing ability of computers that carry out these algorithms, and the cost required to do so [5]. With current technology, it may be challenging to implement them in real-life contexts. Promisingly, several areas of improvement have been identified and are currently being researched.

5.1 Challenges in Open-Ended Evolution (OEE)

One of the challenges within evolutionary computation and the wider artificial life field is simulating open-ended evolution (OEE) [14]. The earliest example of open-ended evolution is John von Neumann's cellular automata, inspired by natural evolution and proposed as a theoretical model for a self-replicating machine. Published in 1966, this remained a theory, and von Neumann could not build an actual machine. However, cellular automata became vital in many fields, including artificial life and open-ended evolution [15]. New models emerged from the cellular automata concept, including John Conway's Game of Life and Christopher Langton's Loop automaton. Concepts including self-replication within cellular automata and emergent OEE are referenced in biological fields such as embryology [15] as a bottom-up approach.

Despite the usefulness of the concept of OEE, it remains a challenge to emulate actual self-replicating and evolving artificial life in the lab. At this time, many evolutionary computational methods often lead to a single stable equilibrium under the laws of thermodynamics, in which evolution virtually stops or becomes predictable. Sometimes, the algorithm can get stuck on an outcome that may not necessarily be the best possible one [14]. POET, mentioned earlier, is a step closer to simulating open-ended evolution. However, it is still directed evolution, since there is a fixed scenario the organisms are selected to adapt to.

One of the best examples of open-ended evolution is biological evolution, where an ancestral single-cell organism gave rise to a large variety of increasingly complex and diverse organisms, including humans. Artificially simulating something like this natural evolution requires an additional understanding of factors impacting how we simulate and observe OEE. For instance, selection pressures in natural selection can act as limitations in open-ended natural evolution, which raises the question of whether constraints from the environment can yield more interesting results from OEE [16]. Other factors include setting more concrete, quantitative criteria for measuring "open-endedness."

One issue with some current research on OEE is that there is much subjectivity and qualitative observations for measuring the "open-endedness" of an evolutionary system, making it challenging to compare discussions and results across the literature [16]. Therefore, evaluating OEE through a more concrete and unbiased lens may benefit the artificial life community by providing a foundation of OEE research that everyone can agree on. One crucial benchmark toward this goal is the definition of different types of OEE listed as a report from the First Workshop on Open-Ended Evolution in York, enumerated below from [17]:

1. Ongoing generation of adaptive novelty:
   (a) Ongoing generation of new adaptations
   (b) Ongoing generation of new kinds of entities
   (c) Emergence of a dynamical hierarchy and major transitions
   (d) Evolution of evolvability
2. Ongoing growth of complexity:
   (a) Entity complexity
   (b) Interaction complexity

Measuring open-endedness may be more definite in the future. Even so, creating a way to implement OEE and simulate realistic evolution has diverse approaches as a rising research topic. Suppose evolutionary computation reaches the stage where it can emulate OEE. In that case, it can provide vital clues for artificially attaining the complexity of the human brain and intelligence, along with other products of biological evolution. Understanding OEE may be the key to developing more valuable and creative artificial intelligence and life [14]. This can also be valuable for developing meta-learning methods for AI systems. By approaching human intelligence not as a machine to reconstruct but as the optimal outcome of open-ended evolution, we could expand its abilities infinitely.

5.2 *What Comes Next?*

The future of evolutionary computation is a vast and uncharted territory, filled with potential and challenges. As the field expands, we must remember that our journey is not just about overcoming technical hurdles but also about understanding the fundamental principles of life and evolution. The challenge of simulating open-ended evolution (OEE) is a testament to this fact. It is not merely a technical problem but a question about the nature of life, complexity, and evolution itself.

As we move forward, we must strive to develop more practical—and as unbiased as possible—methods for evaluating OEE. This will help us compare and contrast different approaches and provide a solid foundation for future research. The definition of different types of OEE, as outlined in a series of Workshops on Open-Ended Evolution in York, Tokyo, Kyoto, and other places [16, 18, 19], is a significant step in this direction.

Nevertheless, while more concrete measures of OEE will prove invaluable, one must also recognize the subjective character inherent to open-endedness. The concept of open-endedness is, to a large extent, in the eye of the beholder. What may seem like a novel or complex outcome to one observer might be perceived as trivial or predictable to another. This may have been somewhat underlying and only implicitly evoked in previous studies, which often focused on a particular (albeit relevant and justified) aspect of evolution. This subjectivity might be better acknowledged, and its weakness may further underscore the importance of developing shared, objective criteria for evaluating OEE, while acknowledging that our understanding and appreciation of open-endedness may evolve as we delve deeper into this fascinating field.

Once the community can successfully emulate OEE, the implications will be profound. It could provide vital clues for the capacity of evolved entities to attain the level of complexity of human brains and other feats of intelligent systems, by capturing the conditions of emergence of such evolutionary paths in artificial models. This could pave the way for the development of more advanced learning and meta-learning methods for AI systems.

For these reasons, the future of evolutionary computation does not solely hinge on developing more advanced algorithms or building more powerful computers. It is directed at understanding the principles of life and evolution that allow for generating heterogeneous, wide-ranging, flexible, open-ended forms of learning and using that understanding to create more intelligent and creative artificial systems. This journey that we have only just started shows promises of an immense potential and remarkable innovations further down the line.

References

1. De Jong K, Fogel D, Schwefel H (1997) History of evolutionary computation. Oxford University Press
2. Hansen N, Arnold D, Auger A (2015) Evolution strategies. In: Springer handbook of computational intelligence. Springer, Berlin, pp 871–898
3. Rozenberg G, Bäck T, Joost N (2012) Handbook of natural computing. Springer, Berlin
4. Kantrowitz M (1997) Genetic algorithms frequently asked questions. CMU Artificial Intelligence Repository
5. Oka M (2022) ALIFE-artificial intelligence to artificial life (originally in Japanese). Tokyo BNN Publishing
6. Munetomo M (2008) Genetic algorithms—its theory and methodologies (originally in Japanese). Morikita Publishing
7. Sims K (1994) Evolving virtual creatures. In: Proceedings of the 21st annual conference on Computer graphics and interactive techniques, pp 15–22
8. Pedersen J, Moult J (1996) Genetic algorithms for protein structure prediction. Curr Opin Struct Biol 6(2):227–231
9. Jumper J, Evans R, Pritzel A, Green T, Figurnov M, Ronneberger O, Tunyasuvunakool K, Bates R, Žídek A, Potapenko A et al (2021) Highly accurate protein structure prediction with alphafold. Nature 596(7873):583–589
10. Callaway E (2022) What's next for the ai protein-folding revolution. Nature 604:234–238
11. Kanoh H (2002) Routed guidance for car navigation using genetic algorithms. Inf Process Soc Jpn (21(2001-ITS-008)):51–58

12. Variengien A, Nichele S, Glover TE, Pontes-Filho S (2021) Towards self-organized control: Using neural cellular automata to robustly control a cart-pole agent. ArXiv:abs/2106.15240
13. Stern A (1999) Ai beyond computer games. In: AAAI AI and interactive entertainment spring symposium 2000
14. Stanley O, Lehman J, Soros L (2017) Open-endedness: the last grand challenge you've never heard of O'Reilly Media
15. Damerow J (2012) John von neumann's cellular automata. Embryo Project Encyclopedia
16. Frans K, Soros L, Witkowski O (2021) Questions for the open-ended evolution community: reflections from the 2021 cross labs innovation science workshop. Artif Life 25(1):4
17. Packard N, Bedau M, Channon A, Ikegami T, Rasmussen S, Stanley K, Taylor T (2019) Open-ended evolution and open-endedness: editorial introduction to the open-ended evolution i special issue. Artif Life 25:1–3
18. Taylor T, Bedau M, Channon A, Ackley D, Banzhaf W, Beslon G, Dolson E, Froese T, Hickinbotham S, Ikegami T et al (2016) Open-ended evolution: perspectives from the oee workshop in york. Artif Life 22(3):408–423
19. Packard N, Bedau MA, Channon A, Ikegami T, Rasmussen S, Stanley KO, Taylor T (2019) An overview of open-ended evolution: editorial introduction to the open-ended evolution ii special issue. Artif Life 25(2):93–103

Chapter 2
Genetic Algorithms: A Technical Implementation of Natural Evolution

Rishabh Duggal

1 Introduction

Now that we have advanced so far technically in this world, finding just a solution to a particular problem is not feasible enough and the situation always demands a more optimized method to tackle it. However, honing these solutions is one of the key characteristics of genetic algorithms, based on their bare ideology of "every generation is smarter than before". This also helps in the development of algorithms in further stages by following similar steps, that is, making the children of the latest generation as parents of the next and creating smarter offsprings and thus solutions for the particular situations. The effortless procedure of using these algorithms, once designed when clubbed with the power of them based on the algorithm's design, coins the term "efficiency" which would be clubbed with GAs most of the time except in some scenarios which we will also analyze in this chapter.

That being said, certain scenarios might appear where these algorithms would not be the best option to choose from. In this chapter, every major aspect of the genetic algorithms would be discussed, reviewed, and on several deciding factors, we'll finally conclude this backed by adequate reasoning.

2 History

The inception of this theory didn't take much time after mankind developed the first machine learning derivative. Alan Turing introduced the world with a "learning machine" which was inspired by evolutionary theories. From that point, several

---

R. Duggal (✉)
Thapar Institute of Engineering and Technology, Prem Nagar, Patiala, Punjab, India
e-mail: rishabhduggal249@gmail.com

M. Khosravy et al. (eds.), *Frontiers in Genetics Algorithm Theory and Applications*,
Springer Tracts in Nature-Inspired Computing,
https://doi.org/10.1007/978-981-99-8107-6_2

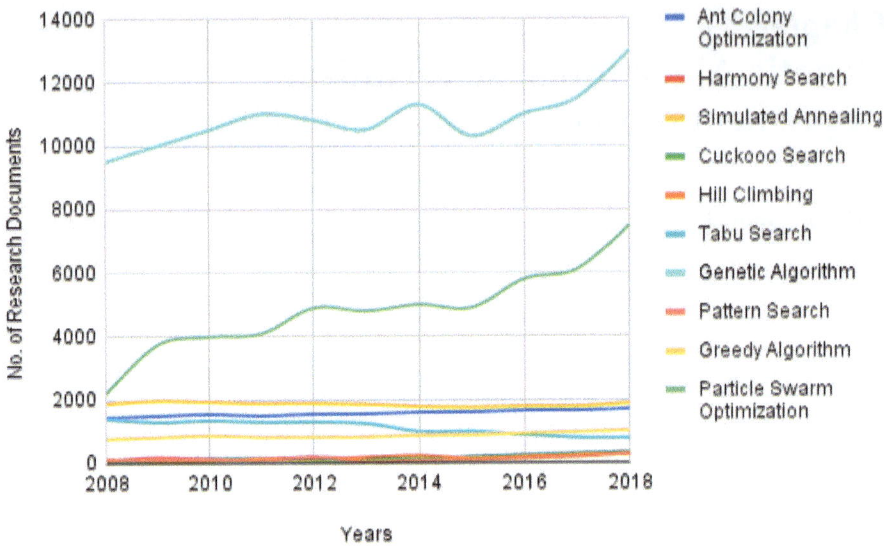

Fig. 1 Increasing popularity of genetic algorithms

experiments were then conducted based on natural selection and how to replicate it to make it a tool for problem-solving. It was John Holland who made this theory work in practical scenarios, his approach being inclined toward the biological aspect of these algorithms.

It was then in the 1990s, a full-fledged technical implementation of these algorithms started to emerge. How evolution has worked since the inception of time and life, especially on this planet, was directly applied to the algorithms which were being used famously back then. The development of this approach has now made genetic algorithms one of the most powerful and efficient ways to work around a specific problem. Figure 1 shows the increasing popularity of GA besides some other state-of-the-art nature-inspired optimization techniques.

In addition to GA, there are other evolutionary algorithms as well. A recent technique inspired from GA is Mendelian Evolutionary Theory Optimizer (METO). While GA deploys Darwin rules, Mendel laws of inheritance are deployed in METO [1]. GA and evolutionary computation are sub-categories of computational intelligence, and all together under the large shelter of artificial intelligence (AI). GA has been deployed in broad range of applications like fake data generation and detection [2], morphological filters [3], power planning [4], Internet of things [5], lightweight AI model design [6, 7], predictive analysis [8], power quality [9, 10], self-organized control [11], maintenance [12], etc.

3 How Algorithms Evolve

The procedure of this approach is not a straightforward single-step solution to a particular problem. Even when you consider nature, when a particular species is present in an environment, its characteristics define how well they are suited for that environment. Luckily, nature always has a hunger to make the residents fitter as time proceeds. Residents make their offspring, which obviously are not all the same. The one which is fitter for the scenario lives on, and others eventually get eliminated. Then the fitter parents again make much fitter offspring, making each generation smarter than the previous one. By this procedure, nature ensures that the overall strength and smartness of its residents evolve positively over time [13].

Since this happens and is being witnessed in real life, we can safely say that each step takes place randomly and naturally. As far as its technical implementation is concerned, we as the algorithm writers have to ensure that the offspring generation aspect is done in the most efficient way. To perform this, we take assistance from the three major operators used for the generation of a genetic algorithm: selection, crossover, and mutation. The block diagram in Fig. 2 shows the GA operation and Fig. 3 gives the GA pseudo-code [14].

In nature, a particular generation mates and creates offspring in a random manner. This makes the whole process of evolution moderately bovine, as is clearly witnessed. To make this aspect efficient, we need to choose the fittest particular individuals, and make them parents, hence ensuring a smarter offspring generation. This procedure is handled by the selection operator.

Selecting the fitter individuals already makes the process efficient, but to boost up the efficiency even more, we take help from the other two operators. Let's take two individuals as parents, now we process their fitness score and based on that, we swap a certain set of genomes (or characteristics) to check if we have a better set of parents. We then create offspring off of these parents to have a much smarter generation. By doing this crossover step, we often skip a repetitive set of parent-offspring generation (redundant computations in technical implementation), making this whole process much more efficient and coherent.

In nature, as evolution proceeds, we often witness some individuals who have completely different characteristics from their fellow mates. This is sometimes beneficial for that particular species or even sometimes a nightmare. Nonetheless, this is a tool that when used lucratively can speed up the evolution process. This is when the mutation operator is used. We alter a particular genome (or characteristic) intentionally to make that particular individual fitter. That individual would then create fitter offspring making the next generation much more intelligent and stronger. Thus, we can see how nature's way of making its residents fitter can be applied directly to algorithm generation, and further by using the above-mentioned tools, we can also speed up this evolution process, making it much more efficient and powerful.

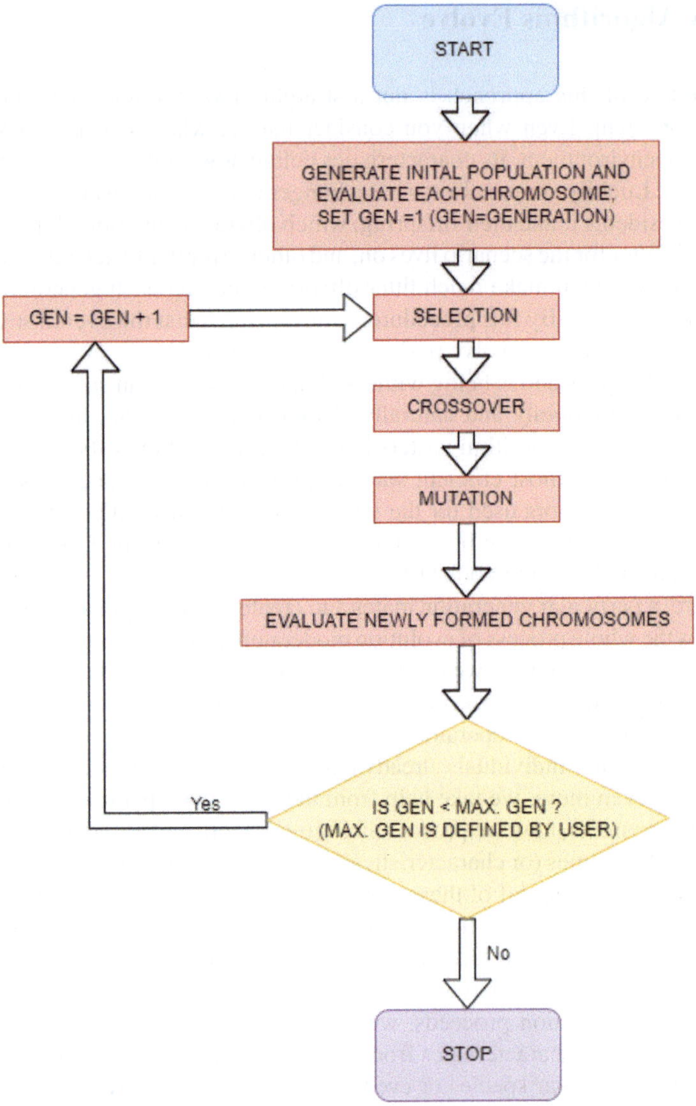

Fig. 2 How a genetic algorithm operates

Fig. 3 Pseudo-code for a genetic algorithm

```
GA()
  initialize population deployment
  Population fitness calculation

  while (saturation/termination approached)
do
    Selection of parent
     crossover operator with probability P(c)
     mutation operator with probability p(m)
     decode and fitness calculation
     selection of survivors and marking them
as fit
     observe fittest
  return fittest
```

4 A Review of Some Research Conducted in this Field

This section considers some important research, particularly the implementation of genetic algorithms in several practical scenarios. Each research is analyzed and a particular verdict is then provided.

4.1 Multi-objective Shape Optimization of an Autonomous Underwater Glider Based on Fast Elitist Non-dominated Sorting Genetic Algorithm

The sample space in this research is building a glider with low drag, high efficiency, low power consumption, and longer voyage. SOLIDWORKS for 3D modeling, ANSYS for meshing and simulation, and MATLAB were the tools applied for data collection [15]. Variables used in the study were altering the shape based on further generations which were offsprings from the proposed GA and comparing it with three more typical predefined shapes. Statistical techniques such as the representation of crowding distance, 3D surface division, sensitivity coefficient for drag and pressure, drag change when velocity is varied, hull surface pressure distribution, different pressure diffusion of all four glider shapes, hull L/D and gliding angle at a different angle of attack were also used.

When designing the NGSA-II, the initial population size and number of iterations are kept at 300 and 180 (fairly huge for performance and efficiency optimization). Hull drag and surface pressure are then mathematically formulated and then the glider shape is optimized. This was the problem-solving approach for this study and as far as the conclusions, perspective, and findings are considered, this research concludes that hull diameter and length are directly proportional to the drag while hull diameter and shape exponents are almost exponentially proportional to the change of hull pressure coefficient. Huge benefits like drag reduction and vortex elimination are presented mathematically, implying that this approach generates the optimum shape every time, overall giving the best hydrodynamic performance and low power consumption of the underwater vessel. This could yield promising results when applying it on lengthy deep sea maneuvers as it would reduce the recharging trips toward the top of the water body, and thus would save a lot of time and energy.

4.2 An Energy Optimization in Wireless Sensor Networks by Using Genetic Algorithm

A sample space in this research is the optimal sensor node design, and energy model for WSNs [16]. Figure 4 shows the diagram of a wireless sensor network. MATLAB simulations conducted after applying the proposed GA and analysis of them using basic graph generators were the tools applied for data collection. Variables used in the study were fitness values, generations, payload size, header size, modulations, number of iterations conducted on the energy models, and energy values. Statistical techniques such as the tabular representation of 20 GA models used for the optimization of energy, energy models' outcomes with respect to the proposed genetic algorithm, and fitness value portrayal of all three energy models using various genetic algorithm models were also used.

Selection of desired energy optimization models of WSNs followed by their optimization using 20 GAs with detailed information stated for selection, crossover, and mutation followed by the analysis of each and every aspect studied and calculated in this research by conducting 60 simulations. This was the problem-solving approach for this study and as far as the conclusions, perspective, and findings are considered, the proposed GA methods in this report promise minimum consumption of energy by delimiting and challenging the limited battery life of WSNs. The analysis results show the positive impact of optimization of the proposed GA which in turn help to stretch the battery life of sensor nodes. This research could have been more refined if we also considered the effect of temperature on both the battery and environment,

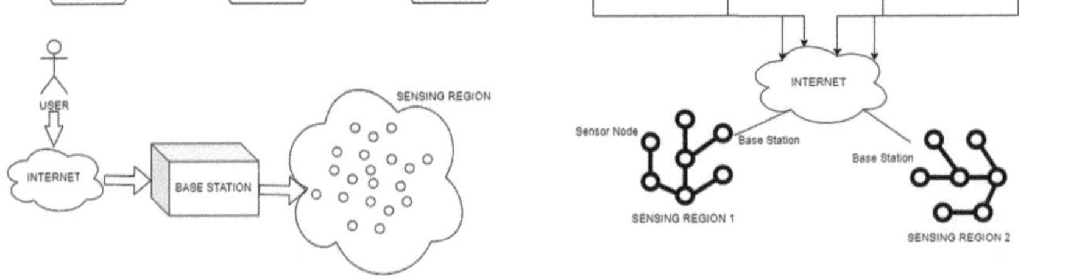

Fig. 4 Wireless sensor networks

applying a unique GA on it to optimize the results which would definitely increase the battery life of WSNs significantly. The inclusion of the effect of different raw materials for creating the battery and its impact on it could have broadened the research to make it more practical.

4.3 An Implementation of Instruction Detection System Using Genetic Algorithm

The sample space in this research is a standard dataset used in the KDD Cup 1999 competition [17], a proposed intrusion detection system [18], and an analysis of similar existing systems. Misuse/signature-based and anomaly/statistical detectors were the tools applied for data collection. Statistical techniques such as bar graph representation of detection rates for each class were also used.

We start by an overview of intrusion detection followed by a discussion of the existing systems and their problems. The principles of intrusion detection systems have been demonstrated in Fig. 5. Then proposing the IDS of the research is built using genetic algorithms followed by its implementation. Experimentations are conducted finally to properly analyze the results of this approach. This was the problem-solving approach for this study and as far as the conclusions, perspective, and findings are considered, the results of the experiment showed a positive response when the proposed GA was applied to the IDS. Researchers also proposed their own IDS which showed better performance during the tests which were conducted. Inclusion of a turning off by a special gesture considering and treating it as a rare case scenario

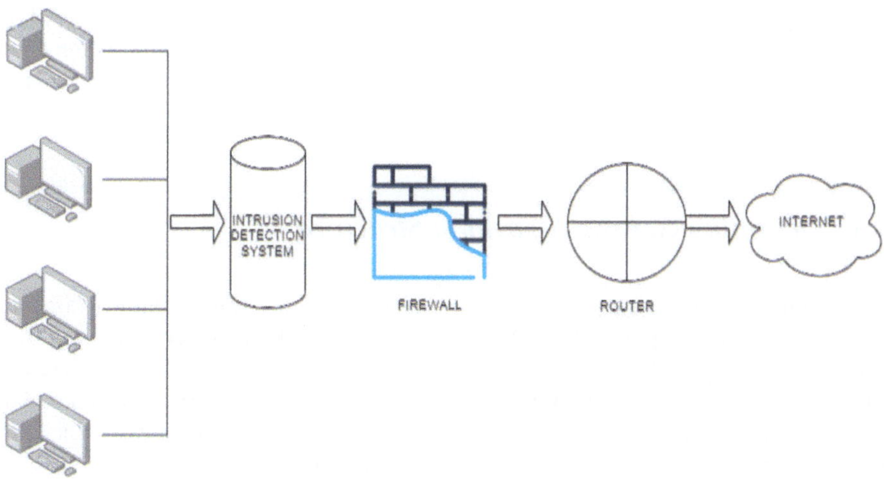

Fig. 5 Intrusion detection systems' principles

would be an amusing gimmick to this research. Real-life applications of such ideas require an expansion of hardware, especially at the output side of the system, which may reduce the practicality of the device.

4.4 Numerical Function Optimization

In application of GA for numerical function optimization [19], the sample space is a numerical function sampled for this research, designing and analysis of the algorithm, and discussions of the effectiveness of various aspects related to the proposed operator. RCGA-rdn requires the input of parameters and an unspecified simulator is the tool applied for data collection. Statistical techniques such as the tabular representation of optimal settings, experimental results of different crossover, replacement, mutation operators and performance comparisons, and performance comparisons. Experiment results of practical engineering optimization problems for different algorithms were also used.

Stating the problem and a static penalty function followed by the construction of operator is by experimentation of the algorithm testing and analysis determines the effectiveness of various aspects of the approach. This was the problem-solving approach for this study and as far as the conclusions, perspective, and findings are considered, optimization of the numerical function using the proposed GA showed better performance in the function. It was able to ignore the local minimas which was not possible before the integration of the proposed GA. Researchers also state that this approach will show promising results even after an increase in population size as this algorithm is capable of maintaining the balance between diversity and convergence. One obvious issue with avoiding the local minimas and maximas is the performance compromises when it comes to corner cases. An approach to this problem would be to integrate a parallel and less powerful variant of the same approach and switch back and forth between the scenarios being tackled depending on which part will provide the optimal performance.

4.5 Designing Genetic Algorithm Structure

For designing GA structure, the work by [20] is over 20 case studies, proposes a framework for designing the GA structures, and analyzes several case studies. Parameters and variables used in the study were population size, crossover, mutation rate, adaptive restarting condition, local solutions which get randomly generated, maximum step size, and reduction rate of it. The Variations in fitness values are achieved after a number of objective function evaluations.

Observing the latest developments of GA structure is followed by the proposed framework for the same. Exploring the case studies and based on all these aspects, analyzing the research. This was the problem-solving approach for this study and as far as the conclusions, perspective, and findings are considered, the framework design showed promising results as the genetic algorithms tuned from this framework were showing respectable performance with quality growth of even 99.97%. Although the GAs designed through this framework were fast, the time taken for their development is a field that could have been explored much more deeply for polishing the research. Ironically, this ETA reduction could be achieved by applying a unique GA on the initial phase of the framework, which would amplify the delivery stage and give much more promising results.

4.6 Selection Techniques in Genetic Algorithm

Theoretical data gathered was manually by the researchers for the analysis in a review work by [21]. Proposing the selection strategies followed by their comparison would be the problem-solving approach for this study and as far as the conclusions, perspective, and findings are considered, tournament selection was the one which presented the highest performance numbers out of all the strategies. All the techniques were covered as far as the selection aspect of a GA is concerned. However, an in-depth comparison of all those techniques and a rigorous use of some statistical techniques would have enhanced the research and made it look much more presentable.

4.7 Face Recognition based on Genetic Algorithm

In the research work by [22], the sample space in this research is CAS-PEAL Face Database for gathering sample faces. The controller requires the input of images consisting of faces, images are collected from a database maintained by the Chinese government wherein the tools applied for data collection. Variables used in the study were error samples, number of features, and number of iterations. Statistical techniques such as bar graph depiction of the correspondence between recognition rate and characteristic number; iterations; and erroneous samples. Plots showing the dependency of error sample with iteration number were also used. Figure 6 illustrates the way that GA can be applied to a face recognition system.

Designing a unique GA and SVM for the given problem and experimenting it for the analysis of this research was the problem-solving approach for this study and as far as the conclusions, perspective, and findings are considered, the research concludes that the proposed technique is successful in detecting the faces when test images are fed into it. The use of PCA assisted in shrinking the image features making the computations much faster. This reduction however comes with a price of limiting the accuracy of the system. Very minute details of a face may get ignored in the computations as the pixels get more merged. A solution to this would be

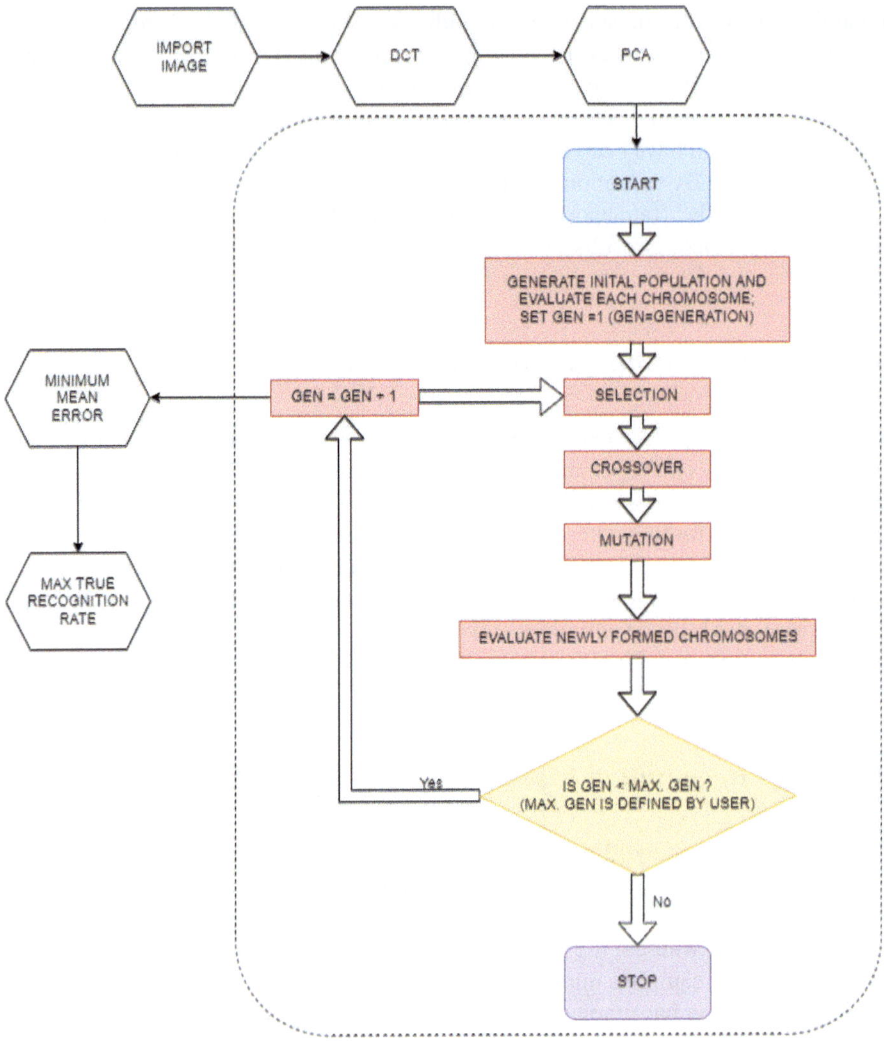

Fig. 6 A genetic algorithm applied on face recognition system

recursively applying the proposed GA to a certain extent which although may reduce the computation time can surely become a much more precise deployment of the idea. Availability of such a huge database provided by the government was visible in the accuracies of the tests performed.

4.8 Genetic Algorithm Performance with Different Selection Strategies in Solving TSP

In research work by [23], the sample space is tested at 8 different TSP instances (0-city, 20-city, 30-city and 40-city, burma14, bay29, dantzig42 and eil51). The tournament size decided for the experiment is two. Since this research is completely theoretical, the data has been gathered manually by the researchers by analyzing other researches. Variables used in the study were distance, number of generations, and deviation. Statistical techniques such as a tabular example of scaled rank with different SP values and results for the best solution for all examples were also used.

Proposing a GA for solving the TSP followed by the selection strategy for reproduction. Conducting experiments to verify the proposal and analysis of results. This was the problem-solving approach for this study and as far as the conclusions, perspective, and findings are considered, researchers conclude that the adoption of rank-based roulette wheel selection method gave significant boosts in the performance of the system with a deviation of 0.9 percent, which is respectable considering the preciseness issues of a genetic algorithm, if not applied to larger generations. Here comes the loophole in implementing GAs for solving the TSPs. Lower generation GAs help in faster calculations which may result in precise outcomes when a TSP is significantly small but if we amplify the TSP, computation speeds may remain the same but the results become much more blunt. Vice versa of this scenario is equally true which makes this approach not the top contender for TSP solving techniques.

4.9 An Improved Localization Algorithm Based on Genetic Algorithm in Wireless Sensor Networks

In research work by [24] that is a localization algorithm for wireless networks, a 100 $\times$ 100 m 2D square area was used for testing, and 100 sensor nodes were deployed. The proposed WSN itself is used for data collection, and MATLAB is used for simulations. Variables used in the study were the number of errors, unknown nodes, and anchor nodes. For variance of communication radius statistical techniques such as plots representing localization errors and average localization errors in different anchor node densities and with change of communication radius were also used.

Proposal of the DV-Hop algorithm followed by stating the WSN localization problem. Designing the GA and an improved DV-Hop algorithm based on GA simulating the whole research and analyzing the results. This was the problem-solving approach for this study and as far as the conclusions, perspective, and findings are considered, the proposed optimization method proved significantly better than the conventional technique when tackling the node localization. This shows how powerful a GA can develop even after fewer generations, which results in the prominent evolution of these algorithms. The location of such nodes was now determined much more accurately and the surprising aspect of the simulation results is that similar performances are achieved even if we ignore the insertion of hardware in the system as far as WSNs are concerned.

4.10 Hybrid Genetic Algorithm for Cloud Computing Applications

In research work by [25] on hybrid GA for cloud computation, the sample space is a grid or population size of 5 and an occupation strategy of 0.25, 0.1 is the execution probability and 10 iterations were performed. For the mutation operator, the population size was 2, and the search radius was 0.2 whereas the mutation rate was 0.05. Variables used in the study were optimizing value, performance difference, and number of failures of two scheduling policies. Statistical techniques such as tabular comparison of the genetic operation between GA and MAGA, bar graph comparison of the performance difference, and various line graph representations of sampling results were also used.

Stating the multiple agent GA followed by the establishment of a load balancing model. Following the algorithmic procedure and simulating the proposed system for better analysis of results. This was the problem-solving approach for this study and as far as the conclusions, perspective, and findings are considered, the research concludes that the proposed hybrid GA shows better performance than the conventional GA as far as higher dimensional OPs are concerned. The CPU utilization of MAGA was also more efficient which directly affects the time consumption in a positive manner. This also ensures that long-run calculations using minimal CPU will hold those performance statistics not making any last-minute compromises. Tweaks in the conventional GA also mean that there are some minimal in the hybrid version of it, where if any corner case falls, then it may result in a lower performance as compared to the conventional GA, but those are rare case scenarios and very unlikely to occur.

4.11 A Genetic Algorithm for Finding a Path Subject to Two Constraints

In [26], GA has been applied to finding a path under two constraints. The sample space is an MCR problem that was chosen to be solved, 15 iterations were performed for the GAs, and simulation studies involved the networks with 20 nodes. MS VC++ is used for all simulation purposes and applied for data collection. Variables used in the study were the number of generations, nodes and requests, success ratio, and computation time. Statistical techniques such as a plot representing convergence property and computation time of each algorithm. Bar graph depiction of the success ratio of each algorithm. A graph representing the dynamic versatility degree of the proposed GA was also used.

Describe the MCR problem followed by the proposal of GA for that problem. Conducting simulations and analyzing the results obtained. This was the problem-solving approach for this study and as far as the conclusions, perspective, and findings are considered, the proposed algorithm was successful in finding paths accurately, which is supported by the simulation results too. As stated by the researchers, the construction of this unique GA is made such that it automatically avoids the illegal chromosomes, enhancing the computation speeds to a much faster level. The mutation operator assists in finding all possible solutions, making this approach much more accurate than other similar problem-solving techniques as this particular scenario is considered to be solved with a unique genetic algorithm.

4.12 Maximizing Influence in a Social Network: Improved Results Using a Genetic Algorithm

The sample space in research work by [27] for maximizing the influence in a social network is two tests were conducted for the analysis of the proposed GA. The theoretical part of data collection in this research is done manually by the experimenters. The simulator used in this research is unspecified wherein the tools applied for data collection. Variables used in the study were influenced number and number of solutions. Statistical techniques such as comparison of influence number with four testing algorithms proposed in the research and bar graph representation of a number of solutions with GA with those four algorithms. A 3D graph with an average influence node number with different parameters was also used.

Proposing the algorithm followed by the conduction of experiments based on them and analysis of results obtained for the research. This was the problem-solving approach for this study and as far as the conclusions, perspective, and findings are considered, the proposed GA showed better results and also improved the efficiency of the system which were respectable outcomes for such scenarios. Simulation results backed these outcomes too. As stated by the researchers, the proposed GA is eminent in providing a diverse set of solutions, this is where a limitation of this approach comes into the picture. A GA can determine a set of results around the point where we want to find the result, but never narrows down to that point. This however makes the GA approach much more safer and reliable than other techniques of finding solutions.

4.13 Feature Selection Based on Hybridization of Genetic Algorithm and Particle Swarm Optimization (PSO)

The work by [28] hybridizes the GA and PSO for feature selection wherein the sample space in this research is two datasets (Indian Pines and Toronto), the first and second scenarios consisted of 10 and 20 inhabitants, respectively. MATLAB was used for the data collection and its computation by the assistance of the datasets used for this research were the tools applied for data collection. Stating the proposed GA, PSO, and HGAPSO + SVM followed by their experimentation and analysis of the results obtained. This was the problem-solving approach for this study and as far as the conclusions, perspective, and findings are considered, one of the major outcomes of this research is that it can be applied to road detection also, which falls perfectly under the study of this state of the art. Along with that, this method was successful in providing even more accurate results in a respectable CPU time. Output feature determination and parameter initialization can also be ignored as it is handled by this technique.

4.14 Modeling of Road-Traffic Noise with the Use of Genetic Algorithm

The traffic noise on the roads has been modeled by [29] using GA. The sample space in this research, which is noise evaluation, was conducted at three stations, three kinds of vehicles (LC, MT, and HT) were used. Mics were located at a height of 3 m on both sides of the test road, 50 km/hr was the reference space for measurements. MATLAB was used for the development of a simulator used in this research and applied for data collection. Variables used in the study were alterations in noise levels of the three stations, measurement hours. Statistical techniques such as tabular representation of noise levels measured at different times at different stations were used. Comparison of results with computed data at all three stations followed by a proposed model's scatter plot was also done.

Proposal of the GAs used in the research followed by the experimentation of them. A mathematical model is then created which is simulated for the analysis of this research. This was the problem-solving approach for this study and as far as the conclusions, perspective, and findings are considered, the model was successful in building a scenario whose topmost priority was to keep the noise pollution levels as low as possible and it showed some respectable results. The idea itself is a unique solution to the noise pollution problem and would be very useful in areas or countries where there is a lot of traffic and traffic light following is not much respected. The models described here to assist with the problem proposed help the calculations to become much simpler which in turn delivers a smoother execution of the system which is to be made to solve this problem. But with the increasing population of electric vehicles, this technique may become redundant in the upcoming future as the electric vehicles are already silent and the only noise they produce is from their horns.

5 Limitations

While this whole concept of applying real-life evolution into the technical aspects of problem-solving sounds quite promising and bulletproof at first glance, sadly there are some shortcomings that are evident when analyzed in depth. In this section, we will focus on such characteristics which might present this technique as not the most favorable to choose for some selected scenarios.

As of now, the genetic algorithms do not show promising results when thrown in scenarios that are not that straightforward. If the number of variables and sample space considered reach some humongous figures, GAs demand some high computational power to tackle them. This often results in either the expensive nature of problem-solving or compromises on the reliability aspects of the solutions presented by the algorithms.

A sub-problem of these types of scenarios or even an independent problem in some cases occurs when sometimes the solutions presented by these algorithms might not have considered the bigger picture when calculated, that is, for example, in a 2D graph, there are always not a single extrema. There are local extremas and one global extrema. The nature of these algorithms might fool them into focusing on one of such local extremes, thereby completely ignoring the global.

Although theoretically there is a proposed way to halt at an optimum solution, it is not a crystal clear approach and always depends on the nature of the problem which when paired with the time limitations and computational power provided to apply the algorithms might not throw the desired results.

These algorithms work perfectly when the data provided at the start is the final set of sample space. Imagine during the run time of an application that uses genetic algorithms, the dataset keeps updating. This would seriously hinder the performance and efficiency of the algorithm and in most of the real-world scenarios, the data is dynamic.

These are some of the limitations that are currently being faced in this field. However, we also have to consider the fact that this field is fairly new and most of these limitations, similar if not exactly the same, are also faced by other approaches, most of which are being practiced for a longer period of time. There are some aspects that are completely unique to genetic algorithms, but at this point, we have to trust the robustness of this approach and hope that we overcome these limitations one by one as time proceeds and this approach gets further developed.

6 Conclusion

The methodology practiced while implementing genetic algorithms holds a promising future. The theory and its several applications discussed in this chapter infer that in most of the scenarios, this method of problem-solving turns out to be the most lucrative one, which is strongly backed by the most prominent evidence witnessed in nature. It has been more than two decades since the first technical application of this algorithm was conducted, which is also a positive, considering its swift success in the software world.

Undeniably, there are limitations that hinder the use case of this approach, but the silver lining to this is that as time proceeds, advancements in the development of GAs assist in the depreciation of these limitations. We have to accept the fact that, at the end of the day, there would not be a single approach or solution that would be accepted for a set of unrelated problems, especially in a bigger picture where the nature of such problems becomes much more diverse. Nonetheless, we can safely conclude that the applications and positive response of this approach are dominated heavily by the fewer limitations it possesses, which often makes the genetic algorithms the finest choice for a particular scenario.

References

1. Gupta N, Khosravy M, Patel N, Dey N, Mahela OP (2020) Mendelian evolutionary theory optimization algorithm. Soft Comput 24:14345–14390
2. Khosravy M, Gupta N, Patel N (2022) Frontiers in nature-inspired industrial optimization. Springer, Berlin
3. Khosravy M, Gupta N, Witkowski O, Pasquali A (2021) Neighborhood base matched morphological filters: cross-fertilization with linear lowpass filtering. In: 2021 international conference on computational science and computational intelligence (CSCI). IEEE, pp 1623–1628
4. Dehghani M, Taghipour M, Sadeghi Gougheri S, Nikoofard A, Gharehpetian GB, Khosravy M (2021) A deep learning-based approach for generation expansion planning considering power plants lifetime. Energies 14(23):8035
5. Khosravy M, Nakamura K, Nitta N, Dey N, Crespo RG, Herrera-Viedma E, Babaguchi N (2022) Social iot approach to cyber defense of a deep-learning-based recognition system in front of media clones generated by model inversion attack. IEEE Trans Syst Man Cybern: Syst 53(5):2694–2704
6. Gupta N, Khosravy M, Patel N, Dey N, Gupta S, Darbari H, Crespo RG (2020) Economic data analytic ai technique on iot edge devices for health monitoring of agriculture machines. Appl Intell 50:3990–4016
7. Gupta N, Khosravy M, Patel N, Dey N, Crespo RG (2021) Lightweight computational intelligence for iot health monitoring of off-road vehicles: Enhanced selection log-scaled mutation ga structured ann. IEEE Trans Ind Inform 18(1):611–619
8. Joshi A, Khosravy M, Gupta N (2021) Machine learning for predictive analysis: proceedings of ICTIS 2020. Springer, Berlin (2021)
9. Jalalzad SH, Yektamoghadam H, Haghighi R, Dehghani M, Nikoofard A, Khosravy M, Senjyu T (2022) A game theory approach using the tlbo algorithm for generation expansion planning by applying carbon curtailment policy. Energies 15(3):1172
10. Senjyu T, Khosravy M (2022) Power system planning and quality control

11. Variengien A, Pontes-Filho S, Glover T, Nichele S (2021) Towards self-organized control: using neural cellular automata to robustly control a cart-pole agent. Innov Mach Intell (IMI) 1:1–14. https://doi.org/10.54854/imi2021.01
12. Takano H, Iwase N, Nakayama N, Asano H (2022) Towards self-organized control: using neural cellular automata to robustly control a cart-pole agent. Innov Mach Intell (IMI) 2:1–11. https://doi.org/10.54854/imi2022.001
13. Khosravy M, Gupta N, Patel N, Senjyu T (2020) Frontier applications of nature inspired computation. Springer, Berlin
14. tutorialspoint: Genetic algorithms—quick guide. https://www.tutorialspoint.com/genetic_algorithms/genetic_algorithms_quick_guide.htm. Accessed 27 July 2023
15. Fu X, Lei L, Yang G, Li B (2018) Multi-objective shape optimization of autonomous underwater glider based on fast elitist non-dominated sorting genetic algorithm. Ocean Eng 157:339–349
16. Jha SK, Eyong EM (2018) An energy optimization in wireless sensor networks by using genetic algorithm. Telecommun Syst 67:113–121
17. KC (1999) Computer network intrusion detection. https://kdd.org/kdd-cup/view/kdd-cup-1999/Intro
18. Hoque MS, Mukit MA, Bikas MAN (2012) An implementation of intrusion detection system using genetic algorithm. arXiv:1204.1336
19. Song Y, Wang F, Chen X (2019) An improved genetic algorithm for numerical function optimization. Appl Intell 49:1880–1902
20. Dao SD, Abhary K, Marian R (2017) An innovative framework for designing genetic algorithm structures. Expert Syst Appl 90:196–208
21. Shukla A, Pandey HM, Mehrotra D (2015) Comparative review of selection techniques in genetic algorithm. In: International conference on futuristic trends on computational analysis and knowledge management (ABLAZE). IEEE, pp 515–519
22. Zhi H, Liu S (2019) Face recognition based on genetic algorithm. J Vis Commun Image Represent 58:495–502
23. Razali NM, Geraghty J et al (2011) Genetic algorithm performance with different selection strategies in solving tsp. In: Proceedings of the world congress on engineering, vol 2. International Association of Engineers Hong Kong, China, pp 1–6
24. Peng B, Li L (2015) An improved localization algorithm based on genetic algorithm in wireless sensor networks. Cogn Neurodynamics 9:249–256
25. Zhu K, Song H, Liu L, Gao J, Cheng G (2011) Hybrid genetic algorithm for cloud computing applications. In: IEEE Asia-Pacific services computing conference. IEEE, pp 182–187
26. Lu T, Zhu J (2013) A genetic algorithm for finding a path subject to two constraints. Appl Soft Comput 13(2):891–898
27. Zhang K, Du H, Feldman MW (2017) Maximizing influence in a social network: improved results using a genetic algorithm. Phys A: Stat Mech Appl 478:20–30
28. Ghamisi P, Benediktsson JA (2014) Feature selection based on hybridization of genetic algorithm and particle swarm optimization. IEEE Geosci Remote Sens Lett 12(2):309–313
29. Rahmani S, Mousavi SM, Kamali MJ (2011) Modeling of road-traffic noise with the use of genetic algorithm. Appl Soft Comput 11(1):1008–1013

Part II
Applications in Logistics, Transportation, and System Identification

Chapter 3
Mating in Genetic Algorithm with Application in System Identification

Md Fahmi Abd Samad and Farah Ayiesya Zainuddin

Acronyms/Abbreviations

ARX	Autoregressive with exogenous variable/input
GA	Genetic algorithm
NARX	Nonlinear autoregressive with exogenous variable/input
OF	Objective function
SI	System identification

1 Background

Numerous researchers have acknowledged the robustness and longevity of genetic algorithm (GA) in tackling various problems, rendering them a potent optimisation mechanism for multiple applications. As a result, GA is often regarded as a powerful and effective tool for optimising solutions. It belongs to the vast group of artificial intelligence (AI) methods, where AI has been playing a significant role in upgrading industrial systems, e.g. in self-organised control [46], maintenance [43], energy management [6, 16, 22, 40], cyber-security [25, 26] and predictive analysis

M. F. A. Samad (✉) · F. A. Zainuddin
Fakulti Teknologi dan Kejuruteraan Mekanikal, Universiti Teknikal Malaysia Melaka, Hang Tuah Jaya, 76100 Durian Tunggal, Melaka, Malaysia
e-mail: mdfahmi@utem.edu.my

M. F. A. Samad
Center for Advanced Computing Technology, Universiti Teknikal Malaysia Melaka, Hang Tuah Jaya, 76100 Durian Tunggal, Melaka, Malaysia

M. Khosravy et al. (eds.), *Frontiers in Genetics Algorithm Theory and Applications*, Springer Tracts in Nature-Inspired Computing,
https://doi.org/10.1007/978-981-99-8107-6_3

[23]. The use of GA can alleviate the computational workload involved in exhaustively testing all potential solution points. GA has been applied successfully in mechanism synthesis [3], motor control [5], image processing, aeronautics, robotics, laser technology and many other real-world challenges [49].

GA commences by randomly initialising a population consisting of a specific number of chromosomes. From this population, in the hope of achieving better solution, certain manipulations are made to the chromosomes. Prior to any manipulation, an assessment of the chromosomes is conducted, and through the selection scheme, certain favourable chromosomes are duplicated into a mating pool, with some potentially being replicated multiple times based on the chosen selection scheme. This is followed by the crossover and mutation processes. Every time a cycle of manipulation is completed, the new chromosomes are evaluated. The procedure persists until a designated number of iterations, referred to as generation. Within the conventional GA, the exploration for a supreme solution amidst the manipulation ensues through three pivotal stages, namely selection, crossover and mutation (Fig. 1) [4, 30, 42].

Throughout GA development, a variety of these genetic operators have been introduced. They have been shown to be able to perform better in various applications than their predecessors. A significant number of researchers improve crossover in order to increase the performance of GA (e.g. [17, 32]). Zainuddin and Samad [49] presents a variety of crossover choices by researchers applied in many engineering usages.

Pavai and Geetha [32] emphasised that crossover is an important operator in GA and plays a key role in the convergence of GA to a solution. By transmitting genes that significantly enhance the likelihood of survival and reproduction to future generations through crossover, the algorithm is able to improve the quality of the population over time and obtain the best possible solution for the given initial population. However, there is always room for improvement in the creation of crossover operators, and

Fig. 1 Traditional GA

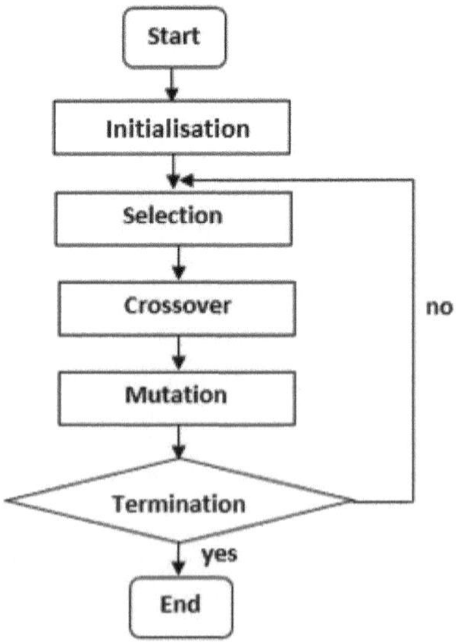

researchers are constantly exploring new approaches to improve the efficiency of GA.

In GA, crossover is an important operator that helps to create new solutions by combining the genetic information of two or more parents. This occurs by the random pick from the mating pool following the selection stage. However, if the selected parents are too similar, the crossover operator may not be effective in generating diverse offspring, thus inhibiting effective evolution. Hence, some other researchers put their attention to mating process to enhance the optimality of GA. A GA incorporated with mating may look as in Fig. 2. These researchers turned into how mating can be better done to allow for a more beneficial crossover process. Mating refers to the process of pairing individuals so that crossover, also known as recombination, can be implemented effectively. Most GA implementations mate their individuals on the basis of first-come-first-served rule, of which the individual that is next selected becomes the individual that is next to be mated. This mating type is also named random mating. With this notion, non-random mating or sometimes referred to as restrictive mating, where a specific rule is applied, should be seen as an area for further development. This chapter explores the mating process as a highly potential area for the development of GA in achieving better performance in optimisations such as in avoiding premature convergence and searching for more accurate system models.

Fig. 2 GA with mating

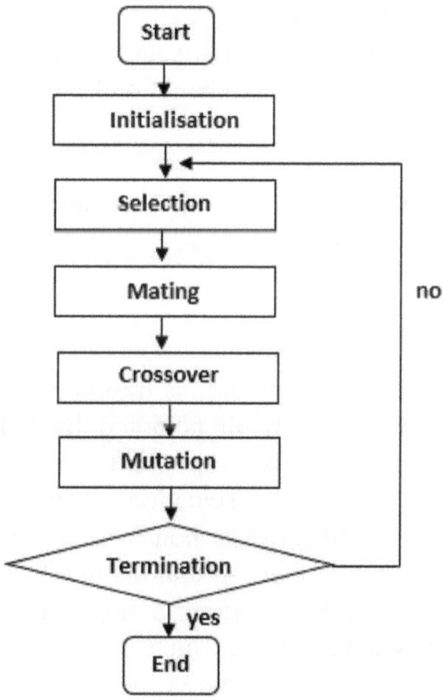

2 Development of Mating as an Operator in GA

According to Rizal and Suyanto [33], a traditional GA incorporates parent selection to enhance its exploration capability and mutation to exploit individuals. The purpose of the parent selection process is to prevent GA from getting stuck too soon at the local minimum. This issue is particularly important to be addressed, especially when the parent's genes do a recombination with alike genes. Both aspects—selection and mutation—are carried out to enhance search capabilities in extensive search spaces. On the other hand, the authors concur more with Fernandes and Rosa [11] that mention recombination (another word for crossover) as more often associated with exploitation while mutation has a crucial impact on exploration as it prevents alleles from dying out. The problem that often occurs with GA is that the algorithm is not able to explore the search space effectively enough to avoid premature convergence while addressing a complicated problem. Premature convergence refers to GA producing a local minimum as a solution, instead of the global one. It is here that the evenness between exploitation (by selective pressure) and exploration (genetic diversity) is being aimed by many researchers in order to make GA avoid the local optimum.

Years back, Hollstien [18] has started to carry out some studies on mating, as cited by Goldberg [15]. Some specific mating preferences were introduced such as random mating, inbreeding, outbreeding and line breeding. Hollstien found out that recurrent inbreeding (mating made between related parents) together with crossbreeding is the most effective method. The combination requires an additional procedure of individual comparison and a comparative evaluation of the effectiveness of one type of mating to the other. Goldberg in his comments on the findings ended with

> … mating tags and templates remain a fertile avenue for future research

On the other hand, in supporting his case for the weakness of not having a proper mating scheme, Huang [21] wrote:

> … a possible deficiency of employing a simple selection scheme that does not permit individuals to actively determine their mates.

Random mating, as traditionally it had been in GA, carries no rule for mating. However, it should be noted that before crossover is performed, these potentials are placed in a mating pool after the selection. The notion of the mating pool may not apply to certain selection schemes as individuals are mated based on what comes next. Restrictive or non-random mating, on the other hand applies a specific rule to these potentials so that a suitable mating is done. There are a variety of strategies for how non-random mating may be applied. One of its classifications is ancestry or kinship-based which deals with the relationship measure between the population members. This brings out terms like 'grandparent', 'parent', 'child', 'sibling' and 'relative' among others. This classification relates to the terms inbreeding and outbreeding or crossbreeding. Another classification is fitness-based which looks into exploiting individuals with high fitness in a fitness-sharing sense. A popular classification is based on assortment or likeness where a similarity measure is made between individuals to see whether they are genotypically or phenotypically the same before a rule of mating is applied.

Delima et al. [7] introduced a new crossover operator called the Inversed Bi-segmented Average Crossover (IBAX). This novel operator was designed to improve the generation of offspring in GA used in numerical variable minimisation and optimisation problems. The researchers suggested a novel mating method for breeding new offsprings utilising the IBAX operator and asserted that it has the potential to result in more effective and refined solutions for variable minimisation. Specifically, it could address the issue of early convergence in GA. Simulation results demonstrated that when combined with a rank-based selection method, the GA using the IBAX operator eliminated 20 variables, which represented 66.66% of the total. Such reduction reduces the dimensionality of the problem, supporting the claim for more efficient GA.

Rizal and Suyanto [33] proposed a new GA called the Human-Like Constrained-Mating Genetic Algorithm (HLCMGA) to address the issue of premature convergence or local optima in complex problems. The HLCMGA incorporates a crossover operation with human-like restricted mating to enhance its exploration capability. It took many analogies of human kinship and ancestry properties and avoided inbreeding. Inbreeding focuses that mating should be done with parents of close relationship whereas outbreeding and crossbreeding look for mates further out of the family ties. According to computer simulations conducted on ten benchmark multimodal functions, HLCMGA outperformed the simple GA. The simulations demonstrated that HLCMGA exhibited greater exploration ability and exhibited superior convergence to the global optimum across multiple search domains.

It is commonly believed that inbreeding reduces genetic diversity within a population, whereas outbreeding increases this diversity. Inbreeding is also known to elevate the frequency of a deleterious allele present in a family [11]. Similarity in genes due to factors like kinship ties has to be avoided as it is one of the factors that causes premature convergence or making solutions trapped in a local minimum [52]. Drezner and Drezner [8] also commented on these mating schemes where they mentioned that more successful reproduction occurs between individuals that are moderately different genetically. They cited on terms such as inbreeding depression and outbreeding depression for researchers to be aware of. Inbreeding depression occurs when individuals that are genetically very similar (e.g. siblings, parents) tend to produce inferior offsprings. This is due to the fact that individuals who share kinship ties, such as parents and their children or siblings, tend to exhibit genetic similarity. The mating reduces biological fitness. On the other hand, outbreeding depression, i.e. breeding between individuals that are very different, may also be disadvantageous.

The terms assortative mating were also used frequently when discussing mating [10, 18, 21]. It is divided into positive assortative (or also referred to as just assortative mating) and negative assortative mating (or also just disassortative mating). Positive assortative mating refers to the process in which the pair is made up of phenotypically similar individuals while negative refers to the opposite. Fernandes and Rosa [10] assert that negative assortative mating, in which unlike individuals are to be mated before crossover, performs better than positive assortative mating. Their work supports that such type of mating leads to better exploration in GA. For complex optimisation problems, variation for exploration in GA is essential. High exploration helps GA to avoid premature convergence [14].

Another study that investigates assortative mating was done by Fernandes et al. [12] who used a statistical measure between individuals to evaluate similarity. Huang [21], on the other hand, in evaluation of the effectiveness of assortative mating, used Hamming distance when evaluating similarity. In Huang [20], he concluded that dissimilar mating (negative assortative mating) is more beneficial in Royal Road optimisation problem. Galán et al. [13] introduced a self-adaptive mating based on similarity or fitness. Their approach was rather a combination between the two aspects to achieve effective mating.

Matsui [29] used the term correlative tournament selection to indicate that their work is an extension of traditional tournament selection. In his method, he added a kind of parent selection method that analyses the correlation between parents in expectation of a more beneficial crossover result. Ting and Büning [44] incorporate the tabu search method when coming up with a mating strategy to allow for a better balance between selective pressure and genetic diversity.

Krauss et al. [28] evolved a neural network with the utilisation of GA where his application includes 'speciation'—another term commonly used when discussing mating preference. Speciation is a process where groups of individuals are made in isolation to allow independent reproduction [21]. Krauss et al. [28] implemented speciation to enable better formation of candidate solutions in a large-scale game problem.

There is a growing awareness that the mating process in GAs has considerable scope for improvement. Enhancements to the mating technique could lead to significant performance gains; however, it is preferable to avoid the requirement to check for parent similarity or introducing additional computational complexity to the process. This calls for an alternative mating technique that can still boost GA performance.

3 Single Parent Mating

Single parent mating was developed as another ambition to enable GA to perform better. The area that was focused was on how the crossover between similar parents may be avoided. The following sections explain the motivation behind its development and how its effectiveness was investigated, particularly in the application area of system identification (SI).

3.1 Motivation

Due to the motivation of having a mating type that allows good exploration yet with minimal need for additional computation of neither similarity nor kinship, this study was done in the proposal of a new type of mating. The study began with an argument for the need to have mating in GA that overcomes having crossover between similar parents. Figure 3 illustrates the possible offsprings using a common crossover, specifically, uniform crossover, onto two presumably random parents from a mating pool after a selection process. It is seen that without any mating strategy, there can only be 4 possible offsprings.

In this study, it is defended that by introducing a suitable mating technique, named single parent mating (SPM), after the selection routine, it is possible to preserve high levels of diversity in the resulting crossover outcomes. A flow chart of GA with the mating technique is shown in Fig. 4.

In SPM, the mating process involves selecting chromosomes or parents from a mating pool that has been constructed using high-quality chromosomes following the selection process. The chromosomes are represented in binary form and are copied and inverted to generate a new set of parents. In the inversion process, all bits that have a value of 1 are transformed to 0, and 0 to 1. Hereby, the mating occurs by pairing the original parent with its inverted version, which is why it is called single parent mating. After the completion of this process, the standard crossover procedure is executed within the pair. A variety of crossover is at our disposal.

Several traditional crossover types are [45, 49]:

1. Single point

The single point crossover involves dividing the parents into fragments and combining them at a single crossover point to produce offspring.

2. Double point

The double point crossover uses two crossover points to exchange chromosome sections between the parents.

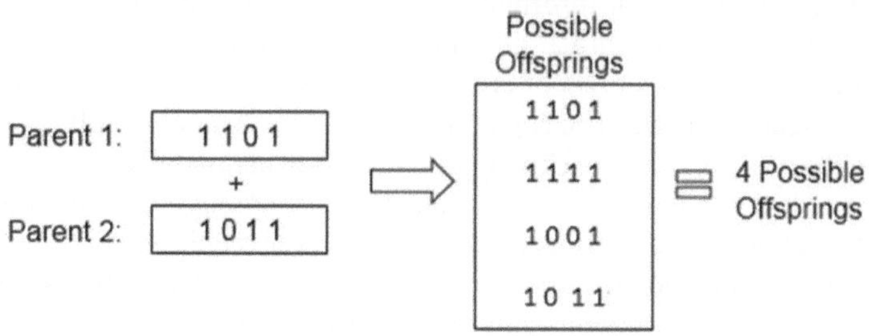

Fig. 3 Possible offsprings without mating

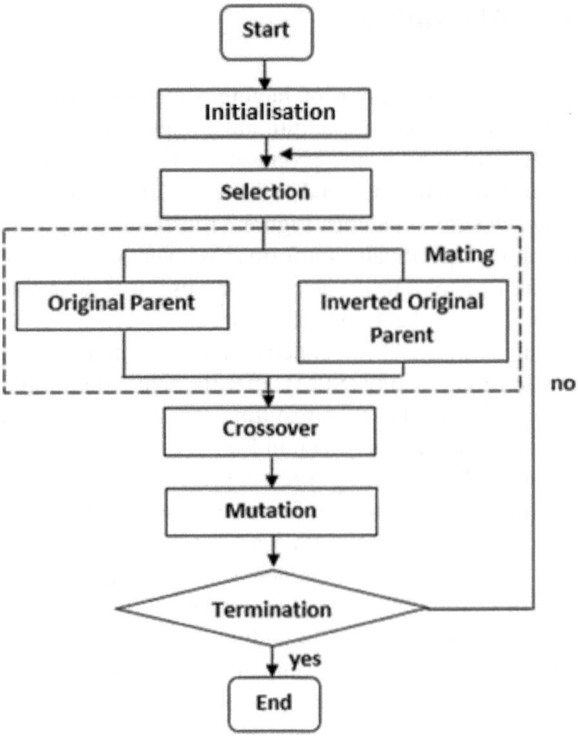

Fig. 4 GA with single parent mating

3. Multiple point

The multiple point crossover selects a random number of crossover points.

4. Uniform

The uniform crossover ensures that the bits of both parents are combined uniformly to create offspring.

The mating emphasises how the pair of chromosomes are made, not how their information are exchanged, as information exchange is carried out through crossover. Figure 5 illustrates that there are 2^4 possible offsprings with SPM, assuming uniform crossover (number and positions of crossover points are random). By incorporating SPM, a larger search space can be explored, and a higher diversity of offsprings can be achieved in the next generation. This approach helps to maintain population diversity, thereby increasing the likelihood of finding better solutions. In terms of computational cost, implementing mating does add to it as it requires additional process. However, since SPM does not require any complex calculation but a few simple processes, i.e. creating inversions and rearranging chromosomes in the population, the additional cost is little.

However, implementing mating to all the individuals in the population is not always advantageous. So, in identifying the suitable percentage of the population in

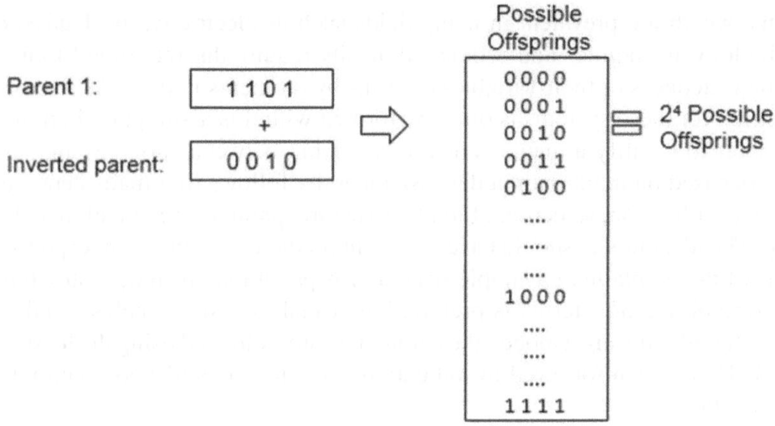

Fig. 5 Possible offsprings with mating

GA to be applied with SPM, the mating was carried out with a specific percentage of the parents in the mating pool. Note that the mating pool is the same size as the population. Several percentages were tested and they were 0, 10, 20, 30, 40, and 50%. Single point crossover was used as it was found to provide the best balance between accuracy and parsimony for the application [50].

As an example, for a 30% SPM, 30% of the parents in the mating pool are replicated and subjected to inversion. This results in another group of parents, which also constitutes 30% of the overall population. Mating was done between the parents and their inverted selves (comprising 60% of the population). The remaining 40% comes from the initial mating pool. With this setting, 0% represents GA without mating.

SPM in GA had been tested where the application area was discrete-time SI. Tests using input–output data of simulated discrete-time models enabled the identified model to be compared directly to the original model. Furthermore, it had also been tested on real data whereas in the simulated data, the identified models were built in the form of linear or nonlinear autoregressive with exogenous input (ARX or NARX) models. The real data came from a flexible robot arm system which has 1024 data points of input (measured reaction torque of the robotic structure) and output (the acceleration of the flexible arm).

3.2 Application of GA with SPM in SI

The process of SI involves establishing a mathematical relationship between variables and terms of a given system based on observed input–output data [9, 24]. The ultimate goal of SI is to enable better control of the system. There are two main types of system modelling: continuous-time and discrete-time modelling. Dynamic systems, which are prevalent in many fields such as electricity, mechanics, chemistry, biology, economics and ecology, typically require discrete-time SI due to the interconnectedness of their variables or terms by moments in time. As a result, data acquisition for these systems is often performed with this assumption in mind.

In order to identify a mathematical relationship between variables and terms of a process based on input–output data, SI generally follows four main steps, sequentially: data collection, selection of model structure, parameter estimation and model testing. The ideal model should have sufficient predictive accuracy in response to the system while maintaining a simple structure. A parsimonious model structure with fewer variables and/or terms is preferred as it makes system analysis and control easier. After identifying a model, the parameters are estimated using the least squares method. This is then followed by validation using methods like correlation tests or error statistics.

Efficiency plays a critical role in the process of identifying systems. Achieving a high level of efficiency in modelling facilitates quick and cost-effective determination of the most appropriate model for a given system. Consequently, metaheuristic approaches, such as evolutionary computation, have been adopted by researchers to enable optimal search for system models. More specifically, as a method known for optimisation and able to reduce much computational burden, GA has been used as a powerful method for SI [9, 34]. Shin et al. [41] successfully employed a real coded GA to identify various time-delay systems. GA was also utilised by Nowaková and Pokorný [31] to identify parameters for second-order systems in the form of difference equations. Additionally, GA was found to be effective in identifying the structure and parameters for various benchmark problems and a liquid level system in neural network form by Zhang and Jili [51]. Samad and Nasir [38] demonstrated that GA can identify systems in the form of discrete-time systems by employing a selection of operator options. Despite the utilisation of GA in SI, its efficiency remains a concern, as evidenced by prolonged processing times and premature convergence. Premature convergence may lead to the selection of a suboptimal solution as the final solution, and this becomes evident when using a simulated (known) model.

The investigation on the effectiveness of SPM in GA had been carried out to see if the mating technique is able to perform well. As a crucial step in SI, optimal model structure selection ensures that an adequate model, i.e. a model with optimal balance between accuracy and parsimony is selected in approximating a system. So, GA with SPM was used for selecting a model structure.

3.3 SI Problem Representation

For discrete-time SI, the chromosomes in GA represented specific model structures where the parameters were calculated using the least squares method. These models were evaluated based on specific optimality measurements (objective functions or OFs) of the models in identifying the system at hand, both accurately and parsimoniously.

In the case of ARX or NARX model structure, by using binary representation, 1 means presence of the variable while 0 means the omission of the variable [36]. As an example, a system that is represented as a NARX model with a degree of nonlinearity = 2, maximum order of input lag = 2, and maximum order of output lag = 2, the maximum possible regressor in the model may be calculated to be 15 [27]. Based on the provided information, a binary model representation with a chromosome length of *lchrom* = 15 can be generated. To search for the optimal model, a search space of size 2^{lchrom}—1 = 32,767 points is available to identify all possible models. The system's output $y(t)$ can be expressed as a linear-in-the-parameter equation using the following formulation:

$$
\begin{aligned}
y(t) &= a_1 + a_2 y(t-1) + a_3 y(t-2) + a_4 u(t-1) + a_5 u(t-2) \\
&+ a_6 y^2(t-1) + a_7 y(t-1)y(t-2) + a_8 y(t-1)u(t-1) + a_9 y(t-1)u(t-2) \\
&+ a_{10} y^2(t-2) + a_{11} y(t-2)u(t-1) + a_{12} y(t-2)u(t-2) + a_{13} u^2(t-1) \\
&+ a_{14} u(t-1)u(t-2) + a_{15} u^2(t-2)
\end{aligned}
\tag{1}
$$

where a_i was the parameter value of regressor i ($1 \leq i \leq$ maximum number of possible regressor) with a_1 designated as a parameter of the system's constant output level.

The chromosome [110 100 001 000 100] based on the model provided in Eq. (1) and specific regressor coding, corresponds to the following model:

$$y(t) = a_1 + a_2 y(t-1) + a_4 u(t-1) + a_9 y(t-1)u(t-2) + a_{13} u^2(t-1) \tag{2}$$

3.4 Evaluation

In order to evaluate the performance of using GA in SI, evaluation of the final models represented by the chromosomes was carried out based on three different OFs. The OFs used were Akaike information criterion (AIC), Bayesian information criterion (BIC), and parameter magnitude–based information criterion 2 (PMIC2). The model that achieves the lowest criterion score receives the highest fitness score, while the one with the highest criterion score receives a fitness score of 0. The formula used for the assignment of fitness value to individual m in each generation was as follows:

$$fitness_m = max(\mathrm{OF}_m) - \mathrm{OF}_m \tag{3}$$

where $max(\mathrm{OF}_m)$ was the highest value of OF calculated from the respective generation and $1 \leq m \leq$ populationsize.

AIC is extensively used for selecting model structures and it is written as [1, 2]:

$$\text{AIC} = N\ln\frac{RSS}{N} + 2k \tag{4}$$

where N represents the number of observations or data points, RSS (residual sum of squares) represents the maximum value of the likelihood function for the assumed model, and k represents the quantity of parameters in the model. The maximum value of the likelihood function for the assumed model (RSS) is defined as follows:

$$\text{RSS} = \sum_{t=N} \varepsilon^2(t) = \sum_{t=N} (y(t) - \hat{y}(t))^2 \tag{5}$$

where $\varepsilon(t)$ is the residual, $y(t)$ and $\hat{y}(t)$ are the correct and predicted output at time t, respectively.

BIC is another popular information criterion. BIC has a stronger penalty than AIC for extra parameters and it is written as [19, 37]:

$$\text{BIC} = N\ln\frac{RSS}{N} + k\ln N \tag{6}$$

An alternative OF utilised to assess the optimality of a model is parameter magnitude-based information criterion 2, expressed as follows [35, 37]:

$$\text{PMIC2} = \sum_{t=N} (y(t) - \hat{y}(t))^2 + \sum_k \frac{1}{\theta_k}; \theta \neq 0 \tag{7}$$

where θ is the value of the parameter.

Each of this information criterion has its unique emphasis on the balance between accuracy and parsimony. The balance between both aspects is considered fundamental in guaranteeing the optimality of a model in modern SI. Their values were later considered as the performance indicator when comparing various outcomes of the study. Low OF indicates high performing approach and vice versa.

3.5 Results and Discussion

Figure 6 shows the average result of one of the evolution found using GA implemented with different percentages of SPM. In this study, an optimal model was pursued using PMIC2 as the OF. The graph is labelled from 0.0 to 0.5 in 0.1 increments, denoting the range from 0 to 50% of mating that occurs through the single point crossover. It can be seen that different percentages of SPM implementation produce different convergence. Here, convergence refers to how GA accepts the

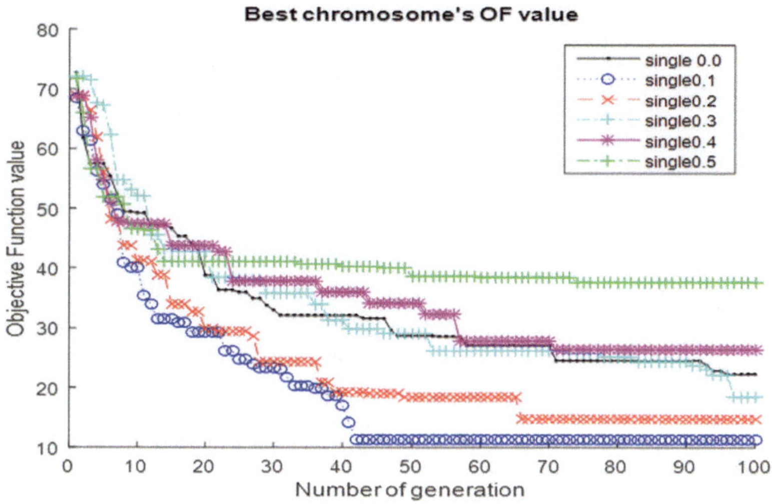

Fig. 6 Average best chromosome's OF value using different SPM percentages

final model in the evolution that carries the lowest OF value in the final generation. This final model is conceived to be the most optimal thus far, based on the set maximum generation. In the figure, using 10% SPM in GA gave the best final model. It is affirmed that in many instances, the final model that carries the lowest OF has a very strong resemblance to the original model and has high accuracy.

Figure 7 shows the overall result for 4 simulated models while Table 1 shows the result of the best mating percentage based on curve fitting of the numerical data from Fig. 7. The curve fitting was based on a polynomial of 2° to the maximum possible degree (5), using intervals of 2.5%. The study concluded that the best SPM percentage is between 10 and 20%, inclusively. First of all, 50% mating is the highest percentage possible since increasing that percentage meant increasing the size of the population. The reason that a high percentage of SPM caused a deficiency in the run was due to the disruptive aspect of the mating. Since the mating requires generating new parents of unexplored fitness, many of these new parents were actually carrying weak genes. The presence of these weak genes did not contribute to the effectiveness of the search and wasted some amount of time, in terms of generation count. Using a high percentage of SPM caused the population to be filled with low-fitness individuals again and again.

A good mating technique should also be applicable to different types of crossover. Therefore, upon finding the suitable percentage, a study had been carried out on SPM in GA by Zainuddin and Samad [50] to investigate the robustness. The effectiveness of using a specific percentage of SPM in GA with various crossover types had been made on a different set of discrete-time simulated models in the form of ARX and NARX models. In compliance with the outcome of an earlier study, the percentage of SPM tested was 15%. The crossover types tested were the traditional ones—single point, double point, multiple point and uniform.

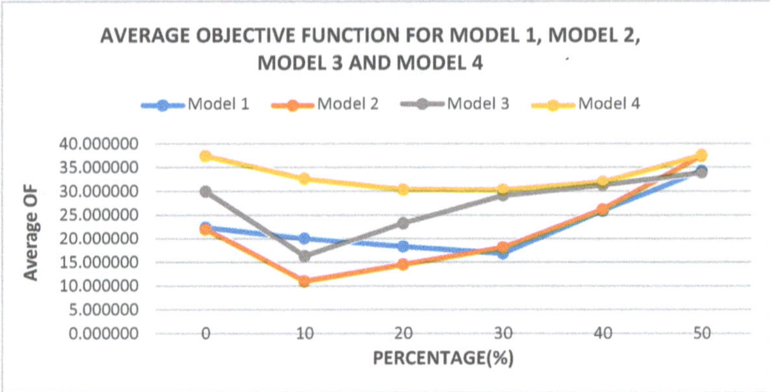

Fig. 7 Average best chromosome's OF value for 4 simulated models across different SPM percentages (the lower the better)

Table 1 Best mating percentage estimated using polynomial curve fitting

	Polynomial 2 (%)	Polynomial 3 (%)	Polynomial 4 (%)	Polynomial 5 (%)
Average A*	20	17.5	12.5	10
Average B[#]	20	20	17.5	17.5

*Average A is based on fitting of average OF value of all simulated models
[#]Average B is based on the average of the best percentage from the fitting of all simulated models, individually

As depicted in Fig. 8, the utilisation of SPM led to lower average OF values for all simulated models when employing single point and double point crossovers, compared to not using SPM. This highlights the effectiveness of the mating procedure in combination with these crossover schemes. Single point and double point crossovers are recognised to strictly maintain their building blocks, owing to the small number of crossover points. By introducing a mating partner with dissimilar structural building blocks, more diverse offsprings can be produced, thereby enabling the exploration of a larger search space.

In contrast, when employing multiple point and uniform crossovers, three out of four cases yielded better results with SPM. Among all 20 scenarios, 16 (80%) were found to perform better with 15% SPM. Zainuddin and Samad [50] noted that the effectiveness of mating was slightly lower compared to single point and double point crossovers. The number of crossover points for multiple point crossover was set randomly, hence in some cases, the number of crossover points was small, such as 1 or 2, resulting in similar behaviour to single point and double point crossovers, allowing the mating to become effective. Likewise, the uniform crossover may demonstrate the same characteristics. In uniform crossover, the building block of an offspring primarily depended on whether adjacent bits were taken together or not. This then constituted to the possibility that only 1 or 2 crossover points were present. In certain

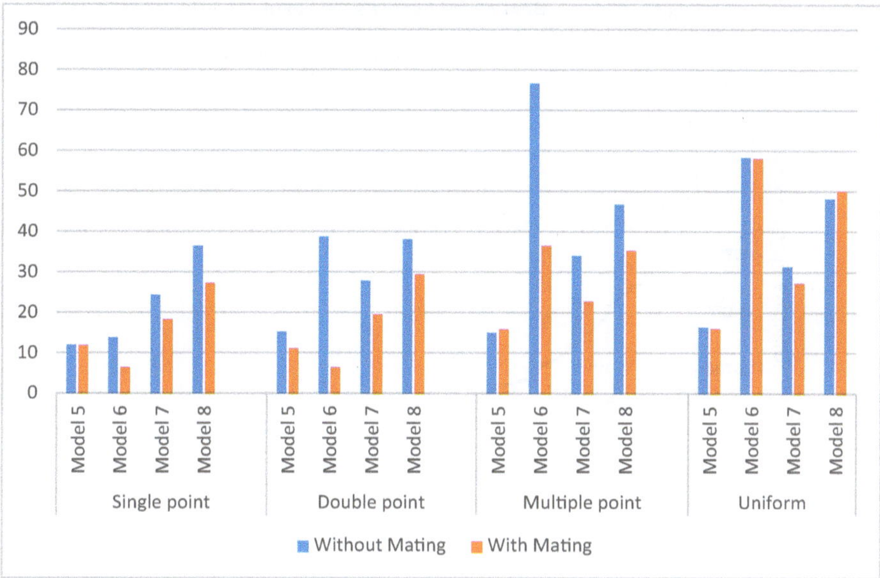

Fig. 8 Average final best chromosome's OF using different crossover (the lower the better)

cases, it behaved like single point or double point crossovers and carried partially the same building blocks as the parents. With proper mating, a larger search space was explored, and a higher variability of offsprings was achieved in the next generation.

An implementation of 15% SPM in GA had also been carried out on real data. The search space for the models was set to include ARX and NARX model choices up until the degree of nonlinearity = 2. In a binary string, 1 represented the presence of a term while 0 represented the omission of a term. There were 44 terms to be selected for model structure and the number of possible models to choose from was $2^{44}—1 > 10^{13}$.

Samad and Zainuddin [39] revealed that when used for real SI, the OF value of 15% SPM was better (lower) than 0% SPM throughout the evolution. Closer investigation revealed that using AIC as OF, GA with 15% SPM managed to find the same, presumably, the most optimal model (a linear model with 8 regressors) in all its runs. Meanwhile, with 0% SPM, the very same model was found in only some of the runs. Some of the other models in 0% SPM gave higher OF which indicated less optimality.

Figure 9 represents a reflection of the population evolution in real SI using AIC. In the graph, 0% SPM technique represents no mating and is labelled '0' in legend while the 15% SPM technique is labelled '0.15'. The graph of 15% SPM started at a high value but then decreased rapidly until the 100th generation. It then settled rather consistently until the 300th generation. The decrement when using 0% SPM was rather slow until the maximum generation. This meant that 15% SPM was better than 0% SPM, noted by achieving a lower OF average of the best chromosomes, and,

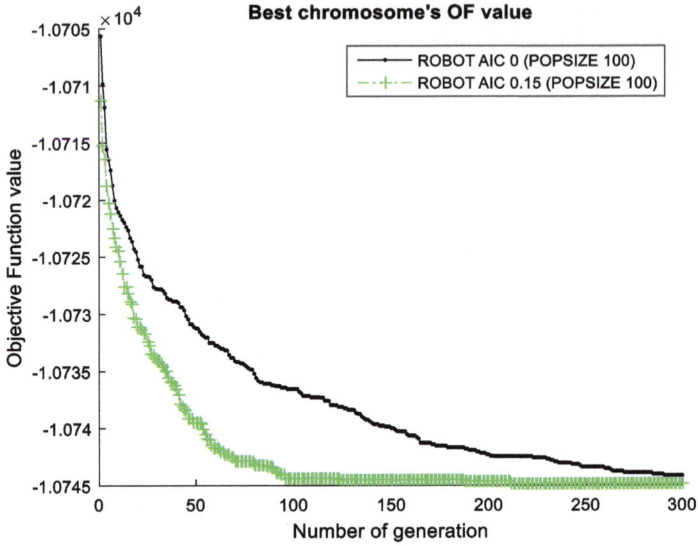

Fig. 9 Best chromosome's OF value for robot arm

additionally, by shorter time. Similar to the observation when using AIC, with BIC, GA with 15% SPM found the same most optimal model (according to BIC)—this time, with only 7 regressors—in all runs whilst 0% SPM found it occasionally. Some of the models in 0% SPM gave higher OF. The selection of the linear models as the optimal ones within the space of nonlinear model choices agrees with Yassin et al. [47]. When using PMIC2, models with lower OF value were also able to be found several times with 15% SPM than the lowest of 0% SPM.

Validation stage of the model with 15% SPM selected using AIC and BIC showed that the models were valid and acceptable [39]. A cross validation where the predicted output of the model from AIC was superimposed to the real output and the error (difference of value between the two) is shown in Fig. 10. The percentage of the highest error in relation to the range of output value was 0.93% while the percentage of average error in relation to the range of output value was 0.27%. With the model selected using BIC, the percentages were 0.96% for the former and also 0.27% for the latter. Furthermore, the mean square error of the models was 2.620×10^{-5} and 2.632×10^{-5} for AIC and BIC models, respectively. These data were comparable to the ones from Zabidi et al. [48] who obtained 2.69×10^{-5} and 2.72×10^{-5} for training data set and testing data set, respectively. It was essential that by incorporating SPM, the models identified for a real data exhibit a balance between accuracy (indicated by a low error) and parsimony (indicated by reasonable number of model terms).

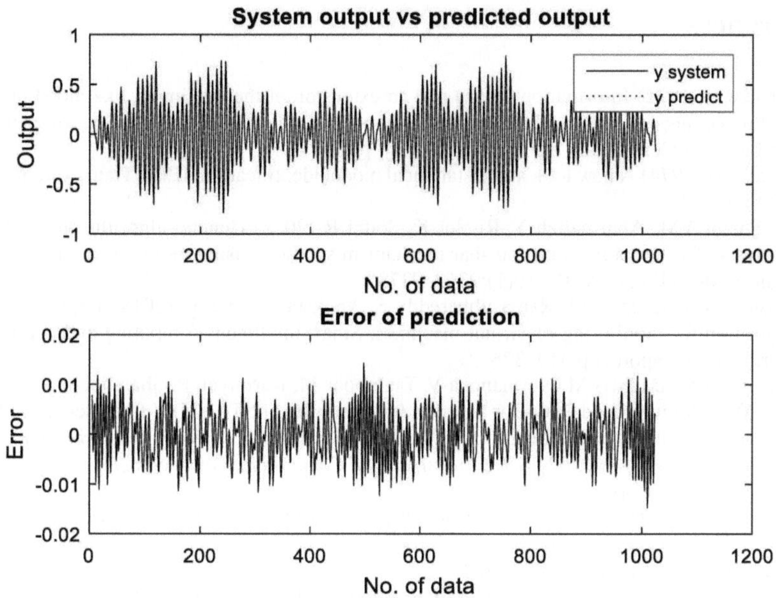

Fig. 10 Superimposition of the predicted output of the model using AIC with 15% SPM onto the real output

4 Conclusion

The development of GA had shown various great improvements. Aside from modifying traditional operators, incorporating local search or hybridising it to other techniques, the mating process is one of the areas that should be looked at further. Researchers have shown the superiority of GA with mating in various cases. Single parent mating avoided the need for complex calculation such as in fitness, similarity or kinship check, yet, had been shown to be able to enhance GA ability for more optimal identification of simulated and real system. It had to be noted that applying too much of the process deteriorates the performance as the diversity of the population makes GA operates like a random walk. A mating technique like SPM may further be investigated and developed so that the outcome reaches the problem's global optimum. Improvements that do not add to further computation complexity are preferable.

References

1. Akaike H (1972) Information theory and an extension of the maximum likelihood principle. In Proceedings of the 2nd International Symposium on Information Theory, Supplement to Problems of Control and Information Theory, pp 267–281
2. Akaike H (1974) A new look at the statistical model identification. IEEE Trans Autom Control 19(6):716–723
3. Al-Smadi YM, Aburmaileh Y, Russel K, Sodhi R (2022) Genetic algorithm for sensitivity analysis of automobile hood four-bar mechanism synthesis using motion generation. Intern J Autom Mech Eng (IJAME) 19(1):9362–9378
4. Basu S, Kannayaram G, Ramasubbareddy S, Venkatasubbaiah C (2019) Improved genetic algorithm for monitoring of virtual machines. Smart intelligent computing and applications. Springer, Singapore, pp 319–326
5. Dehghani M, Bagheri M, Nurmanova V, Taghipour M, Karimyan P, Gharehpetian GB, Abedi M (2020) An improved PSO-GA based fuzzy controller of STATCOM for cascaded induction motors of drilling system. In: *2020 IEEE International Conference on Environment and Electrical Engineering and 2020 IEEE Industrial and Commercial Power Systems Europe (EEEIC/I&CPS Europe)*, pp 1–6. IEEE
6. Dehghani M, Taghipour M, Sadeghi Gougheri S, Nikoofard A, Gharehpetian GB, Khosravy M (2021) A deep learning-based approach for generation expansion planning considering power plants lifetime. Energies 14(23):8035
7. Delima AJP, Sison AM, Medina RP (2019) A modified genetic algorithm with a new crossover mating scheme. Indonesion J Elect Eng Inform (IJEEI) 7(2):165–181
8. Drezner Z, Drezner TD (2020) Biologically inspired parent selection in genetic algorithms. Ann Oper Res 287(1):161–183
9. Duong V, Stubberud AR (2002) System identification by genetic algorithm. Proceedings, IEEE Aerospace Conference, vol 5
10. Fernandes C, Rosa AC (2008b) Self-adjusting the intensity of assortative mating of genetic algorithms. J Soft Comp 12(10):955–979
11. Fernandes CM, Rosa AC (2008a) Evolutionary algorithms with dissortative mating on static and dynamic environments. In: Kosiński W (Ed.) Advances in Evolutionary Algorithms, I-Tech Education and Publishing, Vienna, Austria
12. Fernandes C, Tavares R, Munteanu C, Rosa A (2001) Using assortative mating in genetic algorithms for vector quantization problems. In: *Proceedings of the 2001 ACM Symposium on Applied Computing (ACM SAC'2001)*, Las Vegas, NV, pp 361–365
13. Galán SF, Mengshoel OJ, Pinter R (2013) A novel mating approach for genetic algorithms. Evol Comput 21(2):197–229
14. Ghareb AS, Bakar AA, Hamdan AR (2016) Hybrid feature selection based on enhanced genetic algorithm for text categorization. Expert Syst Appl 49:31–47
15. Goldberg DE (1989) Genetic algorithms in search, optimization and machine learning. Addison-Wesley Publishing Company Inc., Reading, Massachusetts
16. Gupta N, Khosravy M, Patel N, Senjyu T (2018) A Bi-Level evolutionary optimization for coordinated transmission expansion planning. IEEE Access 6:48455–48477
17. Hassanat AB, Alkafaween EA (2017) On enhancing genetic algorithms using new crossovers. Int J Comput Appl Technol 55(3):202–212
18. Hollstien RB (1971) Artificial genetic adaptation in computer control systems, PhD Thesis, University of Michigan
19. Hosseini S, Saberali SM, Sabahi MF (2014) Model order selection based on different information criteria for PDF estimation using maximum entropy method and application in cognitive radio systems, In: *7th International Symposium on Telecommunications*, IEEE, pp 1239–1244
20. Huang CF (2001) An analysis of mate selection in genetic algorithms, technical report CSCS-2001-002. University of Michigan, Center for the Study of Complex Systems
21. Huang CF (2002) A study of mate selection in genetic algorithms, PhD Dissertation, University of Michigan

22. Jalalzad SH, Yektamoghadam H, Haghighi R, Dehghani M, Nikoofard A, Khosravy M, Senjyu T (2022) A game theory approach using the TLBO algorithm for generation expansion planning by applying carbon curtailment policy. Energies 15(3):1172
23. Joshi A, Khosravy M, Gupta N (Eds) (2021) Machine learning for predictive analysis: proceedings of ICTIS 2020, Singapore: Springer
24. Keesman KJ (2011) System identification: an introduction, Springer Science & Business Media
25. Khosravy M, Nakamura K, Hirose Y, Nitta N, Babaguchi N (2022a) Model inversion attack by integration of deep generative models: privacy-sensitive face generation from a face recognition system. IEEE Trans Inf Forensics Secur 17:357–372
26. Khosravy M, Nakamura K, Nitta N, Dey N, Crespo RG, Herrera-Viedma E, Babaguchi N (2022b) Social IoT approach to cyber defense of a deep-learning-based recognition system in front of media clones generated by model inversion attack. IEEE Trans Syst Man Cybern Syst 53(5):2694–2704
27. Korenberg M, Billings SA, Liu YP, McIlroy PJ (1988) Orthogonal parameter estimation algorithm for non-linear stochastic systems. Int J Control 48(1):193–210
28. Krauss R, Merten M, Bockholt M, Drechsler R (2021) ALF: a fitness-based artificial life form for evolving large-scale neural networks. In: Proceedings of the genetic and evolutionary computation conference companion, pp 225–226
29. Matsui K (1999) New selection method to improve the population diversity in genetic algorithms. In: *Proceedings of the 1999 IEEE International Conference on Systems, Man and Cybernetics (IEEE SMC'99)*, Tokyo, Japan, pp 625–630
30. Mirjalili S, Dong AS, Sadiq AS, Faris H (2020) Genetic algorithm: theory, literature review, and application in image reconstruction. In: Mirjalili S, Dong JS, Lewis A (eds) Nature-inspired optimizers: theories, literature reviews and applications, pp 69–85
31. Nowaková J, Pokorný M (2014) System identification using genetic algorithms, Proceedings of the Fifth International Conference on Innovations in Bio-Inspired Computing and Applications IBICA, pp 413–418
32. Pavai G, Geetha TV (2019) New crossover operators using dominance and co-dominance principles for faster convergence of genetic algorithms. Soft Comput 23(11):3661–3686
33. Rizal AC, Suyanto S (2020) Human-like constrained-mating to make genetic algorithm more explorative. In: 2020 8th International Conference on Information and Communication Technology (ICoICT), pp 1–5
34. Samad MFA (2014) Evolutionary computation in system identification: review and recommendations. Intern Rev Autom Cont 7(2):208–216
35. Samad MFA, Nasir ARM (2018b) Implementation of parameter magnitude-based information criterion in identification of a real system. Def Sci Techn Bull 11(1):99–106
36. Samad MFA (2017) Applications of evolutionary computation in system identification. Penerbit UTeM Press.
37. Samad MFA, Nasir ARM (2017) Parameter magnitude-based information criterion in identification of discrete-time dynamic system. J Mech Eng (JMechE) SI4(1):119–128
38. Samad MFA, Nasir ARM (2018a) Comparison of information criterion on identification of discrete-time dynamic system. J Eng Appl Sci 12(Special Issue 1):5660–5665
39. Samad MFA, Zainuddin FA (2022) Single parent mating in genetic algorithm for real robotic system identification. IAES Intern J Artif Intell (IJ-AI) 12(1):201–208
40. Senjyu T, Khosravy M (2022) Power system planning and quality control. Energies 15(14):4995
41. Shin GW, Song YJ, Lee TB, Choi HK (2007) Genetic algorithm for identification of time delay systems from step responses. Int J Control Autom Syst 5(1):79–85
42. Sivanandam SN, Deepa SN (2008) Genetic algorithms, In: Sivanandam SN, Deepa SN (eds) Introduction to genetic algorithms. Springer, Berlin, Heidelberg, pp 15–37
43. Takano H, Iwase N, Nakayama N, Asano H (2021) Decision support in maintenance for electrical components of distribution networks. Innov Mach Intell (IMI) 2:1–11
44. Ting CK, Büning HK (2003) A mating strategy for multi-parent genetic algorithms by integrating tabu search. In: *The 2003 Congress on Evolutionary Computation, CEC'03*, vol 2, pp 1259–1266

45. Umbarkar AJ, Sheth PD (2015) Crossover operators in genetic algorithms: a review. ICTACT J Soft Comp 6(1):1083–1092
46. Variengien A, Pontes-Filho S, Glover TE, Nichele S (2021) Towards self-organized control: using neural cellular automata to robustly control a cart-pole agent. Innov Mach Intell (IMI) 1:1–14
47. Yassin IM, Zabidi A, Amin Megat Ali MS, Md Tahir N, Zainol Abidin H, Rizman ZI (2016) Binary particle swarm optimization structure selection of nonlinear autoregressive moving average with exogenous inputs (NARMAX) model of a flexible robot arm. Intern J Adv Sci Eng Inform Technol 6(5):630-637
48. Zabidi A, Md Tahir N, Yassin IM, Rizman ZI (2017) The performance of binary artificial bee colony (BABC) in structure selection of polynomial NARX and NARMAX models. Intern J Adv Sci Eng Inform Techn 7(2):373–379
49. Zainuddin FA, Samad MFA (2020) A review of crossover methods and problem representation of genetic algorithm in recent engineering applications. Intern J Adv Sci Technol 29(6s):759–769
50. Zainuddin FA, Samad MFA (2021) A mating technique for various crossover in genetic algorithm for optimum system identification. Intern Rev Mechan Eng (I.R.E.M.E.) 15(11):574–581
51. Zhang R, Jili T (2017) A nonlinear fuzzy neural network modeling approach using an improved genetic algorithm. IEEE Trans Industr Electron 65(7):5882–5892
52. Zhao Y, Cai Y, Cheng D (2017) A novel local exploitation scheme for conditionally breeding real-coded genetic algorithm. Multim Tools Applic 76(17):17955–17969

Chapter 4
A Hybrid Biased Random-Key Genetic Algorithm for the Container Relocation Problem

Andresson da Silva Firmino and Valéria Cesário Times

1 Introduction

In recent years, the need to increase terminal efficiency and reduce operational costs has become more pressing as container traffic has increased. According to the World Bank, world container port traffic has grown by 230% since 2000, with 758.7 million TEUs (20-foot Equivalent Units) handled in 2020 [1]. Furthermore, the United Nations Conference on Trade and Development (UNCTAD) predicts an annual average growth rate of 2.4% in world maritime trade for the 2022–2026 period [2]. The COVID-19 pandemic has further intensified the cost pressures on maritime trade and container terminals. Given the significant impacts of the pandemic on the global supply chain, reducing the operational costs of container terminal systems has become even more crucial.

One of the primary measures of port competitiveness and efficiency is the time spent on operations [3]. This time can be evaluated by examining container dwell time and ship turnaround time at container terminals. The former represents the average duration for a container to remain at the terminal, while the latter is the average time required to load and unload a docked ship. Both of these measures are heavily influenced by the container-stacking process, making optimization of this process a critical research area for container terminal process improvement.

Different modes of transportation, such as land or maritime transport, at container terminals are often unsynchronized. As a result, containers are temporarily stored in a container yard after arriving at the terminal until they are needed. Due to limited storage space, the containers are stacked vertically in tiers, creating blocks of stacks

A. da Silva Firmino (✉) · V. Cesário Times
Center for Informatics, Federal University of Pernambuco, Recife, PE, Brazil
e-mail: asf2@cin.ufpe.br

V. Cesário Times
e-mail: vct@cin.ufpe.br

M. Khosravy et al. (eds.), *Frontiers in Genetics Algorithm Theory and Applications*, Springer Tracts in Nature-Inspired Computing,
https://doi.org/10.1007/978-981-99-8107-6_4

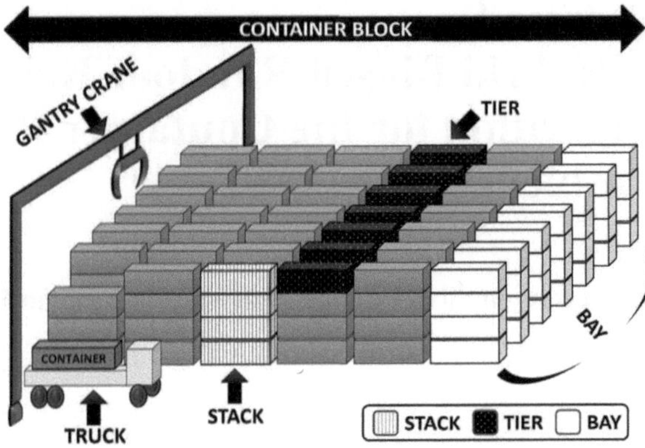

Fig. 1 A single container block divided into tiers, stacks, and bays

delimited by terminal traffic lanes, as shown in Fig. 1. Each block consists of a set of parallel bays, and each bay contains an array of stacks. The number of tiers in a bay is limited by the maximum stack height. Rail-mounted gantry cranes (RMGCs) typically handle containers within a block, accessing the containers from the sides of the bay and moving them between stacks within the same bay. This is because, for various reasons, including safety concerns, RMGCs are generally not allowed to travel across block bays while carrying a container [4]. The movement of containers within blocks is critical for the efficient operation of container terminals.

Regrettably, the target container, which has the highest priority, is not always located at the top of its stack. The containers are generally stacked on a first-come, first-served basis because the removal priority of a new container in the container yard is not always known in advance. Therefore, retrieving the target container when it is not at the top of its stack requires relocating containers above it to other stacks in the bay. These additional operations result in extra costs and time waste, which consequently diminishes the productivity of container-stacking operations and the operational efficiency of container terminals. The Container Relocation Problem (CRP) refers to finding an optimal sequence of operations for the crane to retrieve all the containers from the bay based on their retrieval priorities. The optimal sequence of operations is determined by minimizing the crane's operating time, thereby maximizing the containers' removal efficiency.

As previously noted, improving stacking operation procedures is a critical area of study for container terminal processes. Furthermore, given the impact of COVID-19 on the global supply chain, minimizing operation time in container terminals has become even more essential, particularly with the cost pressures on maritime trade. In this chapter, we address the CRP to optimize the crane's operation time, which can help to decrease operational costs and enhance operational efficiency in container terminals. To this end, we propose a hybrid *Biased Random-Key Genetic Algorithm* (BRKGA) to solve the CRP, as biased random-key genetic algorithms have garnered significant attention from researchers studying complex optimization issues [5].

The algorithm presented in this paper is based on the BRKGA metaheuristic, which was first introduced in [6]. BRKGA is a type of genetic algorithm that offers several advantages over traditional genetic algorithms, including faster prototyping and rapid convergence to high-quality solutions [5]. The hybrid BRKGA proposed in this study combines the constructive exploration abilities of an intelligent metaheuristic to generate good initial solutions with the exploitation abilities of a biased random-key genetic algorithm. This hybrid approach allows for the initialization of the population with promising solutions, leading to a more efficient search for optimal solutions using the BRKGA method.

The following is the structure of the remainder of this chapter. Section 2 presents an overview of the CRP and introduces its fundamental terminology. Section 3 explores prior research and discusses the primary approaches to the CRP, establishing connections between them and the proposed method. In Sect. 4, the hybrid BRKGA method for resolving the CRP is outlined. Subsequently, Sect. 5 presents the computational experiments carried out. Finally, Sect. 6 summarizes the conclusions reached and offers suggestions for future research.

2 The CRP Definition

The CRP is a crucial challenge for container terminal systems worldwide, particularly given the heightened time and cost constraints for port stacking operations. The problem directly impacts the time spent on crane operations, and it involves finding a sequence of operations that enables the crane to retrieve all the containers from the bay according to their priorities while minimizing operating time. This is a combinatorial optimization problem that falls under the NP-hard optimization problem class. In the literature, it is also referred to as the container retrieval problem or the block problem.

The CRP takes as input the initial configuration of a bay, which consists of H tiers and W stacks defining the maximum height and width of the bay, respectively. N uniformly shaped containers are initially located in the bay and occupy specific slots. A slot is defined by its coordinates (i, j), where $i \in 1, \ldots, W$ and $j \in 1, \ldots, H$. The problem's solution is the sequence of operations performed by the crane, which includes retrieval and relocation. These operations are time-consuming and subject to specific constraints.

- The relocation operation carries the container to stack k and tier l, such that $k \leq W$ and $l \leq H$, i.e., complying with the maximum width and height.
- The relocation operation always moves the container from the top of one stack to the top of a different stack, and when a stack is empty, the top is the ground (i.e., tier 1).
- The retrieval operation extracts the container with the highest priority out of the bay when this container is located at the top of a stack.
- When the crane performs any operation, the bay is changed, producing a new bay configuration. Consequently, the crane must recognize this new configuration to conduct the next operation.
- The container with priority N is only retrieved in the crane's last operation.

In [7], a mathematical model was introduced for the CRP with the objective of minimizing the operating time of the crane. When solving the CRP, if a retrieval operation cannot be executed for a given bay configuration, the crane must choose a relocation operation from a list of eligible relocations that satisfy the constraints mentioned earlier. The set of all eligible relocations for a given bay configuration is denoted as the *LCR* notation in this chapter.

There are two versions of the CRP based on the relocation strategy: the *restricted* CRP and the *unrestricted* CRP. In the restricted version, the crane can only relocate containers positioned above the container with the highest priority, while this limitation is not present in the unrestricted version. This restriction significantly reduces the problem's search space and simplifies the development of optimization techniques. For instance, Figure 2 illustrates a restricted CRP solution with four restricted relocations obtained from an instance comprising eight containers, four stacks, and three tiers.

In this example, for the sake of simplicity, the container bay is represented as a two-dimensional storage structure, where the black-bordered boxes indicate the bay slots. The white boxes with numbers indicate the slots occupied by containers (numbered according to their priorities), while the others represent empty slots. The presented solution consists of 12 operations, denoted from A to L, including eight retrievals and four relocations of containers 8, 4, 6, and 8 (again). The relocations

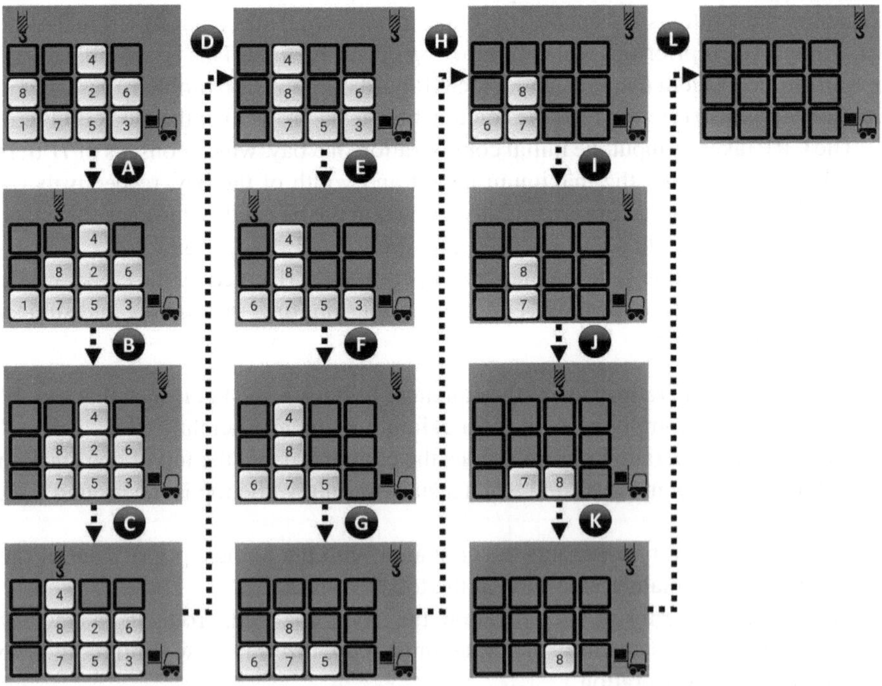

Fig. 2 An illustration of a solution for the container relocation problem

are, respectively, represented by (A), (C), (E), and (J). These relocations enable all containers to be retrieved at the upper right corner of the bay. It is noteworthy that this solution corresponds to a restricted CRP instance with four restricted relocations.

This research investigates both the unrestricted and restricted versions of the CRP, and the proposed algorithms are suitable for solving both versions. To achieve this, the LCR, which is utilized by all algorithms, should consist of either restricted or unrestricted relocations exclusively, depending on the CRP version being considered. For example, for the bay configuration presented in Fig. 2, the LCR list will comprise nine eligible relocations for the unrestricted version: move container 8 to the second or fourth stack (i.e., relocations 1 and 2); move container 7 to the first or fourth stack (i.e., relocations 3 and 4); move container 4 to the first or second stack (i.e., relocations 5 and 6); move container 4 to the fourth stack (i.e., relocation 7); and move container 6 to the first or second stack (i.e., relocations 8 and 9). On the other hand, the LCR for the restricted version will consist of only two eligible relocations: relocations (1) and (2).

The crane's operations are known to be time-consuming. The time taken by the crane to empty the bay is referred to as *the crane's operating time* [8, 9]. In solving the CRP, this work employs the computation of the crane's operating time, which is formally defined in [8]. To the best of our knowledge, this is the most accurate time computation method applied to the CRP. This method accounts for the crane movement (either horizontal or vertical) as a uniformly accelerated rectilinear movement; it considers the twistlock time and includes several parameters of storage distance to calculate the time taken for travel.

3 Related Work

The *Container Relocation Problem* (CRP) is an NP-hard problem [10] that was first investigated by [11]. Since its inception, numerous techniques have been proposed to tackle the CRP. Traditionally, minimizing the number of relocation operations has been the primary optimization objective for the CRP. Optimal methods that provide the exact solution can be found in [12–19], while near-optimal solutions are presented in [20–28] to handle larger problem instances. While optimal methods guarantee the exact solution, the NP-hardness of the CRP significantly limits the size of the problem instances that these methods can address compared to near-optimal techniques that aim to find near-optimal solutions.

One of the potential reasons for the widespread focus on minimizing the number of relocations in CRP research is the belief that reducing the number of relocations leads to lower operational costs for retrieving containers from the bay. However, it is worth noting that minimizing relocations may not always result in the optimal solution for minimizing operating costs, as demonstrated in [29]. Although relocation operations can be time-consuming and should be minimized, it is not necessarily the case that minimizing the number of relocations leads to the minimum crane operating time, as has been observed in previous works [4, 7, 30].

Taking into account the aforementioned information, some disruptive research has explored alternative optimization objectives that deviate from minimizing the number of relocations and instead aim to achieve the most precise computation of operational costs. For instance, in [29, 31, 32], the CRP's objective function is, respectively, to minimize operating costs, the crane's trajectory, and fuel consumption, while [4, 7–9, 25, 30, 33–36] address the crane's operating time. Notably, in response to the divergent approaches used in the CRP literature to compute the crane's operating time, [8] proposes an improved method to accurately determine the crane's operating time. To the best of our knowledge, this definition is more accurate and comprehensive than previous CRP calculations since it considers the (horizontal or vertical) crane movement as a uniformly accelerated rectilinear movement (u.a.r.m), incorporates the twistlock time, and accounts for various storage distance parameters to calculate the traveled time. Therefore, the present study adopts the exact computation of the crane's operating time specified in [8].

As previously mentioned in Sect. 2, the CRP literature explores two versions of the problem based on the relocation type: the *restricted* CRP and the *unrestricted* CRP. The literature predominantly focuses on the *restricted* CRP, possibly due to the unrestricted version's higher complexity compared to the restricted version. The restricted CRP only allows containers placed above the highest priority container to be relocated, whereas the unrestricted CRP has no such restrictions, leading to a more complex search space and making optimization methods more challenging. The unrestricted CRP is more prone to fall into local optimal solutions [21], whereas the restricted CRP simplifies the search space and optimization method design. While most optimization studies concentrate on the restricted CRP, the unrestricted CRP presents more optimization opportunities than the restricted version [28], leading to better solutions. In the unrestricted CRP, the crane can perform promising relocations, moving a container that requires relocation in a way that never necessitates relocation again [8]. Therefore, the present study deals with both the restricted and unrestricted versions of the CRP.

The literature discusses various works that aim to minimize the crane's operating time. In [7], a Reactive GRASP algorithm is proposed for the restricted CRP, while [9] presents the Triad heuristic, which outperforms the heuristics suggested in [4, 35]. Moreover, PILOT and ACO algorithms have been developed based on the Triad heuristic, with better solutions found than TRIAD within short runtimes. da Silva Firmino and Times [8] introduce the Coronavirus Optimization Algorithm (CVOA), which simulates the coronavirus's spread in solving the CRP. CVOA is the first application of a bioinspired optimization algorithm to an NP-hard problem, and it has been shown to outperform PILOT and ACO algorithms when used alone. However, it is important to note that the experiments were only conducted in an unrestricted CRP scenario, even though the algorithms can be applied to both the restricted and unrestricted versions.

Bioinspired optimization methods have attracted significant attention in the field of optimization [37]. Additionally, genetic algorithms have proven to be effective in various industrial applications [38–40]. However, few studies have explored the potential of bioinspired methods in addressing the CRP [8, 9, 20, 22, 25]. As mentioned previously, [8] proposed the CVOA approach, while ant colony optimization was used in [9, 25]. The studies in [20, 22] employed genetic algorithms to solve the restricted CRP. Biased random-key genetic algorithms (BRKGAs) have been shown to be effective in solving complex optimization problems [5], and there has been a growing trend toward hybrid metaheuristics that combine different methods to achieve better performance. In the case of BRKGAs, these hybridizations often involve the inclusion of metaheuristics as local search procedures [41]. For example, [5] used path-relinking, [42] adopted a variable neighborhood search, and local search heuristics were utilized in [43].

This study stands out from prior research due to several factors. Primarily, no prior study has been employed BRKGA to tackle the CRP, to our knowledge. Additionally, unlike most research, this study aims to optimize the crane's operating time in both the restricted and unrestricted CRPs. Further, our approach to hybridizing BRKGA differs from previous methods, which often involve using metaheuristics as local search techniques. Instead, we combine the exploratory qualities of an intelligent, constructive metaheuristic to generate promising initial solutions with the exploitative capabilities of a biased random-key genetic metaheuristic. As a result, the hybrid BRKGA strategy can initialize the population with more promising solutions, leading to a more efficient search and better BRKGA solutions. We provide further details about the implementation of the hybrid BRKGA approach in the next section.

4 The Hybrid Biased Random-Key Genetic Algorithm

The Biased Random-Key Genetic Algorithm (BRKGA) [6] is a metaheuristic that encodes solutions as vectors with n elements of real numbers between zero and one (i.e., the interval [0, 1]). Each vector is called a *chromosome*, and each component c_l is called a *gene*, with its value being the *allele*. The BRKGA maintains a population of such vectors, and a *fitness function* is used to assess each individual chromosome's fitness. This function creates a solution from the chromosome and computes its value. The resulting values indicate the fitness of each individual in the population.

The BRKGA metaheuristic follows an iterative population-based process, which begins by creating an initial population that is typically randomized. The second step involves repeating the BRKGA's evolution process until a specified stopping criterion is met, with the goal of improving the population over successive generations. During this step, four tasks are performed. First, the population P is divided into two groups: the *elite* set (P_e) and the *non-elite* set ($P \setminus P_e$), based on the individuals' fitness. The P_e set consists of the best solutions for the entire population, while the non-elite set contains the remaining individuals, which are usually the worst elements of the current population.

The second step of the BRKGA metaheuristic involves three additional tasks to evolve the population and generate the next population P'. Firstly, all the individuals in the elite set are directly copied from P to P'. Secondly, two parent individuals are randomly selected from population P, with one parent coming from the elite set and

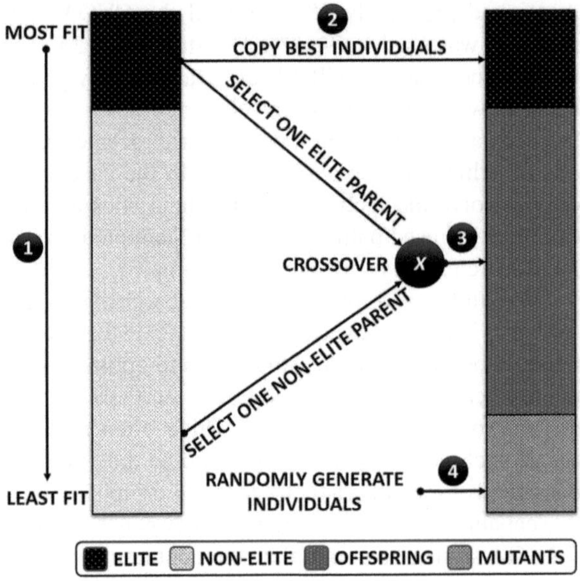

Fig. 3 Overview of BRKGA's evolution process. Task (1) sorts the current population and partitions the population into elite and non-elite sets. Task (2) copies the elite individuals to the next population. Task (3) fills up the next population with offspring. Task (4) populates the next population with mutant individuals

the other from the non-elite set. The resulting offsprings are generated using a biased uniform crossover, where the gene is selected from the elite parent with a probability of ρ, and from the non-elite parent otherwise. Thirdly, a small percentage of random individuals (called mutants, P'_m) are added to generate the remaining individuals in population P'. Figure 3 depicts all the tasks involved in the population evolution process between consecutive generations of a BRKGA.

In order to enhance the robustness of the standard BRKGA's iterative process, additional steps can be incorporated, such as a local search procedure and/or reset procedure [41]. The reset procedure is typically used to prevent the BRKGA process from getting trapped in local optima, where the process cannot improve the best solution after many generations. However, this approach may be too extreme as it eliminates well-formed parts of genomes that have been found in promising solutions during evolution. An alternative approach to avoid this problem is the multi-population strategy proposed in [44]. In this strategy, a number of independent populations, denoted by I_p, are evolved simultaneously over generations. After a predetermined number of generations, denoted by I_g, the populations exchange information by swapping their best I_i individuals. This strategy can help prevent local optima and mitigate the issue of diversity loss during the evolutionary process by improving the variability of individuals [5, 45]. More information about the implementation of this strategy can be found in the following section.

The three distinct phases stated earlier, which are the population initialization, evolution, and exchange of elite populations, serve as the fundamental elements for our BRKGA technique. However, in order to surpass the effectiveness of this conventional approach, our chapter introduces a hybrid BRKGA method. The hybridization is introduced during the initial population setup step. In this regard, a constructive metaheuristic that is intelligent is incorporated into the BRKGA to generate optimal starting solutions and speed up the search for improved solutions in the subsequent evolution step. The forthcoming sections will elaborate on the particulars of the proposed hybrid BRKGA for the CRP.

4.1 Implementation Details

For providing details on the implementation of the proposed HBRKGA, initially, an explanation of the association between a BRKGA individual and a CRP solution is necessary since a BRKGA individual represents a CRP solution. The sequence of crane operations needed for building a CRP solution defines an individual's genetic code (i.e., a chromosome), such that each chromosome gene corresponds sequentially to each relocation operation of a CRP solution. Consequently, the crane's operating time in the solution denotes the individual's fitness. This association between an individual's genetic code and a CRP solution is proposed in [8] for the CVOA approach previously mentioned. In our BRKGA approach, this association is extended so that each gene allele is a real number, indicating the relocation operation's position from within the LCR. For example, the chromosome vector $[0; \frac{1}{3}; 0; \frac{1}{3}]$ can be seen as a restricted CRP solution for the instance illustrated in Fig. 2.

In the presented figure, Fig. 2, there are a total of eight retrievals and four relocations in the solution. These relocations, denoted by (A), (C), (E), and (J), are selected after examining the current bay configuration to obtain an LCR. Each eligible relocation in the LCR is indexed by a position n, representing its position within the LCR. The indexation of relocations in the LCR is illustrated in Fig. 4, where $RL(i, k)$ denotes the relocation of the container at the top of stack i to the top of stack k. The figure shows the bay configuration before each relocation (i.e., A, C, E, and J) and the resulting LCR. For example, relocation A has two possible options in the LCR: moving container eight from the first stack to the second stack (i.e., $RL(1, 2)$) and moving container eight from the first stack to the fourth stack (i.e., $RL(1, 4)$). These relocations are, respectively, located at positions 0 and 1 within the LCR.

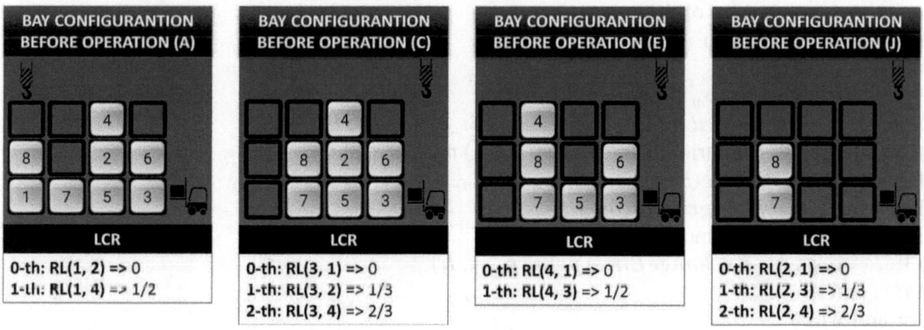

Fig. 4 Numeric representation of relocation operations into the LCR

It is important to note that every relocation in the LCR has an associated real number allele. In this case, $RL(1,2)$ and $RL(1,4)$ are linked to the values 0 and $\frac{1}{2}$, respectively, which indicate the position of the relocation within the LCR. This position is determined by multiplying the allele value by the number of elements in the LCR. For example, position 1 of $RL(1,4)$ is obtained by multiplying the allele value ($\frac{1}{2}$) by the number of elements in the LCR (2). Also, note that the relocations A, C, E, and J are represented by the alleles 0, $\frac{1}{3}$, 0, $\frac{1}{3}$, respectively, which correspond to $RL(1,2)$, $RL(3,2)$, $RL(4,1)$, and $RL(2,3)$. Therefore, the chromosome vector $[0; \frac{1}{3}; 0; \frac{1}{3}]$ is a restricted CRP solution for the instance depicted in Fig. 2. Further information on this decoding process is provided in the next section.

This innovative genetic coding approach, associating a gene allele to the relocation operation's position from within the LCR, indeed presents advantages over existing genetic coding approaches for the CRP. As you have mentioned, the proposed approach solves the issue of invalid genes that may appear within chromosomes in previous genetic coding approaches. By pointing to LCR elements, all genes in the proposed approach are valid, making the encoding process more efficient and accurate. Additionally, the proposed approach is not limited only to the restricted CRP version, as the LCR can be employed to solve both restricted and unrestricted versions of the CRP. This expands the potential applications of the proposed approach, making it a more versatile solution for various real-world scenarios.

Algorithm 1 Pseudocode of the HBRKGA for solving the CRP

Require: bay: initial bay configuration, I_p: the number of independent populations
Require: I_g: the number of generations for exchange
Require: I_i: the number of best individuals in the exchange
Require: ρ: the probability of an offspring inheriting the allele of its elite parent
Require: α: the adjustment factor for chromosome size
Require: $p_e\%$: the percentage to define the elite set size
Require: $p_m\%$: the percentage to define the mutant set size
Require: $limit_{brkga}$: HBRKGA runtime limit
Ensure: best solution found

$generation \leftarrow 0$
$init\_sols \leftarrow GetInitialSolutions(bay, limit_{brkga})$
$num\_rel \leftarrow$ get the number of relocations of the best solution in $init\_sols$.
$size \leftarrow$ get the size of the $init\_sols$.
$n \leftarrow (num\_rel * \alpha)$
$p \leftarrow (size/I_p) + 1$
$p_e \leftarrow (p * p_e\%)$; $p_m \leftarrow (p * p_m\%)$
$pops \leftarrow GenInitialPopulations(bay, init\_sols, n, p, I_p)$
while stopping criteria (including $limit_{brkga}$) not satisfied **do**
  $pops \leftarrow Evolve(bay, pops, n, p, p_e, p_m, \rho)$
  $generation \leftarrow generation + 1$
  if ($generation$ mod I_g) $= 0$ **then**
    $pops \leftarrow ExchangeElite(pops, p, p_e, I_i)$
  end if
end while
$sol_{best} \leftarrow$ get the solution associated with the best individual in $pops$
return sol_{best}

The HBRKGA for the CRP is outlined in Algorithm 1. Initially, the algorithm initializes the number of generations ($generation$), chromosome size (n), population size (p), elite set size (p_e), and mutant set size (p_m) (lines 1-7). The values for the last four variables are computed using $init\_sols$, which is described in more detail in Sect. 4.3.

The initial solutions set ($init\_sols$) is fundamental for estimating more accurately the chromosome length that is needed (i.e., n). The best solution in $init\_sols$ indicates the number of relocations necessary to solve an instance of the problem (line 3). This number (i.e., num_rel) is an excellent reference to estimate the chromosome size since the number of genes in a chromosome corresponds to the number of relocations in a CRP solution. However, there may be a solution that is better than the best solution of $init\_sols$, and this better solution has a number of relocations higher than num_rel. Thus, in these cases, the algorithm applies an adjustment factor (α) in a way that solutions with a higher number of relocations can be produced (line 5).

Similarly, the population size p is also estimated using $init\_sols$: specifically, the length of $init\_sols$ (i.e., $size$) is used. This estimated value results from the distribution of all solutions of $init\_sols$ between the independent populations (quantified by I_p), i.e., $(size/I_p)$. Also, to ensure more diversity in individuals, the value one is added to the estimated value such that at least one population will have random individuals (i.e., $(size/I_p) + 1$). Once the population size p is defined, p_e and p_m are computed as percentage values of p (line 7). Next, the initial populations ($pops$) are generated utilizing $init\_sols$ (line 8).

The HBRKGA iterations are performed until some stopping criterion is met (lines 9–15). One of these criteria is the runtime limit ($limit_{brkga}$). The evolution procedure (i.e., $Evolve$) is conducted at the beginning of each iteration to improve the independent populations $pops$ during successive generations. At each HBRKGA iteration, the number of generations is incremented (line 11). Next, the algorithm checks whether it is time to perform the exchange of populations (line 12). If it is, the algorithm conducts the procedure $ExchangeElite$ to swap the best I_i individuals between the populations (line 13). Finally, the algorithm returns the best solution found through the best individual discovered among the populations (lines 16–17).

The decoding process is the most crucial part of the BRKGA approach since it maps (translates) a chromosome vector to a valid solution in the problem's solution space. This process is necessary for the execution of the step of initializing, evolving, and elite-exchanging of populations. In addition to the decoding process, the proposed hybrid BRKGA approach requires an encoding process. In our approach, the initial populations are not fully randomly generated, but they are created through existing CRP solutions. Thus, the encoding process is essential to map CRP solutions for chromosome vectors. Further details about these decoding and encoding processes are given in the next section. After that, in the subsequent sections, the procedures $GetInitialSolutions$, $GenInitialPopulations$, $Evolve$, and $ExchangeElite$ are detailed.

4.2 Encoding and Decoding

The encoding process is employed in the proposed HBRKGA to encode a CRP solution into a chromosome vector. Algorithm 2 details the proposed procedure for encoding a CRP solution. All solution operations (i.e., relocation and retrieval) are examined (lines 2–11). If the examined operation is a relocation, the LCR is created considering the current bay configuration, denoted by *bay* (line 4). Next (lines 5–7), the allele value (*allele*) is computed, dividing the relocation position of the LCR (p) by the number of LCR's elements (*size*). After the allele value corresponding to the operation has been computed, it is added to the chromosome vector *ch* (line 8). At the end of each examination, the current bay configuration is updated to the new bay configuration generated through the execution of the operation *operation* (line 10).

Consider now that the chromosome size n is greater than the number of elements added to the vector *ch* (i.e., *size*). In this case, the remaining alleles are populated with the value zero (lines 13–15). This situation can occur when the number of relocations in solution *sol* is lower than the provided chromosome size n. Finally, the procedure returns the chromosome *ch* fulfilled, which genetically symbolizes the CRP solution *sol*.

To solve the CRP, the HBRKGA implementation requires a decoder procedure, which is outlined in Algorithm 3. Initially, the solution (sol) and the indexer i for the chromosome vector ch are initialized (line 1). The procedure's iterations occur continually while the bay (bay) is not empty (lines 2–19). At the start of each procedure iteration (line 3), the procedure verifies whether the container with the highest retrieval priority in the bay is currently located at the top of the stack (i.e., if the target container is available for retrieval). If so, the container is retrieved, meaning that a retrieval operation is produced (line 4). Otherwise, the relocation operation

---

Algorithm 2 Pseudocode of the Encode Procedure

Require: *bay*: initial bay configuration, *sol*: solution, n: the chromosome size
Ensure: *ch* chromosome vector
1: $ch \leftarrow \emptyset$
2: **for all** *operation* in *sol* **do**
3:   **if** *operation* is a relocation **then**
4:     $LCR \leftarrow$ create the LCR using the bay configuration *bay*
5:     $size \leftarrow$ get the size of the LCR
6:     $p \leftarrow$ get the position of *operation* within the LCR.
7:     $allele \leftarrow (p/size)$
8:     add *allele* to *ch*
9:   **end if**
10:   $bay \leftarrow$ get the new bay configuration executing *operation* in *bay*
11: **end for**
12: $size \leftarrow$ get the number of elements in ch; $r \leftarrow (n - size)$
13: **for** r remaining **do**
14:   add 0.00 to *ch*
15: **end for**
16: **return** *ch*

Algorithm 3 Pseudocode of the Decode Procedure

```
Require: bay: initial bay configuration, ch chromosome vector
Require: n: the chromosome size
Ensure: sol: solution
1: sol ← ∅; i ← 0
2: while the bay is not empty do
3:   if the target container is available for retrieval then
4:     operation ← execute the retrieval operation in bay
5:   else
6:     if i ≤ n then
7:       LCR ← create the LCR using the bay configuration bay
8:       size ← get the size of the LCR
9:       allele ← get the i-th allele in ch
10:      p ← round (allele * size); p ← min (size − 1, p)
11:      operation ← execute, in bay, the p-th relocation in LCR
12:      i ← i + 1
13:    else
14:      solution ← ∅
15:      return solution
16:    end if
17:  end if
18:  add operation to solution
19: end while
20: return sol
```

located in position p of the LRC (i.e., p-th relocation) is executed, and the indexer i is incremented (lines 11–12). The position p is computed by multiplying the allele value ($allele$) and the number of LCR's elements ($size$), rounding the result to the nearest integer and limiting it to $size - 1$ to ensure that the relocation index is within the LRC (line 10). After executing the operation, the current operation ($operation$) is added to the solution (line 18). Finally, the procedure ends when the bay becomes empty, and a decoded solution is returned (line 20).

It is worth mentioning that the chromosome vector ch can encode totally or partially a solution whose number of relocations is greater than the chromosome size n. In this case, it is assumed that this is an unworthy solution because it presents a higher number of relocations than is necessary to solve the problem. Therefore, the procedure returns an empty solution when the indexer i is greater than the chromosome size n (lines 14–15). In the HBRKGA proposed here, the crane's operating time of an empty solution is initialized with a maximum value, denoting the worst solution. Thus, a chromosome vector that has been decoded to an empty solution presents the worst fitness.

4.3 Population Initialization

In the proposed HBRKGA approach, the initial populations are not fully randomly generated. BRKGA assumes that initializing populations with predominately non-random chromosomes provide a more efficient search for better solutions by the genetic approach when compared to initializing in an entirely randomized way. As a strong indication of this, in [44], the computational results showed that inserting only four non-random chromosomes in the initial population significantly improved the solutions' quality. Accordingly, the HBRKGA proposed in this chapter employs the exploration aspects of an intelligent, constructive metaheuristic to produce initial promising solutions.

The goal of using a metaheuristic to generate initial populations is to explore diverse and promising solutions in a limited time, allowing the BRKGA to have a more effective search for better solutions. However, since the generation of initial solutions should not consume too much of the overall HBRKGA runtime, a time limit is imposed on the metaheuristic. This time limit is set to half of the HBRKGA runtime limit. This is based on the approach taken in a previous study [8], which used a similar approach to determine the first individual of the CVOA algorithm. The runtime limit ensures that the metaheuristic can generate diverse initial solutions while not impeding the BRKGA's search for better solutions.

This study employs the Pilot Method as the metaheuristic to generate the initial solutions set, specifically adopting the PILOT algorithm proposed in [9]. The Pilot Method provides an intelligent, constructive mechanism to explore the solution space and build solutions [46]. In its processing, the Pilot Method metaheuristic evaluates every possible construction step using a solution produced by its pilot heuristic. Thus, when this metaheuristic ends, many diverse solutions have been created in the exploration process. The distinct set obtained with these many solutions is used by the proposed HBRKGA approach to initialize the populations. Other favorable reasons for the adoption of the PILOT algorithm in the proposed HBRKGA are given as follows: (1) it can be used to solve both versions: restricted and unrestricted CRPs; (2) it has generated the best results in the experiments detailed in [9]; and (3) the PILOT algorithm embedded inside the CVOA algorithm has produced the best results as reported in [8].

Algorithm 4 describes the proposed procedure for generating the initial solutions set. The set of initial solutions is obtained by the metaheuristic, which searches for

Algorithm 4 Pseudocode of the GetInitialSolutions Procedure

Require: bay: initial bay configuration
Require: $limit_{brkga}$: HBRKGA runtime limit
Ensure: $init\_sols$: initial solutions set
1: $lim_m \leftarrow limit_{brkga}/2$
2: $init\_sols \leftarrow$ get the distinct solution set obtained by metaheuristic having bay and lim_m as inputs.
3: **return** $init\_sols$

diverse solutions while utilizing lim_m as a stopping criterion, and bay as the initial configuration. The metaheuristic operates for a maximum time limit of lim_m units. After obtaining the initial solutions set, denoted by $init\_sols$, the initial populations can be generated with $init\_sols$.

Algorithm 5 generates the initial population for HBRKGA. Initially, it creates an empty population with a specific size for each independent population I_p (line 2). The population is sorted by fitness in ascending order. The procedure then iterates p times to add new individuals to the population (lines 3–11). At the beginning of each iteration, the procedure extracts the next solution of $init\_sols$ (line 4). Then, it encodes the extracted solution and generates the chromosome vector ch (line 6). If no more solutions are found in $init\_sols$, ch is generated randomly (line 8). Once ch has been generated, the chromosome vector ch is added to the population pop (line 10). After pop has been entirely populated, the population pop is added to the independent populations set $pops$. Finally, $pops$ is returned when all independent populations have been initialized (line 14).

4.4 Evolutionary Process

This procedure improves populations according to the BRKGA's evolution process mentioned in Sect. 4. Algorithm 6 presents the pseudocode of the evolutionary procedure in the HBRKGA. All populations of the independent population set ($pops$) undergo the evolution process (lines 1–22), producing a new (next) population. Initially (lines 2–3), the individuals in each population (pop) are divided into two groups: the elite set (pop_e) and the non-elite set (pop_{ne}). Then, the procedure

Algorithm 5 Pseudocode of the GenInitialPopulations Procedure

Require: bay: initial bay configuration, $init\_sols$: initial solutions set
Require: n: the chromosome size, p: the population size
Require: I_p: the number of independent populations
Ensure: $pops$: the initial populations

1: **for** I_p populalions **do**
2: $pop \leftarrow \emptyset$;
3: **for** p individuals **do**
4: $sol \leftarrow$ get the next solution in $init\_sols$.
5: **if** sol is not nil **then**
6: $ch \leftarrow Encode(bay, sol, n)$
7: **else**
8: $ch \leftarrow$ get a random chromosome vector.
9: **end if**
10: add ch to pop
11: **end for**
12: add pop to $pops$
13: **end for**
14: **return** $pops$

Algorithm 6 Pseudocode of the Evolve Procedure

Require: bay: initial bay configuration, $pops$: the populations
Require: n: the chromosome size, p: the population size
Require: p_e: the elite set size, p_m: the mutant set size
Require: ρ: the probability of that an offspring inherits the allele of its elite parent
Ensure: $next\_pops$: the improved populations
1: **for all** pop in $pops$ **do**
2: $pop_e \leftarrow$ get the elite set in pop according to p_e.
3: $pop_{ne} \leftarrow$ get the non-elite set in pop according to p_e.
4: $next\_pop \leftarrow \emptyset$
5: $next\_pop \leftarrow$ copy pop_e to the elite set of $next\_pop$.
6: **for** $p - p_e - p_m$ individuals **do**
7: $offspring \leftarrow \emptyset$
8: $ind_e \leftarrow$ select an individual randomly in pop_e.
9: $ind_{ne} \leftarrow$ select an individual randomly in pop_{ne}.
10: **for** $i \leftarrow 0$ to $n - 1$ **do**
11: $is\_elite\_parent \leftarrow$ define whether ind_e is a parent with probability ρ.
12: **if** is_elite_parent **then**
13: $offspring[i] \leftarrow ind_e[i]$
14: **else**
15: $offspring[i] \leftarrow ind_{ne}[i]$
16: **end if**
17: **end for**
18: add $offspring$ to $next\_pop$
19: **end for**
20: $next\_pop \leftarrow$ create the mutants set in $next\_pop$ according to p_m.
21: add $next\_pop$ to $next\_pops$
22: **end for**
23: **return** $next\_pops$

initializes the next population with the entire elite set copied from pop (lines 4–5). Following this, the next population is populated with $(p - p_e - p_m)$ new individuals (offspring) derived from a genetic combination (lines 6–19). During the mating process, two parent individuals are randomly selected from the population pop. One parent is selected from the elite group (pop_e) and the other from the non-elite group (pop_{ne}) (lines 8–9).

Subsequently, each gene of the new individual $(offspring)$ is derived from a biased uniform crossover. The offspring gene $(offspring[i])$ is determined by selecting it from the elite parent with a probability of ρ, or if not selected, it is taken from the non-elite parent (lines 10–17). When the new individual's genes are completed, this offspring is added to the next population (line 18). After that (line 20), the remaining individuals in the new population $next_pops$ are generated by randomly initializing them (i.e., the mutant individuals) and adding them to the population until the total population size is reached (lines 20–24). This process ensures that the population maintains its diversity, which is essential for the success of the evolutionary algorithm.

Once the evolution process is completed, the next population $next\_pop$ is inserted into the new independent population set (line 21). Lastly, the procedure returns the new independent improved population set (i.e., *next_pops*) when all populations have evolved. It is worth pointing out that, at the end of this procedure, the returned populations are ordered by fitness in ascending order. The fitness computation of new individuals (i.e., offsprings or mutants) is performed through the procedure *Decode* (Algorithm 3) using the bay configuration, indicated by *bay*, and their respective chromosome vectors that are given as input data.

4.5 Exchanging Elite Individuals

This procedure controls the exchanging of information among the independent populations. Algorithm 7 outlines this procedure. Each population *pop* receives the I_i best individuals from each one of the remaining solutions (pop_b) and swaps its worst individuals with the newcomers (lines 1–15). The maximum number of individuals that a population is allowed to receive is equal to the number of non-elite individuals of this population (i.e., $(p - p_e)$). This constraint is imposed so that the elite set of populations remains unchanged during the swappings. In other words, the exchanging process should only change non-elite individuals of a population. The individuals participating in the exchange are identified by the w and i indexers, where w indexes the current worst individual (i.e., a non-elite individual) of *pop* and i indexes the current best individual (i.e., an elite individual) of pop_b. When the constraint mentioned earlier is satisfied (i.e., $w < (p - p_e)$), the elite individual ind_b of pop_b is swapped

---

Algorithm 7 Pseudocode of the ExchangeElite Procedure

Require: *pops*: the populations, p: the population size
Require: p_e: the elite set size, I_i: the number of best individuals in the exchange
Ensure: *pops*: the populations with exchanging
1: **for all** *pop* in *pops* **do**
2: $w \leftarrow 0$
3: **for all** pop_b in *pops* **do**
4: **if** $pop \neq pop_b$ **then**
5: **for** $i \leftarrow 0$ to $I_i - 1$ **do**
6: **if** $w < (p - p_e)$ **then**
7: $ind_b \leftarrow$ get the i-th best individual in pop_b.
8: $ind \leftarrow$ get the w-th worst individual in *pop*.
9: $pop \leftarrow$ swap *ind* in *pop* for ind_b
10: $w \leftarrow w + 1$
11: **end if**
12: **end for**
13: **end if**
14: **end for**
15: **end for**
16: **for all** *pop* in *pops* **do**
17: sort *pop* by fitness in ascending order.
18: **end for**
19: **return** *pops*

---

with the non-elite individual *ind* of *pop* (lines 7–9). Note that the swapping may change the population's ordering according to fitness. Consequently, each population is sorted by fitness in ascending order after all exchanges have been performed (lines 16–18). Finally, the populations derived from this exchanging process are returned (line 19).

5 Computational Experiments

This section discusses the experimental methodology used to assess the performance of the proposed biased random-key genetic algorithm on CRP instances of different sizes, and to compare its results with those of other existing algorithms for the same problem. The experimental setup was executed on a computer with 16 GB of RAM, running on *Windows 10 Education 64 bits* operating system, equipped with an *Intel(R) Core(TM) i7-9750H CPU @ 2.60GHz*. The randomness generator used was the *Mersenne Twister* from the Apache Commons Math 3.3 library. Random seeds were generated from decimal places of π (.14159265358979323846264...) and grouped in sets of seven, such as $seed_1 = 1415926$, $seed_2 = 5358979$, $seed_3 = 3238462$, and so forth.

The algorithms under evaluation were coded in Java and tested on a set of 1200 instances obtained from [7]. These instances varied in storage capacity, with a range of 20 to 128 containers. The instances were divided into four categories, namely, small, normal, medium, and large, based on their storage capacities. It should be noted that some instances in this suite are much larger than those commonly used in practical applications, where storage capacities are typically between 24 and 60 containers [9]. The instance suite is available for download at www.cin.ufpe.br/~asf2/csp/instances/.

The crane's operating time used in the experiments was computed using the operating time function defined in [8]. The parameters used to calculate the operating time were the same as those employed in the experiments conducted in [8]. Based on the literature studied, only the works described in [4, 8, 9, 35] can be compared with the proposed HBRKGA as they are the only ones that optimize the crane's operating time for solving the unrestricted CRP. However, among these works, only PILOT [9] and CVOA (coronavirus optimization algorithm) [8] were considered in the experiments presented in the current paper. This decision was made because PILOT and CVOA showed the best results among all algorithms proposed in [4, 8, 9, 35] in the experiments carried out in [8, 9]. Therefore, the proposed HBRKGA was compared to PILOT and CVOA. The input parameter values used for PILOT and CVOA in these experiments were the same as those adopted in [8]. The proposed hybrid BRKGA (HBRKGA) is also compared with the standard BRKGA (simply called BRKGA) to demonstrate the performance improvement achieved by the proposed algorithm. The difference between BRKGA and HBRKGA is related to how populations are initialized. BRKGA initializes the populations with only randomized individuals, whereas HBRKGA uses an intelligent, constructive metaheuristic (described in Sect. 4.3) to obtain initial promising individuals. It is also worth mentioning that the computational experiments were conducted for the restricted and unrestricted versions of CRP, respectively, named throughout this section as *RCRP* and *UCRP*.

In the experiments, both HBRKGA and BRKGA used the same parameter values, except for the adjustment factor for chromosome size, denoted by α. The parameter values used were $I_p = 3$, $I_g = 100$, $I_i = 2$, $\rho = 0.7$, $p_e\% = 0.3$, and $p_m\% = 0.15$. These parameter values were obtained from tuning experiments described in [5]. The values of I_p, I_i, and ρ were suggested by [44]. The adjustment factor α was used to determine the chromosome size in both HBRKGA and BRKGA. HBRKGA and BRKGA used different values of α because BRKGA starts from random individuals and therefore needs a larger chromosome size to avoid generating unworthy solutions within the runtime limit. Additionally, the UCRP demands a larger chromosome size than the RCRP to avoid generating unworthy solutions due to the broader and more complex search space. Thus, four values of α were used: α_{hr} and α_{hu} for HBRKGA, and α_{br} and α_{bu} for BRKGA. The pair $(\alpha_{hr}, \alpha_{hu})$ was used to solve the RCRP with HBRKGA, while $(\alpha_{br}, \alpha_{bu})$ was used to solve the RCRP with BRKGA. The values of $\alpha_{hr} = 1.20$ and $\alpha_{hu} = 1.25$ were used for HBRKGA, while $\alpha_{br} = 2$ and $\alpha_{bu} = 2 * C$ were used for BRKGA. Here, C represents the storage capacity of the evaluated instance. Finally, these parameter values were used in the experiments to evaluate the performance of HBRKGA and BRKGA.

5.1 Analysis of HBRKGA

The goal of the experiments was to evaluate the performance of the proposed HBRGKA algorithm in solving the RCRP and UCRP. Four different algorithms were compared: HBRGKA, CVOA, PILOT, and BRGKA. It is important to note that HBRGKA uses PILOT as a metaheuristic for generating initial populations. The purpose of comparing HBRGKA with PILOT was to determine the effectiveness of a hybrid metaheuristic versus a basic metaheuristic. All algorithms were run for a maximum of 40 seconds and were executed 14 times per instance with a unique random seed for each execution. For instance, execution 1 used $seed_1$ (1415926), execution 2 used $seed_2$ (5358979), and so on. This execution setup was used for both RCRP and UCRP.

The objective of the first assessment of HBRKGA is to determine the consistency of the algorithm's solutions. To achieve this goal, the algorithm's solutions were evaluated for variability over 14 runs on each of the 1200 instances. The results indicate that HBRKGA produces solutions with low dispersion of objective values, indicating a high level of homogeneity in the solutions built. The highest coefficient of variation was 4.11% for RCRP and 2.65% for UCRP, while the mean coefficient of variation was 0.44% and 0.14% for RCRP and UCRP, respectively. Table 1 presents these values along with the mean coefficients of variation for each instance class: small, normal, medium, and large. Despite being a random method, the proposed algorithm shows a tendency to converge, even with different random seeds. It is

Table 1 Coefficient of variation for the solutions produced by HBRKGA over the instances

Instances class	RCRP highest	RCRP mean	UCRP highest	UCRP mean
Small	3.43	0.57	2.65	0.26
Normal	4.11	0.50	1.40	0.12
Medium	1.72	0.35	0.69	0.08
Large	1.54	0.30	0.71	0.09
All	4.11	0.44	2.65	0.14

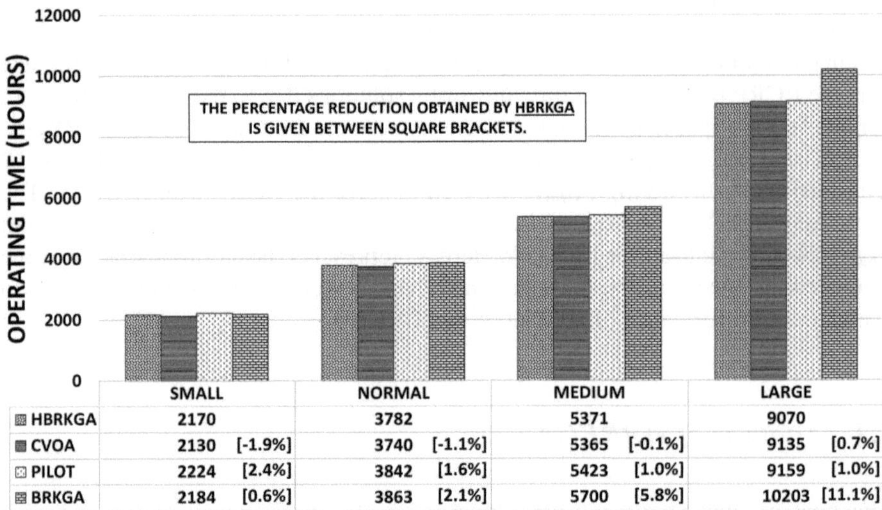

Fig. 5 A comparison of HBRKGA, CVOA, PILOT, and BRKGA algorithms in relation to the crane operating time solving the restricted CRP

worth noting that all coefficients of variation values for the RCRP are higher than those for UCRP, which is due to the increased complexity of the unrestricted version, making it challenging to find better and substantially different solutions in terms of the crane's operating time, even with different random seeds.

Figures 5 and 6 present the results of the performance evaluation of HBRKGA, CVOA, PILOT, and BRKGA, based on 14 runs performed on 1200 instances. These figures present the total crane operating time for solutions obtained by each algorithm on different instance classes, as well as the percentage reduction values achieved by the HBRKGA algorithm. The analysis reveals that HBRKGA outperforms PILOT and BRKGA for both RCRP and UCRP, producing solutions with lower total crane operating time. Moreover, as the instance sizes increase, it appears that the HBRKGA algorithm consistently outperforms the BRKGA algorithm in terms of solution quality, particularly for larger problem instances. The reduction rates achieved by HBRKGA compared to BRKGA were 11.1 and 96.4% for RCRP and UCRP instances, respectively, indicating a significant improvement.

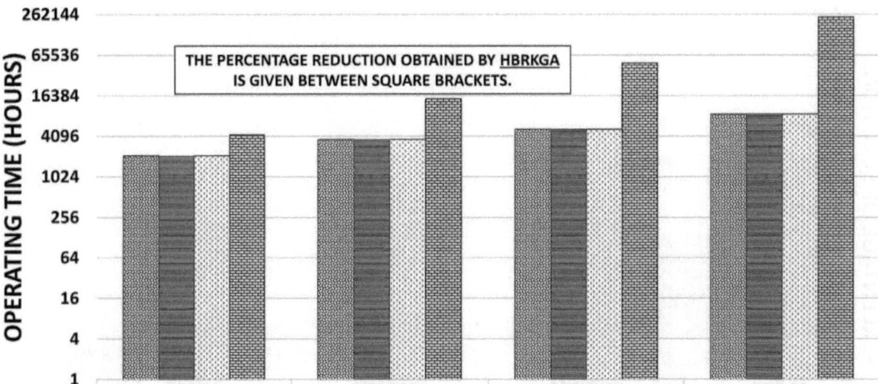

	SMALL	NORMAL	MEDIUM	LARGE
HBRKGA	2095	3660	5155	8648
CVOA	2051 [-2.2%]	3618 [-1.1%]	5130 [-0.5%]	8620 [-0.3%]
PILOT	2106 [0.5%]	3667 [0.2%]	5169 [0.3%]	8662 [0.2%]
BRKGA	4301 [51.3%]	14820 [75.3%]	49868 [89.7%]	238112 [96.4%]

Fig. 6 A comparison of HBRKGA, CVOA, PILOT, and BRKGA algorithms in relation to the crane operating time solving the unrestricted CRP

These performance results demonstrate that the proposed HBRKGA algorithm outperforms the other algorithms for both the RCRP and the UCRP. It is also noteworthy that the HBRKGA algorithm produces more homogeneous solutions than the other algorithms, as discussed previously. Indeed, the results suggest that the proposed algorithm performs well in generating high-quality solutions with low variability among them. The results also suggest that the hybridization of the BRKGA and PILOT algorithms allows the HBRKGA algorithm to exploit the strengths of both algorithms while mitigating their weaknesses. The PILOT algorithm, which has a strong exploration ability, is used to generate diverse initial populations, while the BRKGA algorithm, which has a strong exploitation ability, is used to refine the solutions obtained from the initial populations.

The results obtained from the experiments indicate that the HBRKGA algorithm is a promising solution for the container relocation problem. The algorithm provides better solutions with a lower variability compared to the other evaluated algorithms. This suggests that the HBRKGA algorithm has the potential to be a reliable and effective method for solving similar optimization problems. It has shown a good performance when compared to other metaheuristics, including the standard BRKGA and the state-of-the-art CVOA algorithm. The HBRKGA has proved to be particularly efficient for large-sized instances, where it has produced significantly better solutions than the other algorithms. These results indicate that the proposed hybridization of the BRKGA algorithm is a valuable technique for improving solution quality for complex optimization problems.

The performance comparison of HBRKGA and other algorithms for RCRP and UCRP instances is illustrated in Figs. 7 and 8, respectively. In both figures, the instances are grouped by their size, and each bar represents the percentage of instances

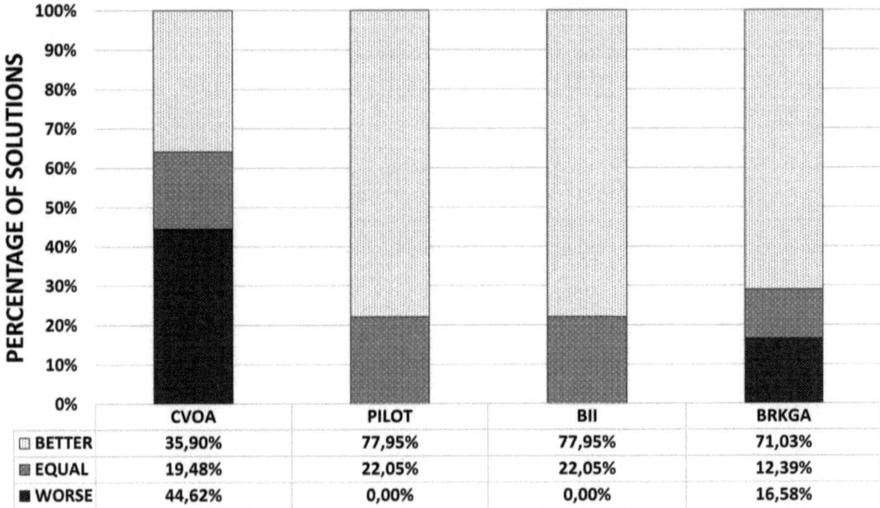

Fig. 7 A comparison of the quality of the solutions generated by HBRKGA solving the restricted CRP

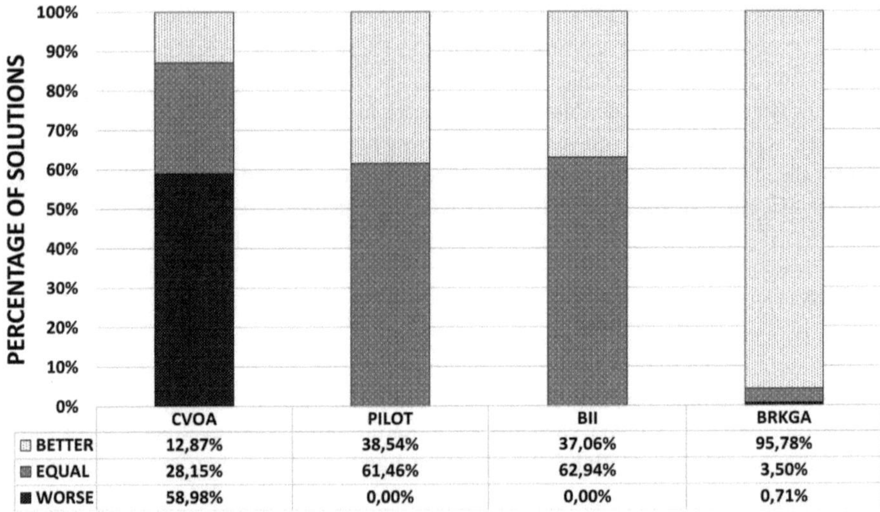

Fig. 8 A comparison of the quality of the solutions generated by HBRKGA solving the unrestricted CRP

in which the HBRKGA solution was classified as *better*, *equal*, or *worse* than the solutions produced by CVOA, PILOT, BRKGA, and BII. BII denotes the best initial individual generated by the HBRKGA procedure, defined in Sect. 4.3, to initialize the populations.

In Fig. 7, it is possible to observe that HBRKGA presented better solutions than CVOA, PILOT, and BRKGA for most of the instances, regardless of their size. Moreover, the comparison with BII shows that the exploitation abilities of HBRKGA indeed contributed to the quality of the solutions produced, as the percentage of instances in which HBRKGA presented better solutions than BII is significantly higher than the percentage of instances in which BII produced better solutions than HBRKGA.

The comparison between the HBRKGA and other algorithms for the UCRP instances is illustrated in Fig. 8. Similar to Fig. 7, HBRKGA presented better solutions than CVOA, PILOT, and BRKGA for most of the instances, especially for the larger ones. However, in this case, the comparison with BII shows that the exploitation abilities of HBRKGA did not contribute significantly to the quality of the solutions produced, as the percentage of instances in which HBRKGA presented better solutions than BII is only slightly higher than the percentage of instances in which BII produced better solutions than HBRKGA.

Overall, the results presented in these figures reinforce the conclusion that the proposed HBRKGA is a competitive algorithm for solving the RCRP and UCRP instances in terms of the crane operating time.

These findings suggest that the proposed HBRKGA method is effective in solving the restricted CRP and can improve the quality of solutions compared to other algorithms. The evolving and elite-exchanging procedures proposed in HBRKGA contributed to the better quality solutions produced compared to the population initialization procedure. Additionally, HBRKGA outperformed the standard BRKGA and PILOT algorithms, indicating the effectiveness of the proposed hybridization. However, the results also suggest that HBRKGA's performance is better for the restricted CRP than the unrestricted CRP. Overall, the study demonstrates the potential of the HBRKGA algorithm for optimizing crane operations and reducing operating time.

6 Conclusion

The results show that HBRKGA can substantially improve solution quality compared to the other algorithms for reducing the crane operating time, as demonstrated by the statistical analysis and the comparison of solution qualities. The proposed evolving and elite-exchanging procedures discovered better solutions than those obtained by the population initialization procedure. Furthermore, the proposed hybrid BRKGA produced better solutions than the standard BRKGA predominantly. The study has also shown that HBRKGA tends to yield better results in solving the restricted CRP than in solving the unrestricted CRP. However, the proposed algorithm has shown great potential for solving both versions of the CRP, making it a promising approach for optimizing container relocation operations in port container terminals.

Moreover, in comparison with another leading hybrid metaheuristic (i.e., CVOA), CVOA presented slightly better results than HBRKGA. However, as the instance size increases, the HBRKGA results tend to improve. In large-sized instances, HBRKGA has produced significantly better solutions than the CVOA while solving the RCRP. In particular, the experiments have reported that the proposed population initialization procedure (with a constructive metaheuristic) can substantially improve solution quality in a BRKGA approach. Besides, the remaining methods of our BRKGA approach have prevalently found better solutions than those produced in the population initialization. It is also worth highlighting that the proposed HBRKGA has presented a better performance while solving the restricted CRP than when solving the unrestricted CRP.

This study can be deepened in several auspicious directions. Optimizing operating time in other stacking problems could definitely be a fruitful direction to explore. It would be interesting to see how the proposed HBRKGA performs in comparison to other metaheuristics when applied to different stacking problems. Incorporating other promising methods, such as *Multi-parent* and *Implicit Path-Relinking*, could also lead to further improvements in solution quality. It would be helpful to evaluate the effectiveness of these modifications and compare them to the original HBRKGA approach. Integrating different optimization problems relevant to container terminal systems, such as *CRP and Wagon Assignment Planning*, could also be a valuable research direction. This would involve solving multiple optimization problems simultaneously, which could be a challenging but worthwhile task. Overall, there are many exciting avenues for future research based on the work presented in this study.

References

1. World Bank: Container port traffic (teu: 20 foot equivalent units) (2020). Accessed 25 Sept 2022
2. UNCTAD: Review of Maritime Transport 2021. United Nations Publication (2021)
3. UNCTAD: Review of Maritime Transport 2019. United Nations Publication (2019)
4. Lin DY, Lee YJ, Lee Y (2015) The container retrieval problem with respect to relocation. Transp Res Part C Emerg Technol 52:132–143
5. Andrade CE, Toso RF, Gonçalves JF, Resende MGC (2021) The multi-parent biased random-key genetic algorithm with implicit path-relinking and its real-world applications. Eur J Oper Res 289(1):17–30
6. Gonçalves JF, Resende MGC (2011) Biased random-key genetic algorithms for combinatorial optimization. J Heuristics 17(5):487–525
7. da Silva Firmino A, de Abreu Silva RM, Times VC (2019) A reactive grasp metaheuristic for the container retrieval problem to reduce crane's working time. J Heuristics 25(2):141–173
8. da Silva Firmino A, Times VC (2022) A coronavirus optimization algorithm for solving the container retrieval problem. In: Khosravy M, Gupta N, Patel N (eds) Frontiers in nature-inspired industrial optimization. Springer Singapore, Singapore, pp 49–76
9. da Silva Firmino A, Times VC, de Abreu Silva RM (2020) In: Optimizing the Crane's operating time with the ant colony optimization and pilot method metaheuristics. Springer, Singapore, pp 364–389
10. Caserta M, Schwarze S, Voß S (2012) A mathematical formulation and complexity considerations for the blocks relocation problem. Eur J Oper Res 219(1):96–104
11. Kim KH, Hong GP (April 2006) A heuristic rule for relocating blocks. Comput Oper Res 33(4):940–954

12. Tanaka S, Voß S (2021) An exact approach to the restricted block relocation problem based on a new integer programming formulation. Eur J Oper Res
13. Bacci T, Mattia S, Ventura P (2020) A branch and cut algorithm for the restricted block relocation problem. Eur J Oper Res
14. Jin B (2020) On the integer programming formulation for the relaxed restricted container relocation problem. Eur J Oper Res 281(2):475–482
15. Lu C, Zeng B, Liu S (2020) A study on the block relocation problem: lower bound derivations and strong formulations. IEEE T Autom Sci Eng
16. Galle V, Barnhart C, Jaillet P (2018) A new binary formulation of the restricted container relocation problem based on a binary encoding of configurations. Eur J Oper Res 267(2): 467–477
17. Quispe KEY, Lintzmayer CN, Xavier EC (2018) An exact algorithm for the Blocks Relocation Problem with new lower bounds. Comput Oper Res 99:206–217
18. de Melo da Silva M, Toulouse S, Wolfler Calvo R (2018) A new effective unified model for solving the pre-marshalling and block relocation problems. Eur J Oper Res 271(1):40–56
19. Tanaka S, Mizuno F (2018) An exact algorithm for the unrestricted block relocation problem. Comput Oper Res 95:12–31
20. Gulić M, Maglić L, Krljan T, Maglić L (2022) Solving the container relocation problem by using a metaheuristic genetic algorithm. Appl Sci 12(15)
21. Jiang T, Zeng B, Wang Y, Yan W (2021) A new heuristic reinforcement learning for container relocation problem. J Phys Conf Ser 1873(1):12050
22. Maglić L, Gulić M, Maglić L (2020) Optimization of container relocation operations in port container terminals. Transport 35:1–11
23. Zhang C, Guan H, Yuan Y, Chen W, Wu T (2020) Machine learning-driven algorithms for the container relocation problem. Transp Res Part B Methodol 139(C):102–131
24. ElWakil M, Gheith M, Eltawil A (2019) A new simulated annealing based method for the container relocation problem. In: 2019 6th international conference on control, decision and information technologies (CoDIT), pp 1432–1437
25. Jovanovic R, Tuba M, Voß S (2019) An efficient ant colony optimization algorithm for the blocks relocation problem. Eur J Oper Res 274(1):78–90
26. Bacci T, Mattia S, Ventura P (2019) The bounded beam search algorithm for the block relocation problem. Comput Oper Res 103:252–264
27. Feillet D, Parragh SN, Tricoire F (2019) A local-search based heuristic for the unrestricted block relocation problem. Comput Oper Res 108:44–56
28. Tricoire F, Scagnetti J, Beham A (2018) New insights on the block relocation problem. Comput Oper Res 89:127–139
29. López-Plata I, Expósito-Izquierdo C, Moreno-Vega JM (2019) Minimizing the operating cost of block retrieval operations in stacking facilities. Comput Ind Eng 136:436–452
30. Voß S, Schwarze S (2019) A note on alternative objectives for the blocks relocation problem. In: Computer logistics. Springer, Berlin, pp 101–121
31. da Silva Firmino A, de Abreu Silva RM, Times VC (2016) An exact approach for the container retrieval problem to reduce crane's trajectory. In: 2016 IEEE 19th international conference on intelligent transportation systems (ITSC), pp 933–938
32. Hussein M, Petering MEH (2012) Genetic algorithm-based simulation optimization of stacking algorithms for yard cranes to reduce fuel consumption at seaport container transshipment terminals. In: 2012 IEEE congress on evolutionary computation (CEC). IEEE, pp 1–8
33. Inaoka Y, Tanaka S (2018) The block relocation problem under a realistic model of crane trajectories. In: 20th international conference on harbor, maritime and multimodal logistics modeling and simulation, HMS 2018, pp 62–66
34. Inaoka Y, Tanaka S (2017) A branch-and-bound algorithm for the block relocation problem to minimize total crane operation time. In: 19th international conference on harbor maritime and multimodal logistics M&S (HMS 2017), pp 98–104
35. Kim Y, Kim T, Lee H (2016) Heuristic algorithm for retrieving containers. Comput Ind Eng

36. Azari E, Eskandari H, Nourmohammadi A (2017) Decreasing the crane working time in retrieving the containers from a bay. Scientia Iranica
37. Helmi AM, Lotfy ME (2020) In: Recent advances of nature-inspired metaheuristic optimization. Springer, Singapore, pp 1–33
38. Variengien A, Pontes-Filho S, Glover TE, Nichele S (2021) Towards self-organized control: using neural cellular automata to robustly control a cart-pole agent. Innov Mach Intell 1:1–14
39. Gupta N, Khosravy M, Patel N, Dey N, Mahela OP (2020) Mendelian evolutionary theory optimization algorithm. Soft Comput 24(19):14345–14390
40. Gupta N, Patel N, Tiwari BN, Khosravy M (2019) Genetic algorithm based on enhanced selection and log-scaled mutation technique. In: Arai K, Bhatia R, Kapoor S (eds) Proceedings of the future technologies conference (FTC) 2018. Springer International Publishing, pp 730–748
41. Prasetyo H, Fauza G, Amer Y, Lee SH (2015) Survey on applications of biased-random key genetic algorithms for solving optimization problems. In: 2015 IEEE international conference on industrial engineering and engineering management (IEEM), pp 863–870
42. Kinast A, Braune R, Doerner KF, Rinderle-Ma S, Weckenborg C (2022) A hybrid metaheuristic solution approach for the cobot assignment and job shop scheduling problem. J Ind Inf Integr 28:100350
43. Chaves AA, Gonçalves JF, Lorena LAN (2018) Adaptive biased random-key genetic algorithm with local search for the capacitated centered clustering problem. Comput Ind Eng 124:331–346
44. Gonçalves JF, Resende MGC (2011) A parallel multi-population genetic algorithm for a constrained two-dimensional orthogonal packing problem. J Comb Optim 22(2):180–201
45. Kummer AF, de Araújo OCB, Buriol LS, Resende MGC (2022) A biased random-key genetic algorithm for the home health care problem
46. Voßs S, Fink A, Duin C (2005) Looking ahead with the pilot method. Ann Oper Res 136(1): 285–302

Part III
Applications in Power Systems

Chapter 5
Genetic Algorithm and Its Applications in Power Systems

Hossein Yektamoghadam, Rouzbeh Haghighi, Majid Dehghani, and Amirhossein Nikoofard

1 Introduction

The electric power industry has undergone sustained expansion in each of its three principal areas: generation, transmission, and distribution. Since electricity is not cost-effectively storable, a constant equilibrium between supply and demand is necessary. The growth in the scale of demand and the intricacy of operations, which include power generation allocation, expansion planning, and unit commitment, has posed substantial challenges in the management and design of power systems [1]. Heuristic optimization techniques, described as methods based on experience, are rapidly growing tools that can effectively and efficiently address complicated optimization challenges. Genetic algorithms, particle swarm optimization (PSO) [2], differential evolution (DE), bacterial foraging algorithms, imperialistic competition algorithms, artificial bee colonies, ant colony optimization, and harmony search are all examples of popular heuristic optimization methods [3].

The vast areas of application of the GA optimization technique to tackle complex problems cannot be handled using conventional methods [4]. A genetic algorithm is

H. Yektamoghadam
University of Tehran, Tehran, Iran
e-mail: Yektamoghadam@ut.ac.ir

R. Haghighi
University of Michigan, Dearborn, MI, USA
e-mail: Haghighi@umich.edu

M. Dehghani (✉)
Oklahoma State University, Stillwater, OK, USA
e-mail: majid.dehghani@okstate.edu

A. Nikoofard
K. N. Toosi University of Technology, Tehran, Iran
e-mail: A.nikoofard@kntu.ac.ir

M. Khosravy et al. (eds.), *Frontiers in Genetics Algorithm Theory and Applications*, Springer Tracts in Nature-Inspired Computing,
https://doi.org/10.1007/978-981-99-8107-6_5

a metaheuristic algorithm based on the concepts of natural selection and the genetics of living organisms. The GA was applied to simulate some of the processes observed during natural evolution using a computer algorithm. Processes that are formed are based on operations on chromosomes (organic systems for coding the genetic structure of living organisms) [5]. A genetic algorithm is a sub-algorithm for search algorithms. However, it is fundamentally different from other search algorithms. The GA is among the most prominent metaheuristic techniques and can yield a global optimum as it operates with a population, i.e., a set of solutions, in contrast to gradient methods that function with a single solution and produce a local optimum.

One of the main applications of GA in power systems is modeling and solving the GEP problem. Due to the integer nature of the decision variables, GEP is an extremely challenging problem that seeks to minimize the investment cost of installing new generating units. According to the literature, researchers have successfully employed GA to address transmission and generation expansion problems by employing a population-based crossover [6]. The power grid infrastructure is optimized via bi-level evolutionary optimization in Gupta et al. [7]. It provides an independent system operator's view on how to plan for the increase of both generation and transmission in a coordinated manner. The proposed research aims to demonstrate how intelligent optimization of generator capacity and placement might lessen expenditures on transmission infrastructure and enhance reliability.

To address the GEP issue with wind power facilities, the authors of Sahragard et al. [8] presented a method based on the GA. The suggested approach intended to maximize wind penetration level into the network at the lowest potential cost to the overall investment. The study demonstrated that maximizing the utilization of wind power generation capacity could lead to greater exploitation of more stable wind regimes. To accurately monitor maximum global power from active power multiple peaks created by Solar-Photo Voltaic, the author of Singh Chawda et al. [9] introduces a Conductance-based PSO technique. The proposed approach modifies the weight component of each particle based on its distance from the global optimum location during tracking. Moreover, it has the advantage of continuously adjusting weight components, which lowers active power fluctuations at the ideal global position. Jalalzad et al. [10] presented a game theory approach to provide an optimal strategy for power plants and results have been compared with different optimization methods such as GA, PSO and TLBO.

In addition, another practical application of the GA is in solving the UC problems presented in Jo and Kim [11]. This study has been done by considering various uncertainties using Monte Carlo simulation. Reserve requirements for power system stability were determined using the produced scenarios for load, wind, solar, and generator failures. The uncertainty integration method (UIM) was used to select the optimal combination of scenarios by minimizing the additional reserve requirements that resulted from the uncertainties associated with power sources. For stochastic unit commitment (SUC) problems, an integration process for uncertainties is formulated and optimized using an enhanced GA. A novel metaheuristic approach for addressing the Security Constrained Hydrothermal Unit Commitment (SCHTUC) problem is provided in a recent study [12], which makes use of GA, an adaptive strategy, and two constraint control repairing methodologies. An enhanced version of a GA-based optimization technique has been proposed to increase its effectiveness in Ponciroli et al. [13]. The authors also defined adaptable power plant operations more precisely and comprehensively. In El-Shorbagy et al. [14], the authors developed a combination of the Manta-Ray Foraging Optimization Algorithm (MRFO) and the GA to prevent the algorithm from getting trapped in local minima. To prove the efficacy of the suggested algorithm in real-world settings, it was applied to the UC problem, one of the most critical issues facing power systems today. To address the OPF problem, KS [15] introduces a novel hybrid optimization method that combines the benefits of both the GA and PSO methods. The suggested method may be used to both single- and multiple-objective optimization situations, with the goal of maximizing environmental, technological, and economic benefits. Generation cost minimization, transmission power loss reduction, voltage stability maximization, voltage profile maximization, and pollution reduction are all examples of goal models that may be employed with this method.

The remainder of this chapter is structured as follows. Following a brief overview in Sect. 1, the GEP definition and its formulation using the GA-based method will be provided in Sect. 2. Afterwards, in Sect. 3, the application and formulation of GA in unit-commitment are presented. Then in Sect. 4, how to implement the GA algorithm in the GEP and UC problem has been presented. After that, a sample GEP problem is solved based on the GA approach, and the results are presented in Sect. 5. In the last section of this investigation, referred to as Sect. 6, the findings and conclusions will be discussed.

2 Generation Expansion Planning

One of the most significant contributions that GA has made to the field of power systems is its ability to model and find a solution to the GEP problem, which seeks to identify the most effective answer to the planning issues that arise [16]. In the GEP problem, GA-based methods attempt to satisfy both technical and financial boundaries for installing new power generation units. Moreover, the extent, lifespan, and nonlinearity of the generating unit size make GEP a difficult challenge to solve. In addition to this, the GEP optimization problem is a non-convex problem due to the existence of binary variables and integer numbers in it. Therefore, the use of non-linear optimization methods such as metaheuristic algorithms is very common in solving the GEP problem. GA is commonly used to generate high-quality solutions for GEP problems. With the generating cost of power plants, operation, and maintenance expenditures, and carbon tax all taken into account, the GEP is calculated by maximizing the government subsidies for energy system management while minimizing the overall investment cost for installing new generators.

In order to fulfill the anticipated demand in the future, the GEP problem entails selecting what kind and how many additional generating units are needed. To calculate the total cost of the system in the scheduled time period for each scenario, the

GEP problem must be solved. In order to address the GEP problem, which will be outlined in the following section, the model seeks to identify the lowest total cost of the system. This includes all expenses associated with investment, generation, and O&M.

2.1 Mathematical Representation of GEP

Following is a formulation of the optimization problem, in which Eq. (1) represents the objective function.

Objective function:

$$\min\left\{\sum_{i=1}^{K} C_i(X_i)\right\},\quad i \in \{1, 2, \ldots, K\} \tag{1}$$

where:

$$C_i(X_i) = CInv_i(X_i) + CGen_i(X_i) + COM_i(X_i) \tag{2}$$

$$CInv_i(X_i) = \sum_{j=1}^{T} Cap_i \cdot Inv_i \cdot X_{i,j} \cdot \frac{1}{(1+r)^j} \tag{3}$$

$$CGen_i(X_i) = \sum_{j=1}^{T} Cap_i \cdot Gen_i \cdot \left(\sum_{j=1}^{T} X_{i,j}\right) \cdot TW \cdot CF \cdot \frac{1}{(1+r)^j} \tag{4}$$

$$COM_i(X_i) = \sum_{j=1}^{T} Cap_i \cdot OM_i \cdot \left(\sum_{j=1}^{T} X_{i,j}\right) \cdot CF \cdot \frac{1}{(1+r)^j} \tag{5}$$

Subject to:

$$(1 + RM_{min}) \cdot D_j \leq XT_j \leq (1 + RM_{max}) \cdot D_j \tag{6}$$

where:

$$RM_{\min} \leq R \leq RM_{\max} \tag{7}$$

$$XT_j = \sum_{i=1}^{K} Cap_i \cdot \sum_{j=1}^{T} X_{i,j} \tag{8}$$

$$0 \leq X_{i,j} \leq \bar{X}_{i,j} \tag{9}$$

Equation (1) as the objective function includes the total costs, which corresponds to the total present value of all investment costs ($CInv_i(X_i)$), generation costs ($CGen_i(X_i)$), operation and maintenance costs ($COM_i(X_i)$) of each i-type power plant as shown in Eq. (2). It is considered that expenses will occur at the start of the respective time periods and will be calculated by Eqs. (3), (4) and (5). Moreover, the discount rate is r and the problem constraint is shown in Eq. (6), which is the reserve margin of the system. Under this restriction, system capacity must be sufficient to meet the peak demand at the end of each period. The minimum and maximum permitted reserve margins are indicated in Eq. (7) as $RM_{\min}$ and $RM_{\max}$. The total capacity of each type of power plant (i) is calculated by Eq. (8). In addition, Eq. (9) includes annual maximum constraints on the number of newly installed units following lower and upper restrictions on the capacity for each fuel type. In these equations, Cap_i is the capacity of each unit in MW, Inv_i is the unit investment cost in \$/MW, Gen_i is the generation variable cost of each unit in \$/MWh, OM_i is O&M expenses of each unit in \$/MW, TW is the theory time working of each unit in hours and CF is the capacity coefficient of each unit [17].

3 Unit-Commitment

Another main application of GA in electricity networks is UC, which is known as a widespread problem. The main objective of the UC is to reduce power generating costs over a given time frame, by establishing an optimal schedule for the system's generators, which must be regulated to many operational restrictions. In addition, UC is a challenging issue in power system engineering, which plays a crucial role in the safe, productive, and cost-effective everyday operation of a network. UC is an optimization issue with the goal of determining when and for how long certain generating units should be operational in order to fulfil future power demand; this demand is generally 24–48 h in the future.

3.1 Mathematical Formulation of UC

The total objective function of the UC problem will be mathematically formulated, which can be written as follows:

Objective function:

$$\min\left\{\sum_{i-1}^{K}\sum_{j-1}^{T} F_i\left(Gen_{i,j}, SU_{i,j}\right)\right\} \tag{10}$$

where:

$$F_i(Gen_{i,j}, SU_{i,j}) = [Gen_{i,j} + SU_{i,j}(1 - U_{i,j-1})]U_{i,j} \tag{11}$$

$$Gen_{i,j} = \sum_{i=1}^{K}\sum_{j=1}^{T}(a_j + b_j PGen_{i,j} + c_j PGen_{i,j}^2) \tag{12}$$

$$SU_{i,j} = \begin{cases} C^H & if\ |X_{i,j}| \leq MDT_j + \tau^C \\ C^C & if\ |X_{i,j}| > MDT_j + \tau^C \end{cases} \tag{13}$$

Subject to:

$$U_{i,j} PGen_{i,j}^{\min} \leq PGen_{i,j} \leq U_{i,j} PGen_{i,j}^{\max} \tag{14}$$

$$\sum_{j=1}^{T} PGen_{i,j}^{\max} = P_D + P_L \tag{15}$$

$$\sum_{j=1}^{T} P_j^{\max} U_{i,j} \geq P_D + SR \tag{16}$$

$$U_{i,j} = \begin{cases} 0 \to 1 & if\ T_{j-1,i}^{off} \geq MDT_j \\ 1 \to 0 & if\ T_{j-1,i}^{on} \geq MUT_j \\ 0\ or\ 1 & otherwise \end{cases} \tag{17}$$

$$R_j^{\min} \leq |x_{j,i} - x_{j-1,i}| \leq R_j^{\max} \tag{18}$$

Total system operational expenses are represented by the objective function in Eq. (10). The total cost is obtained by Eq. (11), where $U_{i,j}$ is the commitment state of generation unit i at interval j (0 for a decommitted unit or 1 for a committed unit). Each generating unit's total operating costs are represented by the quadratic function in Eq. (12), where $Gen_{i,j}$ is the fuel cost function of the power plant i. In this function, a_j, b_j and c_j are constant coefficients and j is the interval index. The start-up cost is the expense incurred upon recommitting a previously decommissioned unit, and it is calculated using the period during which the unit was out of service in Eq. (13). Where C^H is the Hold Start Cost of unit i; C^C is the Cold Start Cost of unit i; MDT_j is the Minimum Down-Time of thermal unit i; τ^C represents the time in hours during which the thermal unit t was decommitted before the interval j [12].

For this particular issue, all of the system limitations that were described must be satisfied, which include the power balance constraint, generation limits, minimum down and uptime constraint, and spinning reserve. Each generator unit's output power must be within a specified range, between its nominal power and its minimum and maximum technical operating levels. In other words, operating constraints must be satisfied, for the upper and lower capacities, which is shown in Eq. (14). In order to maintain a stable power grid, system losses and load must be balanced by the total amount of energy generated by all power plants, which can be expressed as a mathematical representation in Eq. (15). Spinning reserve is defined as the difference between the total available generation from generators and the present load and loss [18]. In Eq. (16), as shown in its mathematical notation, P_D is the load demand (MW) at hour j. $P_j^{\max}$ denotes the maximum power generating capacity. Here SR represents the system spinning reserve in MW at j time interval. Commitment and post-commitment periods must include a brief on/off cycles for the generating units. In the Eq. (17), MDT_j and MUT_j are minimum downtime and minimum uptime of the i-th unit in j-hour, respectively. Here $T_{j-1,i}^{off}$ and $T_{j-1,i}^{on}$ display the on and off times (in hours) of the i-th unit in continuous format up to (t-1) hour. Ramping constraints are sometimes included in unit commitment formulations. Equation (18) denotes the maximum allowed change in generator output between time periods. In the following mathematical formulation $R_j^{\min}$ and $R_j^{\max}$ refer to the ramp-down and ramp-up constraints, respectively.

4 Applying GA in GEP and UC Problem

In this section, the method of implanting a GA on the GEP and UC problem is introduced. The first step of optimization algorithms based on GA is to generate a random chromosome; in a UC problem, the perfect solution is to present a year-long and daily UC schedule that works for the entire units; in the GEP problem, an optimal plan for installing new generation units should be defined. In these problems, the input is the scheduling plan of the UC or the new unit installation plan of the GEP; therefore, the structure of the chromosome will be formed as the problem input. A GEP/UC schedule is defined by an optimal GEP/UC scenario, which is accompanied by either an electricity production cost or an investment cost [13]. A basic GA typically consists of the following sections illustrated in Fig. 1.

Initial Population Generation: To ensure that the generated tentative GEP/UC chromosomes satisfy the constraints related to the operation and commitment of units over time, a specialized procedure has been developed. This procedure aims to efficiently explore the solution space while taking into account the constraints on the GEP/UC schedules of various units, such as their operational time frames, start-up and shut-down periods, and required commitment periods. Due to the difficulty of

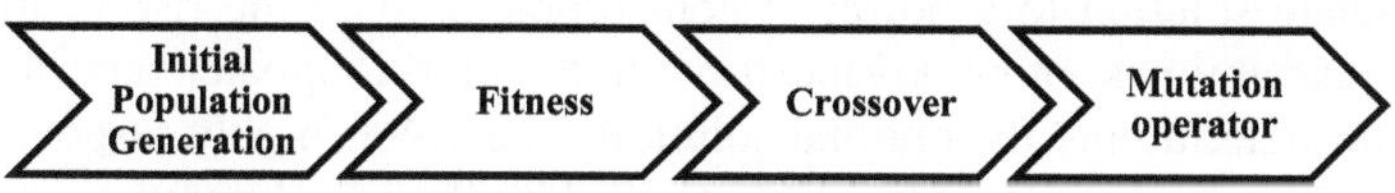

Fig. 1 A basic GA typically consists of the following sections

meeting these requirements with a randomly generated sequence of binary values, a specialized method was established.

Fitness: After generating a complete set of GEP/UC scenarios, the next generation of chromosomes is made by randomly selecting pairs of scenarios. The likelihood of a certain GEP/UC schedule being picked is related to its fitness, as measured against the average fitness of the whole generation. Being the goal of the GEP/UC problem, the minimization of the whole system operation cost or the investment cost, chromosome with lower costs will have higher chances of being selected. The objective of the GEP/UC problem is to minimize the overall system operation cost or investment cost. Hence, the chromosomes that have a lower overall cost will have a greater probability of being chosen.

Crossover: Pairs of chromosomes are chosen as parents for their next generation after their fitness is assessed. In GEP/UC scenarios, two offsprings are produced when the genetic material from both parents was been combined through a crossover procedure, in which some of each parent's chromosomes is swapped with that of the other. This is a vital stage in the GA algorithm. The process of recombination is accomplished by vertically cutting the parent chromosomes.

Mutation operator: Each child experiences a mutation with a predetermined probability when the crossover operator is applied. The purpose of mutations is to introduce new information that is not already present in the population by randomly changing portions of the new chromosome at a mutation spot that is also chosen at random.

Afterward, the next generation of chromosomes will be produced, and the mentioned steps will be repeated until reach the chopping criterion or the assumed minimum of error in the results. Figure 2 shows the flowchart of the GA algorithm (1).

5 Implementation of GA in a GEP Problem

In this section, an example of a GA simulation of the GEP problem as a power system problem is presented. The proposed framework aims to investigate a long-term conventional GEP problem using a GA algorithm. This study considers Iran's power plants as a research issue. The implementation of the GA on other power system problems is similar to what happens in solving the GEP problem, and as a result, the implementation of the GA algorithm in another model has been neglected.

While the GA is straightforward, it may require a relatively long time to arrive at the ultimate solution to a power system problem. With the rise of high-speed computing capabilities, more advanced methods offering greater precision can be the basis for determining the optimal solution. The GA is highly responsive to the initial feasible values and may be characterized by the initial population chosen to initiate the search process.

Fig. 2 GA algorithm flowchart (1)

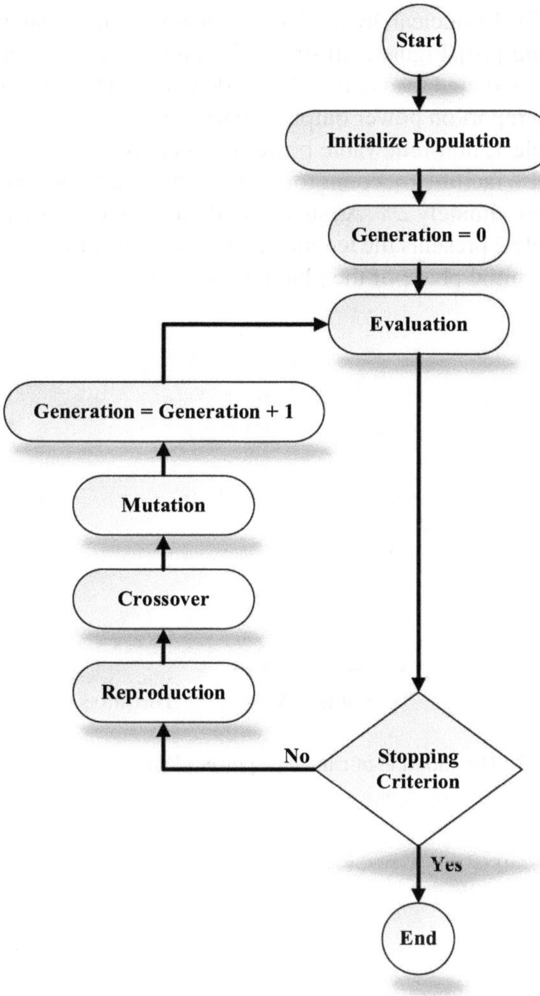

The goal of resolving the GEP optimization problem is to determine the optimal number of units to be installed within the planning horizon. As it has been mentioned that the GEP problem is a nonlinear optimization problem, it cannot be solved using convex optimization methods. The objective function of the problem consist of different costs as presented in Eqs. (1) and (2). This a centralized optimization which can be solved by GA. In doing so, GA attempts to find the best solution using the search technique presented in Sect. 4. In the following the problem definitions have been presented and the results will be discussed.

For this study, nine technologies were identified as the most appropriate options for Iran over a 15-year planning period. The chosen alternatives are natural gas steam turbine (NGST), natural gas combined cycle (NGCC), natural gas combustion turbine

(NGCT), nuclear, hydroelectric, onshore wind, solar, biomass, and geothermal power plants [19]. Figure 3 illustrates the range of potential power plant choices. Meanwhile, the predicted load demands and demand peaks for Iran are depicted in Fig. 4 (as seen in a report on power output capacity by the TAVANIR Company [20]). As shown in Table 1, non-renewable power plants make up over 83% of the total capacity, while hydro facilities account for 14.8%, and renewable and nuclear power plants make up approximately 2%. As such, fossil fuels are the primary energy source in this market. Table 2 presents the technical and economic attributes of the selected alternatives for the initial phase of the planning horizon.

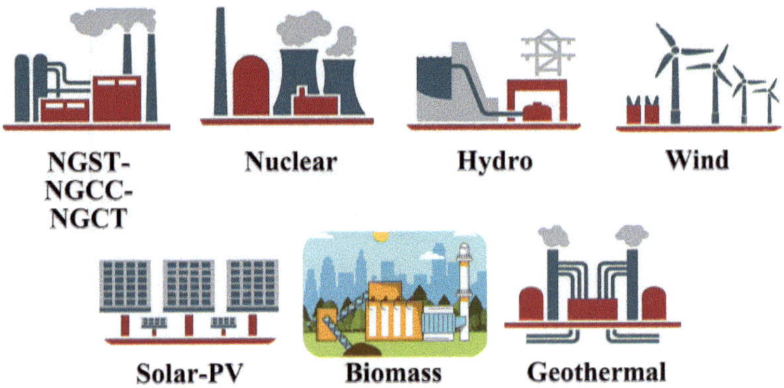

Fig. 3 The variation of candidate power plants

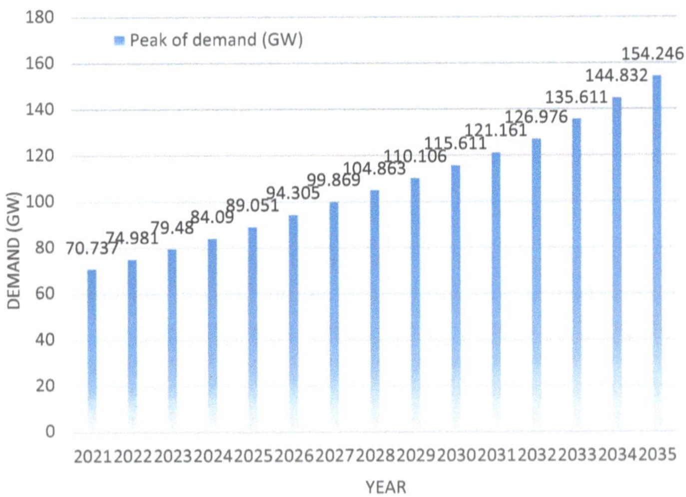

Fig. 4 The predicted peaks of Iran's demand

Table 1 The Power Sector in Iran as it stands right now regarding different technologies [15]

Classification of the unit	NGST	NGCC	NGCT	Nuclear	Hydro	Wind	Solar PV	Biomass	Geothermal
Unit capacity (MW)	160	300	160	1000	200	100	100	100	50
Number of each unit	99	59	155	1	52	2	1	1	2
Total installed capacity (%)	22.5	25.2	35.3	1.4	14.8	0.28	0.14	0.14	0.14
Available capacity (MW)	15840	17700	24800	1000	10400	200	100	100	100

Table 2 The choices, in terms of their technical and economic attributes [13]

Classification of the unit	NGST	NGCC	NGCT	Nuclear	Hydro	Wind	Solar PV	Biomass	Geothermal
Unit capacity (MW)	160	300	160	1000	200	100	100	100	50
Max num. of unit	5	22	55	5	33	39	39	17	55
Investment ($10^3 \times$\$/MW)	1250	1035	676	5615	2936	2213	3950	4114	6243
Fuel cost (\$/MWh)	58.82	47.45	78.43	12.06	74	0	0	41.47	0
O&M cost (\$/MW)	37150	14390	6700	88750	85000	28070	16700	100500	129484
Capacity factor	0.77	0.85	0.85	0.85	0.5	0.3	0.11	0.8	0.75
Theory time working (h)	6500	7000	7000	7000	7000	3000	2640	8000	8000
CO_2 generation rate (ton CO_2/MWh)	0.76	0.344	0.52	0	0	0	0	0	0

The suggested framework has been implemented for Iran's power sector, whereby numerous criteria such as cost components and performance attributes of electricity production technologies serve as vital inputs to the model and have major impacts on the outcomes. In the GA optimization, the inputs to this issue were treated as chromosomes. The system's reserve capacity have been adjusted at a maximum of 0.15 ($RM_{\max}$) and a minimum of 0.05 ($RM_{\min}$). Limits of 10 and 22% of total installed capacity for renewable energy have been defined to ensure the stability of the electricity grid. Furthermore, the upper limits of technologies have been set at 35% of the total installed capacity.

After implementing the GA optimization on the GEP problem the results have been illustrated in the following tables and figures. The number of new units of i-technology to be added during the planning horizon is given in Table 3. Moreover, the total amount of new units that will be added across the planned horizon is presented in Fig. 5.

The trend of the shares of fossil-based and renewable technologies during the planning horizon is depicted in Fig. 5. It can be observed that the share of renewable energies gradually increases over time, while the share of fossil-based technologies decreases. It appears that the key reason to achieve such a result is taking into account the macro requirements in the standard GEP problem. The findings point to a growing movement in the country toward a greener, more sustainable and ecologically friendly electricity source. This transition also presents opportunities for the country, such as improved employment, increased national capabilities, export potential, and lower fossil fuel usage.

Table 3 Numbers of future units throughout the planned horizon

Type of unit/year	NGST	NGCC	NGCT	Nuclear	Hydro	Wind	Solar PV	Biomass	Geothermal
2021	4	4	9	2	16	31	5	5	41
2022	3	3	9	1	13	32	32	12	10
2023	4	11	8	5	13	39	8	16	53
2024	0	5	3	2	0	0	7	8	54
2025	0	7	1	1	7	0	15	3	22
2026	0	2	0	1	13	18	5	4	2
2027	3	1	20	1	0	34	15	0	1
2028	2	0	21	1	9	9	12	3	3
2029	3	1	14	0	11	2	9	6	50
2030	0	8	0	0	28	23	7	3	10
2031	0	3	4	0	0	2	7	0	14
2032	0	4	1	1	10	8	4	1	22
2033	4	2	34	2	26	27	18	13	44
2034	0	8	3	1	0	11	8	1	1
2035	3	15	45	2	12	20	21	10	46

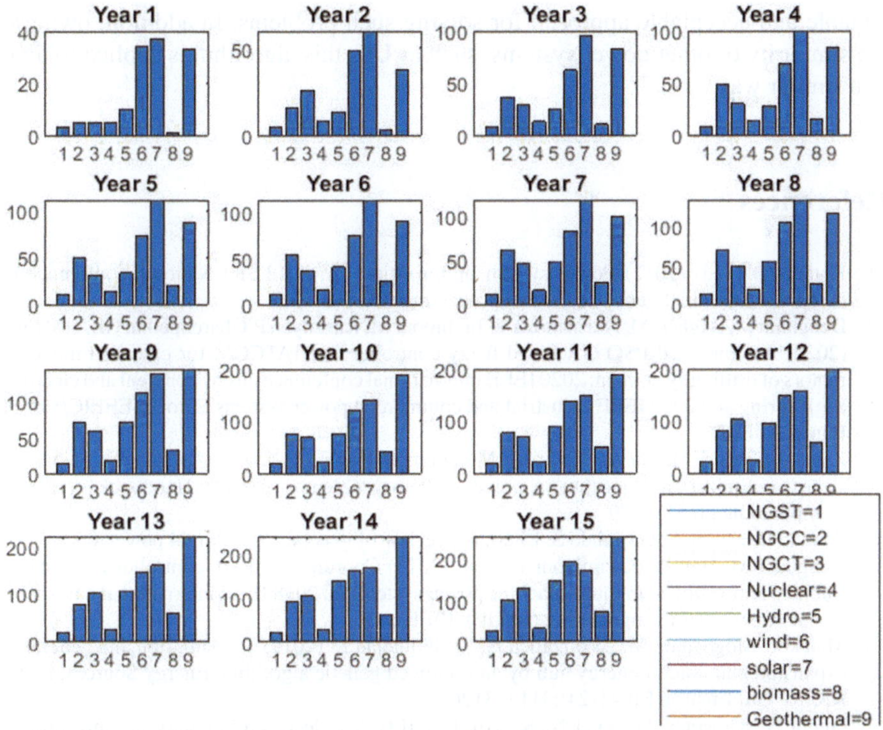

Fig. 5 Accumulated estimation of the number of units to be constructed during the planning period

6 Conclusions

This chapter focuses on the application of genetic algorithms to power systems. In the beginning, two important issues in power systems including GEP and UC have been introduced with the mathematical modelling of these problems. Afterwards, the description of how to apply the GA to GEP and UC problems is given. Following the formulation of the GEP problem for Iran's power industry, the GA was applied to find the optimal number of units of each technology that should be installed during the planning horizon. The GA uses a population of candidate solutions, which evolve and improve through several generations of selection, crossover, and mutation. The objective function used in the GA is the total cost of the system over the planning horizon, which includes investment and operational costs of all technologies. The upper and lower bounds of the renewable energy share, as well as the upper bound of each technology, were taken into account during the optimization process to ensure system stability and reliability. The results indicate a decreasing share of fossil-based technologies and an increasing share of renewable energies, indicating a move towards a more sustainable power supply for Iran. The results obtained from this algorithm showed that the GA implementation on power system problems can be a suitable and acceptable approach for solving such problems. In addition, owing to the similarity of other power systems, such as UC, this algorithm is applied to them in a similar way.

References

1. Kothari DP (ed) (2012) Power system optimization. In: 2012 2nd National conference on computational intelligence and signal processing (CISP). IEEE
2. Dehghani M, Bagheri M, Nurmanova V, Taghipour M, Karimyan P, Gharehpetian GB, et al (eds) (2020) An improved PSO-GA based fuzzy controller of STATCOM for cascaded induction motors of drilling system. In: 2020 IEEE international conference on environment and electrical engineering and 2020 IEEE industrial and commercial power systems Europe (EEEIC/I&CPS Europe). IEEE
3. Zamani-Gargari M, Nazari-Heris M, Mohammadi-Ivatloo B (2017) Application of particle swarm optimization algorithm in power system problems. Elsevier, Handbook of Neural Computation, pp 571–579
4. Shende D, Morey S, Gawande S, Literature review of genetic algorithm in power system
5. Dehghani M, Han W, Karimipour H (eds) (2018) Coordinated fuzzy controller for dynamic stability improvement in multi-machine power system. In: 2018 IEEE international conference on smart energy grid engineering (SEGE). IEEE
6. Malakoti-Moghadam M, Askarzadeh A, Rashidinejad M (2019) Transmission and generation expansion planning of energy hub by an improved genetic algorithm. Energy Sources, Part A Recove Util Environ Eff 41(24):3112–3126
7. Gupta N, Khosravy M, Patel N, Senjyu T (2018) A bi-level evolutionary optimization for coordinated transmission expansion planning. IEEE Access 6:48455–48477
8. Sahragard A, Falaghi H, Farhadi M, Mosavi A, Estebsari A (2020) Generation expansion planning in the presence of wind power plants using a genetic algorithm model. Electronics 9(7):1143
9. Singh Chawda G, Prakash Mahela O, Gupta N, Khosravy M, Senjyu T (2020) Incremental conductance based particle swarm optimization algorithm for global maximum power tracking of solar-pv under nonuniform operating conditions. Appl Sci 10(13):4575
10. Jalalzad SH, Yektamoghadam H, Haghighi R, Dehghani M, Nikoofard A, Khosravy M et al (2022) A game theory approach using the TLBO algorithm for generation expansion planning by applying carbon curtailment policy. Energies 15(3):1172
11. Jo K-H, Kim M-K (2018) Improved genetic algorithm-based unit commitment considering uncertainty integration method. Energies 11(6):1387
12. Postolov B, Iliev A (2022) New metaheuristic methodology for solving security constrained hydrothermal unit commitment based on adaptive genetic algorithm. Int J Electr Power Energy Syst 134:107163
13. Ponciroli R, Stauff NE, Ramsey J, Ganda F, Vilim RB (2020) An improved genetic algorithm approach to the unit commitment/economic dispatch problem. IEEE Trans Power Syst 35(5):4005–4013
14. El-Shorbagy MA, Omar HA, Fetouh T (2022) Hybridization of Manta-ray foraging optimization algorithm with pseudo parameter-based genetic algorithm for dealing optimization problems and unit commitment problem. Mathematics 10(13):2179
15. KS GD (2019) Hybrid genetic algorithm and particle swarm optimization algorithm for optimal power flow in power system. J Comput Mech Power Syst Control 2:31–37
16. Dehghani M, Taghipour M, Sadeghi Gougheri S, Nikoofard A, Gharehpetian GB, Khosravy M (2021) A deep learning-based approach for generation expansion planning considering power plants lifetime. Energies 14(23):8035

17. Haghighi R, Yektamoghadam H, Dehghani M, Nikoofard A (2021) Generation expansion planning using game theory approach to reduce carbon emission: a case study of Iran. Comput Ind Eng 162:107713
18. Singh A, Khamparia A (2020) A hybrid whale optimization-differential evolution and genetic algorithm based approach to solve unit commitment scheduling problem: WODEGA. Sustain Comput Inform Syst 28:100442
19. Noorollahi E, Fadai D, Ghodsipour SH, Shirazi MA (2017) Developing a new optimization framework for power generation expansion planning with the inclusion of renewable energy—a case study of Iran. J Renew Sustain Energy 9(1):015901
20. Iran's energy balance sheet 2014. http://epprjournal.ir/index.php?slc_lang=fa&sid=1

Chapter 6
Modified Genetic Algorithm for the Profit-Based Unit Commitment Problem

Lucas Santiago Nepomuceno, Edimar Jose de Oliveira, Arthur Neves de Paula, and Leonardo Willer de Oliveira

1 Introduction

In the regulated electricity industry, the Unit Commitment (UC) problem refers to the optimization of generation resources over a specified period (usually daily or weekly) to minimize the operational cost of the system. The UC problem considers several constraints, such as technical constraints of the generating units, network constraints, balance demand, etc. [1]. The solution of the UC problem specifies the ideal scheduling of the generating units along the analyzed horizon, i.e., the solution provides the commitment schedule (on/off) and the generation scheduling of committed units [2].

As the process of restructuring and deregulation in the electricity industry became more common, the main objective of the UC problem changed to maximize the profit of electricity generation companies (GENCOs) that participate in competitive markets. This gave rise to the Profit-Based Unit Commitment (PBUC) problem. In deregulated environments, GENCOs define the amount of energy and reserve to be traded in the day-ahead and ancillary markets and program their generating units

This research was supported in part by Brazilian Federal Agency for Support and Evaluation of Graduate Education (CAPES), National Research Council (CNPq), State of Minas Gerais Research Foundation (FAPEMIG) and Brazilian Institute of Science and Technology (INERGE).

L. S. Nepomuceno · E. J. de Oliveira (✉) · A. N. de Paula · L. W. de Oliveira
Department of Energy, Federal University of Juiz de Fora, Juiz de Fora, Brazil
e-mail: edimar.oliveira@ufjf.br

L. S. Nepomuceno
e-mail: lucas.nepomuceno@engenharia.ufjf.br

A. N. de Paula
e-mail: arthur.paula@engenharia.ufjf.br

L. W. de Oliveira
e-mail: leonardo.willer@ufjf.br

M. Khosravy et al. (eds.), *Frontiers in Genetics Algorithm Theory and Applications*, Springer Tracts in Nature-Inspired Computing,
https://doi.org/10.1007/978-981-99-8107-6_6

to maximize profits by respecting operating constraints [6]. The PBUC solution is obtained based on forecasts of the amount and the price of energy, and the reserve requested in each analyzed time interval. The quality of the PBUC solution is significant for GENCOs as it forms the basis of a good strategy in market competition. It is important to emphasize that this problem is not uniform in all regions and countries, as each electricity market has its own structure [7].

Generally, PBUC is formulated as a complex mixed-integer nonlinear programming problem. PBUC can be a single-objective problem when profit maximization is sought or a multi-objective problem when, in addition to profit maximization, is sought to minimize the cost of emission of pollutants or maximize social welfare [9]. The main differences between UC in the regulated environment and PBUC are the objective function, the demand balance, and spinning reserve constraints. In the deregulated environment, these constraints are modeled as inequality equations since GENCOs are not obligated to supply all the demand and all the expected reserve, while in UC in the regulated electricity industry these constraints are modeled as equality equations [8]. In a regulated environment, the generation units are exclusively owned by a single company or government entity. Therefore, the UC problem in regulated environments is intrinsically related to the minimization of operating costs [4]. In contrast, in PBUC, the objective function is established to maximize the profit of a GENCO [8].

The PBUC problem has three solution categories: Numerical methods, Metaheuristics, and Hybrid methods [9]. Numerical methods are purely mathematical optimization techniques, also known as classical methods. These methods include Dynamic Programming (DP), Mixed-Integer Programming (MIP), and Lagrangian Relaxation (LR). However, classical optimization methods have limitations, such as long computational time and the risk of getting trapped in a local optimum due to non-convexity, multimodality, and high-dimensional systems. Therefore, it is not advisable to use classical methods for solving PBUC problems [9].

Metaheuristic techniques are adequate for solving the PBUC problem. They reduce the computational time in comparison to classical methods, do not require the use of derivatives, can avoid stagnation in local minima, and are flexible for algorithmic adaptations and hybridizations with other techniques [9]. Furthermore, although obtaining the global solution is not guaranteed, metaheuristics allow the use of more complex formulations without the need for many simplifications.

The Genetic Algorithm (GA) was one of the first metaheuristic techniques used to solve the PBUC. In [8], the authors adapted the Genetic-Based Unit Commitment Algorithm method described by [17] to maximize the profit of a GENCO in the energy market. In [5], the authors presented a GA-based solution for the PBUC problem. The proposed method utilizes penalties on solutions (chromosomes) that violate constraints, with the value of the penalty smoothly increasing as the number of generations increases. Additionally, the authors incorporated two special operators in the GA structure. In [18], the GA is used to solve the PBUC by considering both the energy and reserve markets.

The studies [5, 8, 18] have adopted penalty functions to handle the violation of constraints in the GA model. However, the use of penalty functions can decrease the efficiency of algorithms in obtaining global best solutions, especially when applied to large-scale test systems [2].

In [19], an improved binary Particle Swarm Optimization (PSO) and a standard value PSO are used iteratively to solve the PBUC problem, and the binary version of PSO is responsible for obtaining on/off units status, and the standard version is used for solving the economic dispatch problem. In [20], the authors also evaluate three other PSO techniques to solve the PBUC problem: Chaotic PSO (CPSO), New PSO (NPSO), and Dispersed PSO (DPSO) and compare the results. The PBUC formulation considers both the energy and reserve markets. The results indicated that CPSO achieved the highest profit, while NPSO and DPSO exhibited a slower convergence.

The PBUC problem has been addressed in various ways in the literature. One approach, as described in [21], involves using the Shuffled Frog Leaping Algorithm (SFLA) to optimize a bi-objective function. The aim is to maximize the profit of the GENCO while minimizing thermal unit emissions. Another study [22] utilizes the Imperialistic Competitive Algorithm (ICA) to solve the PBUC problem. A third approach, outlined in [24], uses a dynamic economic dispatch method and a Binary Fish Swarm Algorithm (BFSA) to solve the problem. This approach considers ramp rate limits and reserve generations during the complete scheduling time horizon in both the day-ahead and ancillary markets.

Several different algorithms have been proposed to solve the PBUC problem, as reported in various studies. In [25], the authors proposed the Binary Coded Fireworks Algorithm (BFWA) to consider both energy and reserve markets. They utilized the tangent hyperbolic transfer function. In [26], a binary version of the Sine Cosine Algorithm (SCA) was used to solve the PBUC problem in both the day-ahead market and reserve market. The authors incorporated various transfer functions, including tangent hyperbolic, sigmoid, modified sigmoid, and arctan. Two other studies, [27, 28], used binary versions of the Whale Optimization Algorithm (WOA) and Grey Wolf Optimizer (GWO), respectively, to solve the PBUC problem. In [29], the Intensify Harris Hawks Optimizer (IHHO) was used to solve the PBUC problem, taking into account the influence of battery electric vehicles. In [30], the Memetic Binary Differential Evolution algorithm (MBDE) was proposed to solve the PBUC problem. The authors also presented a novel unit commitment strategy based on a priority list of units and total profit, along with a straightforward algorithm based on the priority method to allocate power to the committed units.

Hybrid techniques involve the integration of two or more methods mentioned above to improve both solution quality and computational time [9]. One of the first proposed hybrid methods for solving the PBUC was presented in [31], where a hybrid model between LR and GA was used to solve the unit commitment problem, with the GA being used to update the Lagrangian multipliers. Since then, the hybridization of the classical optimization technique Lagrangian Relaxation (LR) and metaheuristic techniques has become popular for PBUC solutions. Hybrid models between LR and Evolutionary Programming (EP) can be cited [32], as well as hybrids between LR and other techniques such as ant colony search algorithm [33], PSO [34], firefly algorithm [35], Artificial Bee Colony (ABC) [36], and Weed Optimization Algorithm (WO) [37].

The specialized literature also presents hybrid methods for solving PBUC, which combine metaheuristics. For instance, a hybrid model between Binary Successive Approach (BSA) and Civilized Swarm Optimization (CSO) has been proposed in [38], as well as a hybrid model between Artificial Immune System (AIS) and GA in [39]. Reference [40] presents a hybrid model between Cuckoo Search (CS) and GWO.

1.1 Contributions

Based on the previous background, the main contribution of this work is to propose suitable adaptations to the GA to improve its performance through the use of special operators, as proposed in [41]. These operators include the Swap Window Operator, Window Mutation Operator, and Swap Window Hill Climb Operator. Additionally, a heuristic is utilized to repair chromosomes (solutions) that violate the minimum up/down time constraints.

The proposed modifications will be tested to solve the Profit-Based Unit Commitment (PBUC) problem, which considers a GENCO seeking to maximize its profits by trading energy and reserves in the day-ahead and ancillary markets, respectively.

The performance of the proposed adaptations will be compared with that of other metaheuristics, namely the GWO, WOA, SCA, and HHO. These metaheuristics were modified to consider binary variables using a modified sigmoid transfer function. To demonstrate the effectiveness of the proposed method, systems of different scales will be considered, including a small-scale system with 10 units and a large-scale system with 100 units.

2 Problem Formulation of PBUC

The PBUC formulation has been widely used in literature as described in reference [9]. This problem is represented by a mixed-integer nonlinear programming (MINLP) optimization problem, with an explosive combinatorial characteristic, seeking to determine the commitment schedule, the generation scheduling of committed units, and reserve allocation. The problem is formulated as follows:

2.1 Objective Function

The profit of the GENCO is calculated as the difference between the total revenue (TR) and the total cost (TC) for all the units over a specified time period, and is formulated in (1).

$$\text{Maximize} \sum_{g}^{NG} \sum_{t}^{NT} (TR\,(g,t) + TC\,(g,t)), \tag{1}$$

where NT and NG are the total time interval and the total number of generating units, respectively.

The total revenue is obtained through the sale of energy in the day-ahead market and reserves in the ancillary market. Therefore, the GENCO coordinates its generating units based on the forecasted energy market price (\$/MWh), forecasted reserve price (\$/MW/h), forecasted load requirement in the energy market (MW), and forecasted reserve requirement in the ancillary market (MW). A good forecast is essential for profit maximization. Additionally, the revenue received depends on the market model used to pay for energy and reserve sales. In this work, two payment methods are considered: Payment for Power Delivered (PPD) and Payment for Reserve Allocated (PRA) [42].

Payment for Power Delivered (PPD): In the PPD method, payment for allocated reserve is made only when it is actually utilized. In this approach, the reserve price is higher than the energy market price [42]. The mathematical model for this approach is presented in (3).

$$\begin{aligned} TR\,(g,t) &= \sum_{g}^{NG} \sum_{t}^{NT} P_{g,t} \cdot FSP_t \cdot UC_{g,t} \\ &+ r \sum_{g}^{NG} \sum_{t}^{NT} R_{g,t} \cdot FRP_t \cdot UC_{g,t}, \end{aligned} \tag{2}$$

where FSP_t is the forecasted energy price at hour t, FRP_t is the forecasted reserve price at hour t and r is the probability for calling and generating the reserve.

Payment for Reserve Allocated (PRA): The PRA approach involves the allocation of reserves to GENCO, which receives a lower reserve price per unit when the reserves are allocated but not used, and an energy market price when the reserves are used. A mathematical model for this approach is presented in (3), as reported in [42].

$$\begin{aligned} TR\,(g,t) &= \sum_{i}^{N} \sum_{t}^{T} P_{g,t} \cdot FSP_t \cdot UC_{g,t} \\ &+ \sum_{i}^{N} \sum_{t}^{T} ((1-r)\,FRP_t + r \cdot FSP_t) \cdot R_{g,t} \cdot UC_{g,t}. \end{aligned} \tag{3}$$

Total Cost: The total operating cost used in this article is obtained by summing the fuel cost, reserve deployment cost, and start-up cost of all generating units, as shown in Eq. (4).

$$TC(g,t) = (1-r)\sum_{i}^{NG}\sum_{t}^{NT} F_i\left(P_{g,t}\right)\cdot UC_{g,t} + r\sum_{i}^{NG}\sum_{t}^{NT} F_i\left(P_{g,t}+R_{g,t}\right)\cdot UC_{g,t} + SUC_{g,t}\cdot UC_{g,t}. \tag{4}$$

The fuel cost and start-up cost of a generator at time t for generated power $P_{g,t}$ and $SUC_{g,t}$, respectively, are represented by $F_g\left(P_{g,t}\right)$ and $SUC_{g,t}$, where the fuel cost function for a thermal generation unit is expressed in a quadratic form as shown in (5).

$$F_i\left(P_{g,t}\right) = a_i + b_i P_{g,t} + c_i P_{g,t}^2. \tag{5}$$

The fuel cost coefficients of unit g are represented by a_g, b_g, and c_g.

The cost to start up a unit at any hour t is dependent on the duration of time that the unit has been idle before start-up, which is mathematically represented in (6).

$$SUC_g = \begin{cases} HSUC_g; & MDT_g \le T_{g,t}^{off} \le MDT_g + TCO_g \\ CSUC_g; & T_{g,t}^{off} > MDT_g + TCO_g. \end{cases} \tag{6}$$

Variables used in the model include the hot start-up cost $HSUC_g$ and the cold start-up cost $CSUC_g$ for a given thermal unit g, as well as the time TCO_g required for the thermal unit to cool down, the minimum down time MDT_g for unit g, and the time interval $T_{g,t}^{off}$ that unit g remained continuously off until hour t.

2.2 Constraint

Demand balance: The demand balance constraint is presented in (7). It is important to note that this constraint is different from the traditional UC problem. The goal of PBUC is to optimize the earnings of a GENCO, and it is not a requirement to fulfill the entire predicted/anticipated demand for a given hour.

$$\sum_{i=1}^{N} P_{g,t}\cdot UC_{g,t} \le PD_t, \tag{7}$$

where PD_t is the forecasted load at hour t.

Spinning reserve: The electric system's requirement for spinning reserve during operation is determined as a percentage of the present hourly demand. Just like the load constraint, the GENCO is not obligated to follow the reserve forecasted for a specific hour. The constraint on the system's rotating reserve service is shown in (8).

$$\sum_{i=1}^{NG} R_{g,t} \cdot UC_{g,t} \geq RD_t, \tag{8}$$

where RD_t is the forecasted reserve at hour t.
Unit minimum up time: A thermal unit should remain committed (on) for a minimum up time. The minimum up time constraints are given in (9).

$$(T_{g,t}^{on} - MUT_i)(UC_{g,t-1} - UC_{g,t}) \geq 0, \tag{9}$$

where MUT_g is the minimum up time period of thermal unit g; $T_{g,t}^{on}$ is the time interval that unit g has been continuously on up to t hours, and it is calculated according to (10).

$$T_{g,t}^{on} = (T_{g,t-1}^{on} + 1)UC_{g,t}. \tag{10}$$

Unit minimum down time: When a thermal unit is de-committed, the unit should remain in an off state for a minimum down time. The minimum down time constraintis given in (11).

$$(T_{g,t}^{off} - MDT_i)(UC_{g,t} - UC_{g,t-1}) \geq 0, \tag{11}$$

where, MDT_g denotes the minimum duration that a thermal unit g must remain offline after being shut down, while $T_{g,t}^{off}$ represents the interval of time that the same unit g has been offline prior to hour t.

$$T_{g,t}^{off} = (T_{g,t-1}^{off} + 1)(1 - UC_{g,t}) \quad g \in [1,\ NG],\ t \in [1,\ NG] \tag{12}$$

Generation power limits: The power produced by each generator operating at hour t must be kept within its designated capacity limit as described in Eq. (13).

$$P_g^{min} \leq P_{g,t} \leq P_g^{max} \quad g \in [1,\ NG],\ t \in [1,\ NG] \tag{13}$$

where P_g^{min} and P_g^{max} represent the minimum and maximum generation capacities of unit g, respectively.
Reserve limits: The reserve limit constraint is present in (14).

$$0 \leq R_{g,t} \leq P_g^{max} - P_{g,t} \quad g \in [1,\ NG],\ t \in [1,\ NG] \tag{14}$$

Initial units status: The value of the Initial Unit Status (IUS) for a unit g is determined based on whether it was committed or de-committed for sub-intervals before the scheduled start time. A positive IUS value indicates that the unit was committed, while a negative IUS value indicates that the unit was de-committed. The calculation of the IUS is given by Eq. (15). This information is used to determine the initial status of units (IUS_g) before the start of the schedule:

$$UC_{g,t=0} = \begin{cases} 0; & IUS_g < 0 \\ 1; & IUS_g > 0 \end{cases} \tag{15}$$

3 Modified Genetic Algorithm

Genetic algorithms, proposed in [43], are global optimization techniques inspired by the Darwinian principle of natural selection. GA can be easily implemented to solve hard optimization problems, providing great modeling flexibility. Algorithms based on GA have demonstrated significant potential in a variety of industrial applications, including self-organized control systems [44], economic load dispatch [45], distributed generator [46], congestion management [47], and many others.

In this work, the modified GA for the PBUC comprises six mechanisms, namely initialization, selection, crossover, mutation, special operators, and constraint handling. The control parameters involved in the GA to solve the PBUC are population size NP (number of chromosomes); the number of participants in the tournament n_{tour} (selection mechanism); crossover probability p_c; mutation probability p_m; special operator probability p_{es}; the number of elite members n_{elite}; and the maximum number of generations NI (maximum number of iterations).

3.1 Initialization

This mechanism is responsible for the random creation of the initial population of candidate solutions. A candidate solution (chromosome), whose length is the product of the number of generating units NG and the scheduling hours NT, is composed of the commitment decision binary variables. A chromosome "X" is represented as follows in (16):

$$\begin{aligned} X = \Big[& UC_{1,1}\; UC_{1,2}\; UC_{1,3}\; \cdots\; UC_{1,NT} \\ & UC_{2,1}\; UC_{2,2}\; UC_{2,3}\; \cdots\; UC_{2,NT}\; \cdots \\ & UC_{NG,1}\; UC_{NG,2}\; UC_{NG,3}\; \cdots\; UC_{NG,NT} \Big] \end{aligned} \tag{16}$$

3.2 Selection

This mechanism is responsible for the selection of the chromosomes that participate in the reproduction process (crossover). In this work, the Tournament Selection process is used, where the chromosomes are selected according to their fitness values from a stochastic roulette wheel in pairs. The selected chromosomes are named parents. After selection, the individuals with better fitness value are used in the crossover process.

3.3 Crossover

This mechanism is a random process of recombination of parent chromosomes. After performing the crossover operation, it gives rise to form two new chromosomes (children) that inherit solution characteristics from both parents.

3.4 Mutation

This mechanism performs change from 0 to 1 and vice versa in randomly selected chromosomes.

3.5 Elitism

This mechanism ensures that the fittest chromosomes are retained in the population across iterations. To achieve this, the n_{elite} top solutions are saved for the next generation in each iteration.

3.6 Constraint Handling

The process of initializing a population randomly and evolving (updating the position) of a metaheuristic can result in solutions that violate the minimum up/down time constraint. Therefore, this work uses a heuristic procedure to avoid the potential violation of these constraints and, thus, improve the quality of the solution. The algorithm is presented in (1). The PBUC problem requires constraint repair only for minimum up/down time but not for spinning reserve, as in the case of conventional UC problems [27].

Algorithm 1 Minimum up/down time constraint handling

Input: UC, MUT, MDT, IUS, NG, NT
Output: UC

```
t ← 0
while t ≤ NT do
  t ← t + 1
  g ← 0
  Compute T^{on}_{g,t} and T^{off}_{g,t}, applying (10) and (12), respectively.
  while g ≤ NG do
    g ← g + 1
    if t = 1 then
      if UC_{g,t=0} = 1 & UC_{g,t} = 0 then
        if T^{on}_{g,t=0} ≥ MUT_g then
          UC_{g,t} ← 1
        else
          UC_{g,t} ← 0
        end if
      else if UC_{g,t=0} = 0 & UC_{g,t} = 1 then
        if T^{off}_{g,t=0} ≥ MDT_g then
          UC_{g,t} ← 1
        else
          UC_{g,t} ← 0
        end if
      end if
    else
      if UC_{g,t-1} = 1 & UC_{g,t} = 0 then
        if T^{on}_{g,t-1} ≥ MUT_g then
          UC_{g,t} ← 1
        else
          UC_{g,t} ← 0
        end if
      else if UC_{g,t-1} = 0 & UC_{g,t} = 1 then
        if T^{off}_{g,t-1} ≥ MDT_g then
          UC_{g,t} ← 1
        else
          UC_{g,t} ← 0
        end if
      end if
    end if
  end while
end while
```

3.7 *Special Operators*

With the objective of improving the convergence of GA in the UC problem, authors in [41] added several operators that are executed if a certain predefined probability occurs.

In the present work, these groups of operators will be called Special Operators, including Swap Window, Window Mutation, and Swap Window Hill Climb. For each of the special Operators, a unique structure is created, giving rise to three algorithms: Genetic

Algorithm with Swap Window (GASW), Genetic Algorithm with Window Mutation (GAWM), and Genetic Algorithm with Swap Window Hill Climb (GASWHC).

The special operators applied in this work are described as follows:

Swap Window Operator:
This operator applies to all individuals in the population.
Step 1: Randomly select two units.
Step 2: Randomly select a "time window" of width w hours between 1 and NT.
Step 3: Randomly select a start period between 1 and $NT - w$ for the "time window".
Step 4: Exchange the scheduled status of the two units included in the window.

Window Mutation:
This operator also applies to all individuals in the population.
Step 1: Randomly select one unit.
Step 2: Randomly select a "time window" of width w hours between 1 and NT.
Step 3: Randomly select a start period between 1 and $NT - w$ for the "time window".
Step 4: All the commitment statuses in the time window are mutated, that is, if its value is "1", it becomes "0", and vice versa.

Swap Window Hill Climb:
This operator is applied only to the best solution of each iteration.
Step 1: Randomly select two units.
Step 2: Randomly select a "time window" of width w hours.
Step 3: Exchange the commitment status of the two units.
Step 4: Evaluate the new solution obtained in the previous step and, if better, it is kept.
Step 5: If the new solution is better, it will be used. Otherwise, it is restored to its initial state.
Step 6: Shift the window by one hour and the
Step 7: Repeat the procedure until the window reaches the last hour of the scheduling horizon.

Flowchart of the proposal:
The flowchart depicting the proposed modifications to the GA for solving the PBUC problem is presented in Fig. 1. The process begins with the random creation of the initial population. The next step is the constraint handling process, which aims to correct solutions that violate the minimum up/down time constraints.

After the correction of the initial population, individuals are evaluated by calculating their fitness value, and the best individuals are stored in the elitism operator, which ensures the preservation of the fittest individuals for the next generation.

The evolutionary process begins with the selection of a random value between 0 and 1. If this value is less than or equal to the special operator probability parameter p_{es}, the entire population will pass through the special operator. Otherwise, the selection operator will be applied, followed by the crossover and mutation operators.

After the application of these operators, the individuals in the population pass through the "constraint handling" again and are evaluated. Finally, the elitism operator is applied before starting a new iteration. This evolutionary process is repeated until the maximum number of iterations is reached.

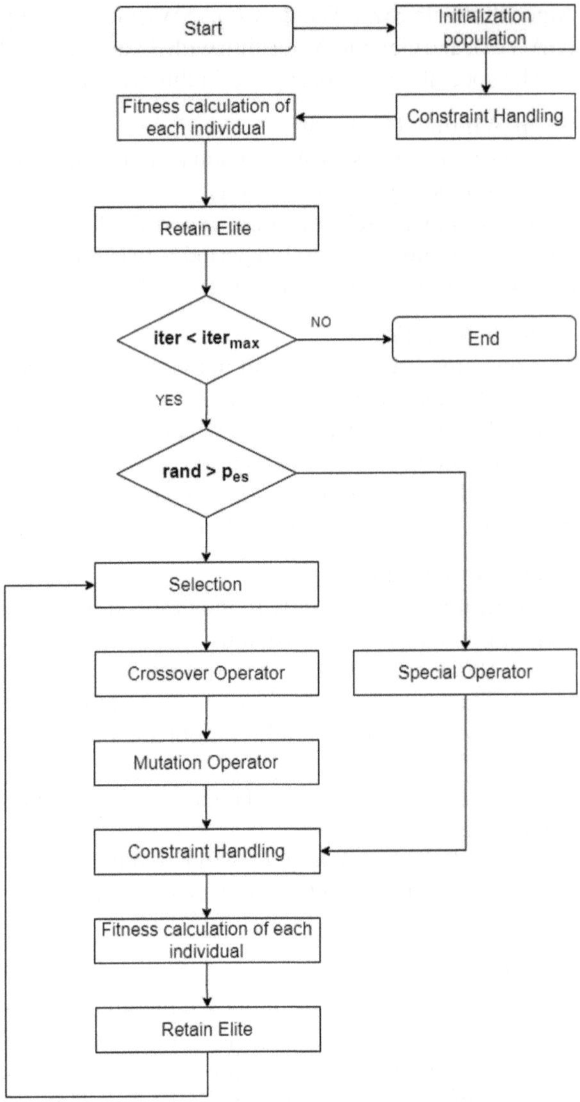

Fig. 1 Modified genetic algorithm for PBUC flowchart

4 Simulation Results and Discussion

In this section, two standard test systems were examined to evaluate the effectiveness of the proposed modifications of the GA to solve the PBUC problem. Additionally, a comparison of its performance was presented with other algorithms such as GWO, WOA, SCA, and HHO. The metaheuristics were adapted for using binary variables via a modified sigmoid transfer function [48], together with the constraint handling mechanism present in Algorithm Minimum up/down time constraint handling.

The first test system is a small-scale system with 10 thermal generation units. The second one is a large-scale system with 100 thermal generation units. For both systems, a period composed of 24 intervals of 1 hour is considered. The 24-h market forecast data is displayed in Table 1. The data for the 10-unit system and 24-h market forecast were obtained from the reference [42]. The 100-unit system is obtained by scaling up the 10-unit system by 10 times, and the hourly market forecast load and reserves are multiplied by a factor of 10. These data sets are widely used in various studies related to PBUC [5, 20, 25, 27].

The simulations were conducted on an Intel Core *i*7 processor with 2.7 GHz using MATLAB®software. Four case studies are conducted as follows:

- Case I—10-unit system with PPD payment method
- Case II—10-unit system with PRA payment method
- Case III—100-unit system with PPD payment method
- Case IV—100-unit system with PRA payment method

In cases I and II, the probability of reserve deployment parameter is set to 0.05 and the reserve price parameter is set to 5 times the energy price of the market, while, in cases III and IV, the probability of reserve deployment parameter is set to 0.005 and the reserve price parameter is set to 0.01 times the energy price of the market. These values are commonly used in case studies in the specialized literature [27].

Fifty simulations were conducted for each case study. The maximum iteration value is set to 100 for cases I and II, while, for cases III and IV, it is set to 500 iterations. The maximum iteration value was used as the stopping criterion for the metaheuristics. Table 2 presents the parameters used for the modified GA for PBUC.

Table 1 Forecasted load, reserve, and energy prices data for 10 thermal generating units

Hour	Load forecasted (MW)	Reserve forecasted (MW)	Energy price forecasted ($/MWh)	Hour	Load forecasted (MW)	Reserve forecasted (MW)	Energy price forecasted ($/MWh)
1	700	70	22.15	13	1400	140	24.6
2	750	75	22	14	1300	130	24.5
3	850	85	23.1	15	1200	120	22.5
4	950	95	22.65	16	1100	110	22.3
5	1000	100	23.25	17	1050	105	22.25
6	1100	110	22.95	18	1000	100	22.05
7	1150	115	22.5	19	1200	120	22.2
8	1200	120	22.15	20	1400	140	22.65
9	1300	130	22.8	21	1300	130	23.1
10	1400	140	29.35	22	1100	110	22.95
11	1450	145	30.15	23	900	90	22.75
12	1500	150	31.65	24	800	80	22.55

Table 2 Parameter Selection for Modified Genetic Algorithm for PBUC

Parameter	Chosen value
population size NP	100
number of participants in the tournament n_{tour}	4
crossover probability p_c	0.8
mutation probability p_m	0.2
special operator probability p_{es}	0.03
number of elite members n_{elite}	10

Table 3 Statistical analysis for Case I

	Best profit ($)	Worst profit ($)	Standard deviation ($)	Mean ($)	Median ($)	Time (s)
bGWO	27,610.90	14,168.44	3,109.06	19,819.81	19,640.57	3.99
bWOA	55,635.89	39,229.79	3,744.27	45,458.84	45,538.57	3.04
bSCA	44,424.23	24,553.47	4,632.95	33,924.00	33,918.05	4.55
bHHO	77,524.66	37,277.63	8,145.66	57,294.25	55,497.05	5.21
GA	102,464.19	58,904.57	8,761.78	84,334.97	83,210.84	3.68
GAWM	105,254.57	73,416.34	7,488.08	87,643.29	88,253.49	3.77
GASW	109,986.38	87,839.30	6,415.14	102,084.62	104,460.88	3.91
GASWHC	102,091.16	73,485.50	7,902.95	89,666.01	89,568.45	3.86

The parameter values used in the genetic algorithm were defined after empirical analyses. For the 100-unit system, this work proposes modifications to the Swap Window and Window Mutation operators. Originally, these operators were executed only once for each chromosome of the population if a random value between 0 and 1 was lower than the pre-specified special operator probability parameter p_{es}. For the 100-unit system, the mechanism was applied 10 times to increase exploration in the large-scale system using a GA. The large number of thermal generating units necessitated the use of special operators multiple times to achieve adequate exploration. However, excessive use of these operators may lead to a reduction in the quality of the population.

4.1 Case I

Table 3 presents a statistical comparison of the metaheuristics used to solve the 10-unit system, considering the PPD payment method. Note that the Genetic Algorithm with the special operator Swap Window (GASW) obtained the best solution with a profit of $ 109,986.38, while the second and third best solutions were obtained by GAWM and GASWHC, respectively.

Table 4 Best power and reserve generation schedule for Case I obtained by GASW

Hour	Power generation of units (MW)										Allocated reserve of units (MW)									
	1	2	3	4	5	6	7	8	9	10	1	2	3	4	5	6	7	8	9	10
1	455	245	0	0	0	0	0	0	0	0	0	70	0	0	0	0	0	0	0	0
2	455	295	0	0	0	0	0	0	0	0	0	75	0	0	0	0	0	0	0	0
3	455	395	0	0	0	0	0	0	0	0	0	60	0	0	0	0	0	0	0	0
4	455	455	40	0	0	0	0	0	0	0	0	0	90	0	0	0	0	0	0	0
5	455	455	65	0	25	0	0	0	0	0	0	0	65	0	35	0	0	0	0	0
6	455	455	130	0	60	0	0	0	0	0	0	0	0	0	102	0	0	0	0	0
7	455	455	130	0	110	0	0	0	0	0	0	0	0	0	52	0	0	0	0	0
8	455	455	130	0	160	0	0	0	0	0	0	0	0	0	2	0	0	0	0	0
9	455	455	130	130	130	0	0	0	0	0	0	0	0	0	32	0	0	0	0	0
10	455	455	130	130	162	0	0	0	0	0	0	0	0	0	0	0	0	0	0	0
11	455	455	130	130	162	0	0	0	0	0	0	0	0	0	0	0	0	0	0	0
12	455	455	130	130	162	0	0	0	0	0	0	0	0	0	0	0	0	0	0	0
13	455	455	130	130	162	0	0	0	0	0	0	0	0	0	0	0	0	0	0	0
14	455	455	130	130	130	0	0	0	0	0	0	0	0	0	32	0	0	0	0	0
15	455	455	130	130	30	0	0	0	0	0	0	0	0	0	120	0	0	0	0	0
16	455	455	130	0	0	0	0	0	0	0	0	0	0	0	0	0	0	0	0	0
17	455	455	130	0	0	0	0	0	0	0	0	0	0	0	0	0	0	0	0	0
18	455	455	90	0	0	0	0	0	0	0	0	0	40	0	0	0	0	0	0	0
19	455	455	130	0	0	0	0	0	0	0	0	0	0	0	0	0	0	0	0	0
20	455	455	130	0	0	0	0	0	0	0	0	0	0	0	0	0	0	0	0	0
21	455	455	130	0	0	0	0	0	0	0	0	0	0	0	0	0	0	0	0	0
22	455	455	0	0	0	0	0	0	0	0	0	0	0	0	0	0	0	0	0	0
23	455	445	0	0	0	0	0	0	0	0	0	10	0	0	0	0	0	0	0	0
24	455	345	0	0	0	0	0	0	0	0	0	80	0	0	0	0	0	0	0	0

In addition to obtaining the best solution, GASW also demonstrated good statistical performance, as shown in Table 3. The three special operators significantly improved the performance of GA, with the Swap Window Operator being the most efficient in obtaining better solutions. Tables 4 and 5 provide the best power and reserve generation schedule found by GASW for Case I and the hourly values of fuel cost, revenue, and profit for the best solution found by GASW, respectively.

4.2 *Case II*

Table 6 presents a statistical comparison of the metaheuristics used to solve the system with 10 units, considering the PRA payment method. Once again, GASW obtained the best solution, which corresponds to a profit of $ 105,079.09. The special operators applied in the GA also yielded the top three performances. Tables 7 and 8

Table 5 Hourly values of fuel cost, revenue, and profit of the best solution found by GASW for Case I

Hour	Fuel cost ($)	Revenue ($)	Profit ($)	Hour	Fuel cost ($)	Revenue ($)	Profit ($)
1	13,744.15	15,892.63	2,148.48	13	26,851.61	32,767.20	5,915.59
2	14,620.00	16,912.50	2,292.50	14	26,217.40	32,046.00	5,828.60
3	16,354.46	19,981.50	3,627.04	15	24,272.84	27,675.00	3,402.16
4	19,346.73	22,027.13	2,680.40	16	20,245.10	23,192.00	2,946.90
5	21,076.02	23,831.25	2,755.23	17	20,245.10	23,140.00	2,894.90
6	21,996.40	25,830.23	3,833.82	18	19,597.58	22,270.50	2,672.92
7	22,964.29	26,167.50	3,203.21	19	20,245.10	23,088.00	2,842.90
8	23,951.09	26,591.08	2,639.99	20	20,245.10	23,556.00	3,310.90
9	27,337.40	29,822.40	2,485.00	21	20,245.10	24,024.00	3,778.90
10	26,851.61	39,094.20	12,242.59	22	17,353.30	20,884.50	3,531.20
11	26,851.61	40,159.80	13,308.19	23	17,186.68	20,531.88	3,345.20
12	26,851.61	42,157.80	15,306.19	24	15,497.41	18,491.00	2,993.59

Table 6 Statistical analysis for Case II

	Best profit ($)	Worst profit ($)	Standard deviation ($)	Mean profit ($)	Median ($)	Time (s)
bGWO	22,618.74	12,492.45	2,710.89	17,226.42	17,012.35	4.23
bWOA	57,480.79	34,993.81	3,355.77	46,619.48	46,911.60	3.22
bSCA	44,277.68	25,070.90	4,068.42	31,516.12	30,919.49	4.15
bHHO	75,539.72	44,008.60	8,166.39	59,255.08	58,354.06	5.64
GA	97,297.24	66,647.27	7,739.67	82,146.42	82,450.98	3.98
GAWM	101,249.54	67,075.24	7,504.65	84,034.46	83,891.39	3.72
GASW	105,079.09	77,076.11	7,235.76	96,424.77	97,600.05	3.42
GASWHC	103,687.61	71,253.52	6,891.79	86,515.66	86,750.62	3.80

show, respectively, the best solution found by GASW and the hourly values of Total Cost, Total Revenue, and Total Profit.

4.3 Case III

Table 9 presents the statistical analysis of the metaheuristics used to solve the 100-unit system using the PPD payment method. The results show that the GA with and without special operators achieved the four best performances, and the GA with the Swap Window operator obtained the highest profit of $ 1,072,859.41. Furthermore,

Table 7 Best power and reserve generation schedule for Case II obtained by GASW

Hour	Power generation of units (MW)										Allocated reserve of units (MW)									
	1	2	3	4	5	6	7	8	9	10	1	2	3	4	5	6	7	8	9	10
1	455	245	0	0	0	0	0	0	0	0	0	70	0	0	0	0	0	0	0	0
2	455	295	0	0	0	0	0	0	0	0	0	75	0	0	0	0	0	0	0	0
3	455	395	0	0	0	0	0	0	0	0	0	60	0	0	0	0	0	0	0	0
4	455	455	0	0	0	0	0	0	0	0	0	0	0	0	0	0	0	0	0	0
5	455	455	0	0	90	0	0	0	0	0	0	0	0	0	72	0	0	0	0	0
6	455	455	0	0	162	0	0	0	0	0	0	0	0	0	0	0	0	0	0	0
7	455	455	0	130	110	0	0	0	0	0	0	0	0	0	52	0	0	0	0	0
8	455	455	0	130	160	0	0	0	0	0	0	0	0	0	2	0	0	0	0	0
9	455	455	0	130	162	0	0	0	0	0	0	0	0	0	0	0	0	0	0	0
10	455	455	130	130	162	0	0	0	0	0	0	0	0	0	0	0	0	0	0	0
11	455	455	130	130	162	0	0	0	0	0	0	0	0	0	0	0	0	0	0	0
12	455	455	130	130	162	0	0	0	0	0	0	0	0	0	0	0	0	0	0	0
13	455	455	130	0	0	0	0	0	0	0	0	0	0	0	0	0	0	0	0	0
14	455	455	130	0	0	0	0	0	0	0	0	0	0	0	0	0	0	0	0	0
15	455	455	0	0	0	0	0	0	0	0	0	0	0	0	0	0	0	0	0	0
16	455	455	0	0	0	0	0	0	0	0	0	0	0	0	0	0	0	0	0	0
17	455	455	0	0	0	0	0	0	0	0	0	0	0	0	0	0	0	0	0	0
18	455	455	0	0	0	0	0	0	0	0	0	0	0	0	0	0	0	0	0	0
19	455	455	0	130	0	0	0	0	0	0	0	0	0	0	0	0	0	0	0	0
20	455	455	0	130	0	0	0	0	0	0	0	0	0	0	0	0	0	0	0	0
21	455	455	0	130	0	0	0	0	0	0	0	0	0	0	0	0	0	0	0	0
22	455	455	0	130	0	0	0	0	0	0	0	0	0	0	0	0	0	0	0	0
23	455	425	0	20	0	0	0	0	0	0	0	30	0	60	0	0	0	0	0	0
24	455	345	0	0	0	0	0	0	0	0	0	80	0	0	0	0	0	0	0	0

Table 10 displays the hourly values of Total Cost, Total Revenue, and Total Profit of the best solution obtained by the GASW after 50 runs for Case III.

4.4 *Case IV*

Table 11 presents a statistical comparison of the metaheuristics used to solve the 100-unit system considering the PPA payment method. The GA with and without special operators achieved the four best performances, and the GA with the Swap Window operator obtained the best solution with a profit of \$ 1,037,659.63. Table 12 presents the hourly values of Total Cost, Total Revenue, and Profit of the best solution found by GASW after 50 runs.

Table 8 Hourly values of fuel cost, revenue, and profit of the best solution found by GASW for Case II

Hour	Fuel cost ($)	Revenue ($)	Profit ($)	Hour	Fuel cost ($)	Revenue ($)	Profit ($)
1	13,689.23	15,528.18	1,838.95	13	20,245.10	25,584.00	5,338.90
2	14,561.05	16,524.67	1,963.62	14	20,245.10	25,480.00	5,234.90
3	16,307.15	19,655.72	3,348.57	15	17,353.30	20,475.00	3,121.70
4	17,353.30	20,611.50	3,258.20	16	17,353.30	20,293.00	2,939.70
5	20,515.99	23,275.03	2,759.04	17	17,353.30	20,247.50	2,894.20
6	21,099.15	24,602.40	3,503.25	18	17,353.30	20,065.50	2,712.20
7	24,004.52	25,892.49	1,887.97	19	20,773.96	23,088.00	2,314.04
8	23,918.06	26,580.66	2,662.61	20	20,213.96	23,556.00	3,342.04
9	23,959.81	27,405.60	3,445.79	21	20,213.96	24,024.00	3,810.04
10	27,951.61	39,094.20	11,142.59	22	20,213.96	23,868.00	3,654.04
11	26,851.61	40,159.80	13,308.19	23	17,845.80	20,505.61	2,659.81
12	26,851.61	42,157.80	15,306.19	24	15,434.42	18,066.97	2,632.55

Table 9 Statistical analysis for Case III

	Best profit ($)	Worst profit ($)	Standard deviation ($)	Mean profit ($)	Median ($)	Time (s)
bGWO	76,989.37	44,006.16	8,655.33	57,142.74	54,349.61	16.45
bWOA	38,693.38	28,730.61	1,860.00	32,944.19	32,988.58	17.18
bSCA	116,130.13	77,034.95	8,288.18	96,536.94	96,122.57	13.32
bHHO	412,447.51	304,329.33	22,642.08	337,736.27	333,217.78	22.38
GA	770,578.93	431,909.65	74,374.23	599,926.80	597,971.18	39.49
GAWM	812,631.97	516,653.53	56,921.79	644,248.67	636,907.72	41.10
GASW	1,072,859.41	789,643.42	64,814.14	990,734.93	1,014,544.54	37.20
GASWHC	1,003,891.45	530,837.48	99,462.15	687,407.59	666,731.96	39.38

5 Conclussion

This chapter introduces modifications to the Genetic Algorithm (GA) aiming to improve its performance in solving the Profit-Based Unit Commitment (PBUC) problem. The PBUC formulation involves a GENCO seeking to maximize its profits by trading energy and reserves in the day-ahead and ancillary markets, respectively. The effectiveness of the proposed modifications was evaluated on two test systems, a small-scale system with 10 units and a large-scale system with 100 units.

Table 10 Hourly values of fuel cost, revenue, and profit of the best solution found by GASW for Case III

Hour	Fuel cost ($)	Revenue ($)	Profit ($)	Hour	Fuel cost ($)	Revenue ($)	Profit ($)
1	144,303.55	158,926.25	14,622.70	13	266,523.23	323,982.00	57,458.77
2	149,071.01	169,125.00	20,053.99	14	259,269.83	315,560.00	56,290.17
3	167,828.68	201,258.75	33,430.07	15	240,694.91	272,576.25	31,881.34
4	190,268.59	219,427.54	29,158.94	16	220,421.49	251,432.50	31,011.01
5	202,829.92	238,295.06	35,465.14	17	209,301.61	239,465.63	30,164.02
6	230,139.81	258,021.11	27,881.30	18	198,267.21	225,902.25	27,635.04
7	234,454.12	263,311.88	28,857.75	19	203,401.39	231,190.80	27,789.41
8	247,732.14	270,301.99	22,569.85	20	210,263.84	241,766.10	31,502.26
9	271,704.19	299,592.00	27,887.81	21	206,957.48	244,721.40	37,763.92
10	275,105.20	397,986.00	122,880.80	22	198,070.00	232,690.05	34,620.05
11	275,105.20	408,834.00	133,728.80	23	178,026.44	209,868.75	31,842.31
12	272,244.54	425,059.50	152,814.96	24	159,361.00	184,910.00	25,549.00

Table 11 Statistical analysis for Case IV

	Best profit ($)	Worst profit ($)	Standard deviation ($)	Mean profit ($)	Median ($)	Time (s)
bGWO	60,047.47	30,923.14	7,568.12	44,932.07	43,902.34	14.59
bWOA	346,139.58	207,719.70	29,120.05	301,485.96	305,616.74	17.17
bSCA	120,415.29	69,231.64	10,678.17	86,865.88	84,225.41	12.64
bHHO	420,733.50	246,407.48	30,892.87	357,746.08	366,536.79	19.53
GA	790,687.94	474,969.37	77,508.18	605,244.45	597,972.47	34.94
GAWM	741,600.96	498,986.59	62,018.81	636,729.90	644,601.64	36.81
GASW	1,037,659.63	823,549.68	54,856.61	961,065.99	968,814.89	34.30
GASWHC	831,808.97	545,176.00	71,665.02	677,022.25	668,600.58	45.03

To enhance the performance of the metaheuristics, three special operators and a constraint handling strategy were employed to prevent violating the minimum up/down-time constraints. The considered special operators were the Swap Window Operator, the Window Mutation Operator, and the Swap Window Hill Climb Operator.

The obtained results demonstrate that the proposed modifications effectively improved the GA performance, yielding the best results in all analyzed cases. Specifically, the special operator Swap Window Operator, in conjunction with the constraint handling strategy, proved to be the most effective operator to enhance the exploration performance of the GA.

Table 12 Hourly values of fuel cost, revenue, and profit of the best solution found by GASW for Case IV

Hour	Fuel cost ($)	Revenue($)	Profit ($)	Hour	Fuel cost ($)	Revenue ($)	Profit ($)
1	141,885.98	155,281.80	13,395.82	13	251,715.85	308,877.60	57,161.75
2	148,410.56	165,246.68	16,836.11	14	247,970.00	303,653.00	55,683.00
3	165,868.45	196,643.54	30,775.10	15	232,560.61	263,925.00	31,364.39
4	188,264.32	215,402.55	27,138.23	16	206,988.62	236,246.20	29,257.58
5	200,690.82	232,807.96	32,117.14	17	198,375.50	227,039.00	28,663.50
6	220,118.47	252,477.45	32,358.98	18	193,959.93	220,524.39	26,564.47
7	234,307.96	258,838.13	24,530.17	19	186,808.30	214,984.80	28,176.50
8	243,898.42	265,887.42	21,989.00	20	190,250.10	222,287.10	32,037.00
9	250,624.05	283,312.80	32,688.75	21	194,020.76	229,706.40	35,685.64
10	268,870.52	383,252.30	114,381.78	22	208,693.78	241,617.60	32,923.82
11	266,852.32	397,618.20	130,765.88	23	176,090.39	204,960.19	28,869.80
12	267,391.18	417,400.20	150,009.02	24	156,144.49	180,430.68	24,286.19

References

1. Montero L, Bello A, Reneses J (2022) A review on the unit commitment problem: approaches, techniques, and resolution methods. Energies 15:1296
2. Kumar V, Naresh R, Singh A (2021) Investigation of solution techniques of unit commitment problems: a review. Wind Eng 45:1689–1713
3. Muralikrishnan N, Jebaraj L, Rajan C (2020) A comprehensive review on evolutionary optimization techniques applied for unit commitment problem. IEEE Access 8:132980–133014
4. Kumar V, Naresh R, Sharma V, Kumar V (2022) Generation scheduling considering emissions in cost-based unit commitment problem. Renew Energy Towards Smart Grid 209–224
5. Georgilakis P (2009) Genetic algorithm model for profit maximization of generating companies in deregulated electricity markets. Appl Artif Intell 23:538–552
6. Derakhshandeh S, Hamedani Golshan M, Masoum M (2013) Profit-based unit commitment with security constraints and fair allocation of cost saving in industrial microgrids. IET Sci Meas Technol 7:315–325
7. Dhanalakshmi S, Kannan S, Mahadevan K (2011) Market modes for deregulated environment-a review. In: 2011 international conference on emerging trends in electrical and computer technology, pp 82–87
8. Richter C, Sheble G (2000) A profit-based unit commitment GA for the competitive environment. IEEE Trans Power Syst 15:715–721
9. Abdi H (2021) Profit-based unit commitment problem: a review of models, methods, challenges, and future directions. Renew Sustain Energy Rev 138:110504
10. Shiina T, Watanabe I (2004) Lagrangian relaxation method for price-based unit commitment problem. Eng Optim 36:705–719
11. Song S, Lee K (2008) A mixed integer linear programming approach for the profit based unit commitment problem under non-linear fuel consumption constraint and maintenance cost. Korean Manag Sci Rev 25:43–53
12. Simoglou C, Biskas P, Bakirtzis A (2010) Optimal self-scheduling of a thermal producer in short-term electricity markets by MILP. IEEE Trans Power Syst 25:1965–1977

13. Lotfi M (2013) Short-term price-based unit commitment of hydrothermal gencos: a pre-emptive goal programming approach. Int J Eng 26:1017–1030
14. Pokharel B, Shrestha G, Lie T, Fleten S (2005) Price based unit commitment for Gencos in deregulated markets. IEEE Power Eng Soc Gen Meet 2005:428–433
15. Chandram K, Subrahmanyam N, Sydulu M (2009) Improved pre-prepared power demand table and Muller's method to solve the profit based unit commitment problem. J Electr Eng Technol 4:159–167
16. Chandram K, Subrahmanyam N, Sydulu M (2008) New approach with muller method for profit based unit commitment. In: 2008 IEEE power and energy society general meeting-conversion and delivery of electrical energy in the 21st century, pp 1–8
17. Maifeld T, Sheble G (1996) Genetic-based unit commitment algorithm. IEEE Trans Power Syst 11:1359–1370
18. Singhal P, Naresh R, Sharma V, Nadakuditi G (2015) Generator scheduling under competitive environment using genetic algorithm. Int J Eng, Sci Technol 7:24–32
19. Xiaohui Y, Yanbin Y, Cheng W, Xiaopan Z (2005) An improved PSO approach for profit-based unit commitment in electricity market. In: 2005 IEEE/PES transmission distribution conference exposition: Asia and Pacific, pp 1–4
20. Raglend I, Raghuveer C, Avinash G, Padhy N, Kothari D (2010) Solution to profit based unit commitment problem using particle swarm optimization. Appl Soft Comput 10:1247–1256
21. Venkatesan T, Sanavullah M (2013) SFLA approach to solve PBUC problem with emission limitation. Int J Electr Power Energy Syst 46:1–9
22. Ghadi M, Baghramian A (2013) A new heuristic method for solving unit commitment problem in competitive environment. Int J Soft Comput Eng 2:2231–2307
23. Columbus C, Simon S (2012) Profit based unit commitment: a parallel ABC approach using a workstation cluster. Comput Electr Eng 38:724–745
24. Singhal P, Naresh R, Sharma V (2015) Binary fish swarm algorithm for profit-based unit commitment problem in competitive electricity market with ramp rate constraints. IET Gener, Transm Distrib 9:1697–1707
25. Reddy K, Panwar L, Kumar R, Panigrahi B (2016) Binary fireworks algorithm for profit based unit commitment (PBUC) problem. Int J Electr Power Energy Syst 83:270–282
26. Reddy K, Panwar L, Panigrahi B, Kumar R (2018) A new binary variant of sine-cosine algorithm: development and application to solve profit-based unit commitment problem. Arab J Sci Eng 43:4041–4056
27. Reddy KS, Panwar L, Panigrahi B, Kumar R (2019) Binary whale optimization algorithm: a new metaheuristic approach for profit-based unit commitment problems in competitive electricity markets. Eng Optim 51:369–389
28. Reddy S, Panwar L, Panigrahi B, Kumar R, Alsumaiti A (2019) Binary grey wolf optimizer models for profit based unit commitment of price-taking GENCO in electricity market. Swarm Evol Comput 44:957–971
29. Nandi A, Kamboj V (2020) A new solution to profit based unit commitment problem considering PEVs/BEVs and renewable energy sources. E3S Web Conf 184:01070
30. Dhaliwal J, Dhillon J (2019) Profit based unit commitment using memetic binary differential evolution algorithm. Appl Soft Comput 81:105502
31. Yamin H, Shahidehpour S (2004) Unit commitment using a hybrid model between Lagrangian relaxation and genetic algorithm in competitive electricity markets. Electr Power Syst Res 68:83–92
32. Yamin H (2004) Profit-based unit commitment for gencos in a deregulated power market. Eur Trans Electr Power 14:321–330
33. Bavafa M, Navidi N, Monsef H (2008) A new approach for profit-based unit commitment using lagrangian relaxation combined with ant colony search algorithm. In: 2008 43rd international universities power engineering conference, pp 1–6
34. Devi S, Rajan C (2009) A hybrid approaches for the profit based unit commitment problem in the deregulated markets. J Electr Eng 9:35–41

35. Rampriya B, Mahadevan K, Kannan S (2010) Unit commitment in deregulated power system using Lagrangian firefly algorithm. In: 2010 international conference on communication control and computing technologies, pp. 389–393
36. Kumar R, Asokan K, Kumar S (2013) Optimal scheduling of generators to maximize GENCOs profit using LR combined with ABC algorithm in deregulated power system. In: 2013 international conference on computation of power, energy, information and communication (ICCPEIC), pp 75–83
37. Sudhakar A, Karri C, Laxmi A (2018) A hybrid LR-secant method-invasive weed optimisation for profit-based unit commitment. Int J Power Energy Convers 9:1–24
38. Anand H, Narang N, Dhillon J (2018) Profit based unit commitment using hybrid optimization technique. Energy 148:701–715
39. Lakshmi K, Vasantharathna S (2013) Hybrid artificial immune system approach for profit based unit commitment problem. J Electr Eng Technol 8:959–968
40. Hooda M, Awasthi Y, Thakur N, Siddiqui A (2020) Profit based unit commitment problem solving using hybrid CUCKOO-GWO algorithm. J Phys: Conf Ser 1478:012015
41. Kazarlis S, Bakirtzis A, Petridis V (1996) A genetic algorithm solution to the unit commitment problem. IEEE Trans Power Syst 11:83–92
42. Attaviriyanupap P, Kita H, Tanaka E, Hasegawa J (2003) A hybrid LR-EP for solving new profit-based UC problem under competitive environment. IEEE Trans Power Syst 18:229–237
43. Holland J (1992) Genetic algorithms. Sci Am 267:66–73
44. Variengien A, Nichele S, Glover T, Pontes-Filho S (2021) Towards self-organized control: Using neural cellular automata to robustly control a cart-pole agent. ArXiv:2106.15240
45. Amjady N, Nasiri-Rad H (2009) Economic dispatch using an efficient real-coded genetic algorithm. IET Gener, Transm Distrib 3:266–278
46. Kashyap M, Mittal A, Kansal S (2019) Optimal placement of distributed generation using genetic algorithm approach. In: Proceeding of the second international conference on microelectronics, computing and communication systems (MCCS 2017), pp 587–597
47. Granelli G, Montagna M, Zanellini F, Bresesti P, Vailati R, Innorta M (2006) Optimal network reconfiguration for congestion management by deterministic and genetic algorithms. Electr Power Syst Res 76:549–556
48. Mirjalili S, Lewis A (2013) S-shaped versus V-shaped transfer functions for binary particle swarm optimization. Swarm Evol Comput 9:1–14

Chapter 7
Optimal Power Flow by Genetic Algorithm

Amir Ebrahimi, Rouzbeh Haghighi, Hossein Yektamoghadam, Majid Dehghani, and Amirhossein Nikoofard

1 Introduction

Power grids are complicated systems which set restrictions on the voltages that can be reached at nodes and the power capacity that can be transmitted along transmission lines. These constraints are present because of voltage and thermal concerns [1]. Supplying the customers consumption load at the lowest possible cost while maintaining network constraints is the most crucial task in the operation of power systems. The OPF program is utilized to run the network. The generated power of each network unit is ascertained by solving the OPF problem, which also allows for the calculation of the voltage of the buses and the lines passing power as well as the observation of any limitations imposed by these factors. It has taken a long time for optimal load distribution to evolve and spread up to this point. Optimal power flow was first introduced by Carpentier in 1962, and as modeling and solution methods

A. Ebrahimi
Amirkabir University of Technology, Tehran, Iran
e-mail: A.ebrahimi1995@aut.ac.ir

R. Haghighi
University of Michigan, Dearborn, USA
e-mail: Haghighi@umich.edu

H. Yektamoghadam
University of Tehran, Tehran, Iran
e-mail: Yektamoghadam@ut.ac.ir

M. Dehghani (✉)
Oklahoma State University, Stillwater, USA
e-mail: majid.dehghani@okstate.edu

A. Nikoofard
K.N.Toosi University of Technology, Tehran, Iran
e-mail: A.nikoofard@kntu.ac.ir

M. Khosravy et al. (eds.), *Frontiers in Genetics Algorithm Theory and Applications*,
Springer Tracts in Nature-Inspired Computing,
https://doi.org/10.1007/978-981-99-8107-6_7

have advanced, it has steadily turned into a fundamental tool for power systems management [2, 3].

Due to its nonlinearity and computational challenges, operating an electric power system is complicated. The optimal power flow issue, often known as optimal scheduling, is one challenge of running a power system safely and cheaply. For a static power system loading scenario, the OPF solution provides the best active and reactive power production schedule as well as the best values for all controllable variables. Due to the huge number and variety of restrictions placed on the power grid, this is an extremely hard nonlinear programming issue. In [4, 5], OPF has been used for reducing fuel expenses, carbon emissions, and active power loss in the network.

A wide variety of optimization strategies have been used for tackling OPF issues. Both classical numerical analysis and more modern metaheuristic approaches may be distinguished among these techniques [6]. Traditional methods [7] frequently resort to approximation in order to reduce intricacy, but this often results in a massive drop in system performance. Search spaces that are very irregular need algorithms that are highly resilient to prevent becoming trapped at a local optimum. The OPF problem is addressed using a wide variety of approaches, including Newton's method [8, 9], interior point method [10], nonlinear programming [10], fuzzy logic [11], and the gradient projection method [12]. To find the best possible answer, these strategies use convexity. In a non-convex solution space, several local minimums may occur unless the connection is simplified.

The immense vastness of power systems and the insufficient domain-dependent expertise of power system engineers are two factors that prevent effective optimal power flow. Numerical optimization approaches based on successive linearization that use first and second derivatives of objective functions and their constraints as the search directions, or linear programming solutions to imprecise models, are effective ways to deal with the first limitation [13]. Such approaches tend to converge to a local minimum [14] and are sensitive to changes in the issue formulation and algorithm choice. However, their mathematical foundations are where their greatest strength lies. Since new restrictions might arise from things like open access, nonconvex solution surfaces, and FACTS devices, it's crucial to create a flexible and trustworthy algorithm. Recently, heuristic methods have been studied for use in power system issues and they have been able to avoid the problems inherent in the more traditional mathematical approaches [15]. Differential evolution (DE) [16], Particle swarm optimization (PSO) [17, 18], tree seed algorithm (TSA) [19], Sine–Cosine algorithm [20], sunflower optimization algorithm [21], Coyote optimization algorithm [22], PSO-GA algorithm [23, 24] and salp swarm algorithm [25] are all examples of nature-inspired, novel metaheuristic methods. These optimization strategies are now being employed to resolve the OPF problem. These metaheuristic-based methods provide a random starting point for the agents and then track their progress to determine which solution is the best.

There are several reasons why evolutionary algorithms (EAs) may be preferred over mixed-integer linear programming (MILP) for OPF: (1) Non-linear constraints: MILP is a linear programming technique that assumes all constraints are linear. However, in power systems, there are many non-linear constraints such as voltage magnitude limits, reactive power limits, and thermal limits on transmission lines. EAs can handle non-linear constraints more easily than MILP; (2) Scalability: MILP becomes computationally expensive as the size of the problem increases. EAs can handle large-scale problems more efficiently than MILP; (3) Robustness: EAs are more robust to uncertainties and disturbances in the system compared to MILP. This is because EAs use a population-based approach that explores multiple solutions simultaneously, whereas MILP relies on a single solution; (4) Flexibility: EAs can be easily modified and adapted to different problem formulations and objectives, whereas MILP requires significant changes to the model formulation for each new problem. Overall, while both techniques have their advantages and disadvantages, EAs may be preferred over MILP for OPF due to their ability to handle non-linear constraints, scalability, robustness, and flexibility.

The GA is an evolutionary strategy that has been successfully applied to a number of power system challenges when more traditional approaches have failed. The decreasing price of powerful computers has made genetic algorithms a viable option for solving optimization challenges like these. These algorithms have recently seen widespread use to find solutions to issues with global optimization where the closed-form optimization method is inapplicable. GA is a kind of global, parallel search method that is based on the idea that natural genetic operators may be used to solve complex problems. Since the GA examines several points in the parameter space simultaneously, it has a better chance of settling on the global answer, in which a continuous or differentiable search space is not required [13]. Authors in [13] developed the GA to tackle the difficulty of OPF in large power distribution systems. Authors in [26] presented a game theory approach to provide an optimal strategy for power plants based on GA and results have been compared with different optimization methods such as GA, PSO and TLBO.

In order to speed up the OPF, the controllable variables are decomposed into active constraints that affect the cost function directly and are included in the GA process, and passive constraints that are updated using a typical load flow program just for once, after the OPF has converged. The aim of the authors in [27] is to create an OPF algorithm for which the answer is based on the application of the GA. The size and complexity of electrical power networks were considered throughout the program's development. A revised OPF solution based on a GA is described in [4]. The approach can deal with nonconvex generating cost curves and is insensitive to initial conditions. To prove the method's worth, they have provided computational implementation results using IEEE standard test systems, and shown that they are equivalent to those obtained using more traditional methods.

According to [28], GA have been utilized to solve the OPF model. In order to enhance the performance of GA, a generating scaling factor (GSF) has also been incorporated into its structure. This integration allows for simultaneous calculations of power flow through transmission lines, thereby providing valuable insights into the capacity of these lines which leads to the congestion management of the grid. A dynamic technique is presented in [29] that is both efficient and trustworthy and whose goal is to reduce generation costs without increasing transmission losses. In order to reliably provide both global and local solutions to OPF at a satisfactory pace, the authors in [30] suggest a hybrid approach that incorporates GA and clustering methods. Osman et al. [31] suggested a novel approach to solve OPF, in which the GA is used to get a workable solution within the constraints of the desired load convergence. Authors in Ref. [32] introduce an enhanced GA that can handle both continuous and discrete control variables. Unit active power outputs and generator-bus voltage magnitudes are treated as continuous control variables, whereas tap settings on transformers and shunt devices are modeled as discrete ones.

The remainder of the current chapter is organized as follows. After presenting the introduction in the first section, Sect. 2 introduces the OPF definition in its mathematical model. Afterward, in Sect. 3, the application of GA in OPF is presented. Then in Sect. 3, a sample OPF problem is solved based on the GA approach, and the results have been presented. Finally, in Section V, the conclusion of this study will be expressed.

2 Optimal Power Flow

In this section, the underlying notion that underpins the OPF as well as its formulation is broken down and explored. The OPF problem seeks to optimize the objective function through suitable adjustments of power system control parameters while adhering to a range of equality and inequality constraints.

2.1 Mathematical Model

The OPF problem can be modelled mathematically, as presented in Eq. (1)

$$\text{Objective function} : \min J(x, u) \tag{1}$$

Subject to:

$$g(x, u) = 0 \tag{2}$$

$$h(x, u) \leq 0 \tag{3}$$

where Eq. (2) and (3) are the equality and inequality constraints, respectively. In this problem J is defined by:

$$J = \sum_{i=1}^{NG} f_1 \tag{4}$$

where f_i is the fuel cost of the i th generator. The generation cost function is represented by a quadratic function as:

$$f_i = a_i + b_i P_{Gi} + c_i P_{Gi}^2 (\$/h) \tag{5}$$

It is noteworthy to say that J can be defined as the other objective such as minimizing power loss, minimizing voltage deviation etc.

2.2 Model Constraints

- Power Flow constrints: Typical load flow equations are represented by the equality constraints.

$$P_{Gi} - P_{Di} - V_i \sum_{J=1}^{NB} V_j \left[G_{ij} \cos(\delta_i - \delta_j) + B_{ij} \sin(\delta_i - \delta_j) \right] = 0 \tag{6}$$

$$Q_{Gi} - Q_{Di} - V \sum_{j=1}^{NB} V_j \left[G_{ij} \sin(\delta_i - \delta_j) - B_{ij} \cos(\delta_i - \delta_j) \right] = 0 \tag{7}$$

  where N_B is the number of buses, P_G is the active power generation, Q_G is the reactive power generation, P_D is the active load demand, Q_D is the reactive load demand, G_{ij} and B_{ij} are the conductance and susceptance between bus i and bus j, respectively.
- Generator constraints: Bus voltages, active power outputs, and reactive power outputs are restricted by their lower and upper limits.

$$V_{Gi}^{\min} \leq V_{Gi} \leq V_{Gi}^{\max}, i = 1, \ldots, N_G \tag{8}$$

$$P_{Gi}^{\min} \leq P_{Gi} \leq P_{Gi}^{\max}, i = 1, \ldots., N_G \tag{9}$$

$$Q_{Gi}^{\min} \leq Q_{Gi} \leq Q_{Gi}^{\max}, i = 1, \ldots, N_G \tag{10}$$

- Transformer constraints: The lower and upper limits of transformer tap.

$$T_i^{\min} \leq T_i \leq T_i^{\max}, i = 1, \ldots, N_T \tag{11}$$

- Shunt reactive power constraints:

$$Q_{Ci}^{\min} \leq Q_{Ci} \leq Q_{Ci}^{\max}, i = 1, \ldots, N_C \tag{12}$$

- Security constraints: The voltages constraints at load buses and transmission line loading define as security constraints:

$$V_{Bi}^{\min} \leq V_{Bi} \leq V_{Bi}^{\max}, i = 1, \ldots, N_b \tag{13}$$

$$S_{li} \leq S_{li}^{\max}, i = 1, \ldots\ldots, N_l \tag{14}$$

It is notable to say that the control variables are generator power except the slack bus.

3 Applying GA to OPF

To deal the OPF problem with GA, the constrained optimization problem should be converted as an unconstrained problem by the penalty function method. A basic GA is a repetitive process that maintains a population of potential solutions of fixed size. Each generation consists of three genetic operators (reproduction, crossover, and mutation) that produce new populations (offspring), and the chromosomes of these populations are ranked according to their fitness, which is a measure of their efficiency relative to a cost function. New better populations of candidate solutions are produced based on these genetic operators and assessments.

Therefore, a basic genetic algorithm shown in Fig. 1 is presented as follows:

1. Loading the initial needed data such as GA parameters and system data .
2. Randomly producing a population with active powers that fall within the allowed range.
3. Running the power flow and evaluate the fitness of every string in the population.
4. Generating offspring sequences by means of reproduction, crossover, and mutation.
5. Analyzing modified strings and determining their fitness levels.
6. Returning the best chromosome as the solution if the search objective is met or a maximum generation size is reached; otherwise, proceeding to Step 4.

4 Implementation of GA in a OPF Problem

An example of a GA simulation of the OPF problem has been conducted and is presented in this chapter. In this regard, the IEEE 30-bus system is considered to perform the OPF problem, as shown in Fig. 2. The system data is given in Table 1 [33]. The test system has a total load that is equal to 189.2 + j107.2 MVA, and it is powered by 6 generators and 41 branches.

All simulations have been done on a PC with an Intel Core i7 CPU @ 3.20 GHz and 12 GB RAM using MATLAB 2021a. Simulation parameters of the GA algorithm

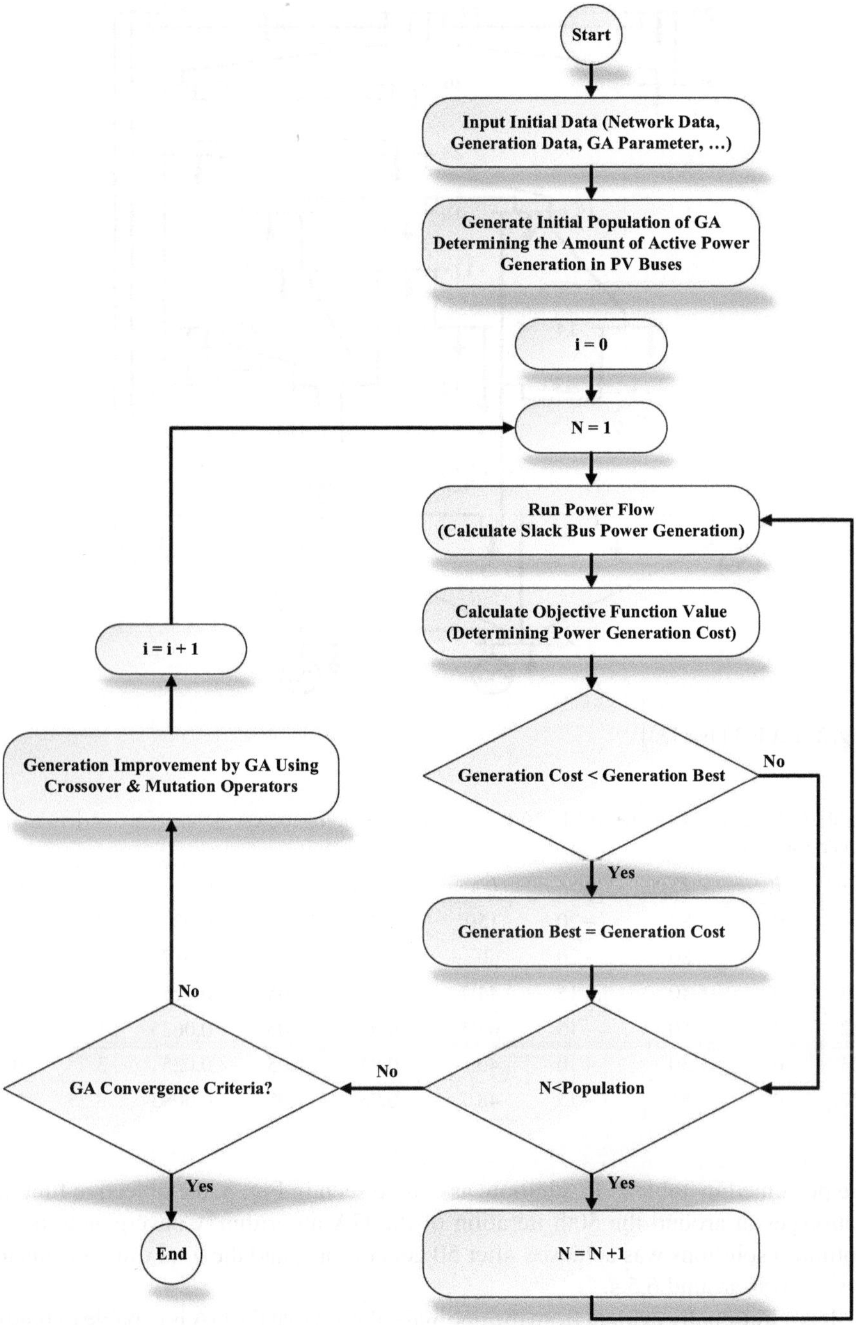

Fig. 1 GA algorithm flowchart

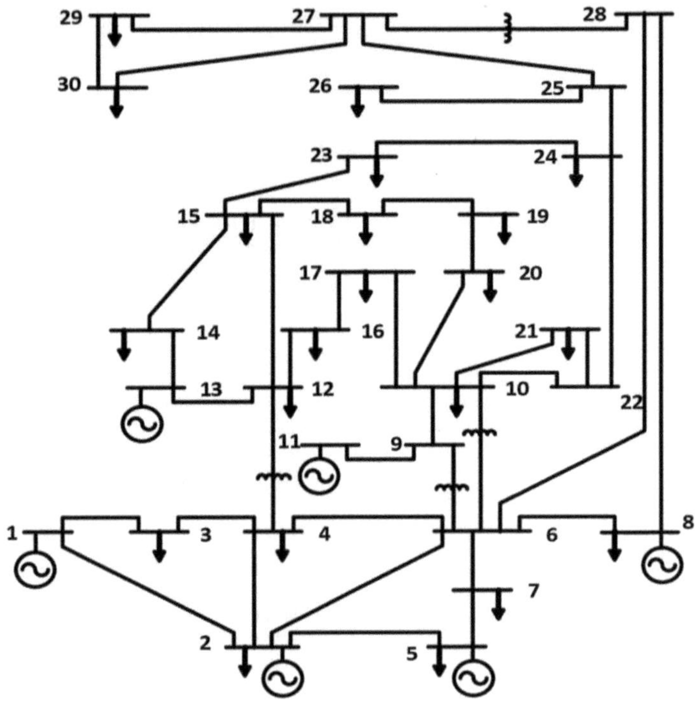

Fig. 2 IEEE 30-bus [33]

Table 1 Parameters of the IEEE 30-Bus System including power generation ristrictions and generator cost

Bus	P_{Min}	P_{Max}	Q_{Min}	Q_{Max}	V_{Min}	V_{Max}	a	b	c
1	0	80	−20	150	0.95	1.05	0.02	2	0
2	0	80	−20	60	0.95	1.05	0.0175	1.75	0
13	0	40	−15	44.7	0.95	1.05	0.025	3	0
22	0	50	−15	62.5	0.95	1.05	0.0625	1	0
23	0	30	−10	40	0.95	1.05	0.025	3	0
27	0	55	−15	48.7	0.95	1.05	0.0083	3.25	0

are presented in Table 2. In addition, as can be seen in Fig. 3, the objective function converges at around the 50th iteration of the GA algorithm. Convergence to very optimum solutions was achieved after 50 generations, and the total time for this test system was around 6.5 s.

It is additionally evident from the optimal solution that the GA is capable of readily avoiding the existing restrictions. Voltage, active and reactive power production are all within acceptable ranges, as are all other control variables and state variables. As can be seen in Fig. 3, fuel costs drop with increasing generations and remain

Table 2 GA Parameters

Iteration	200
Population	10
Crossover rate	0.7
Mutation rate	0.3

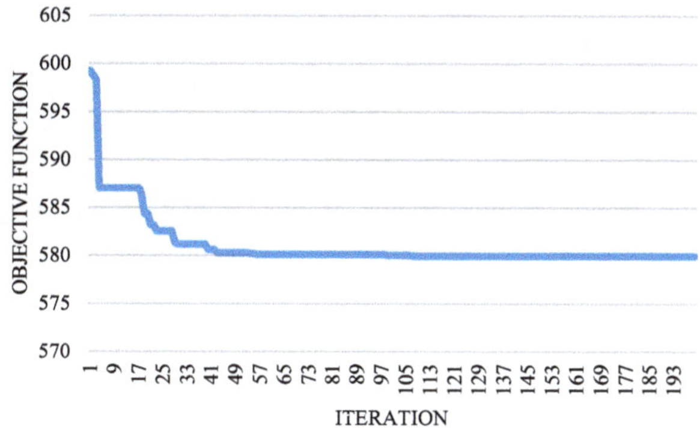

Fig. 3 GA performance

almost unchanged after 50 generations. Figure 4 shows the bus voltage in the test system, which is obtained from performing OPF by the GA algorithm presented in this Sect. 1.3. Moreover, the OPF results have been illustrated in Table 3.

The second test system we use for optimal load flow problem is the IEEE 118-bus system. This system has 411 lines, 54 generators and its load is equal to 4242 + 1338i. The specifications of the generators are in the table below. The figure below shows the IEEE 118-bus system.

By executing the optimal load flow program by genetic algorithm, we understand that it converges in iteration 20 and the figure below shows this. The execution of this program takes about 62 s.

Table 5 shows the amount of load that the generators produce. This table shows that the generators produce power within the permitted range and comply with the

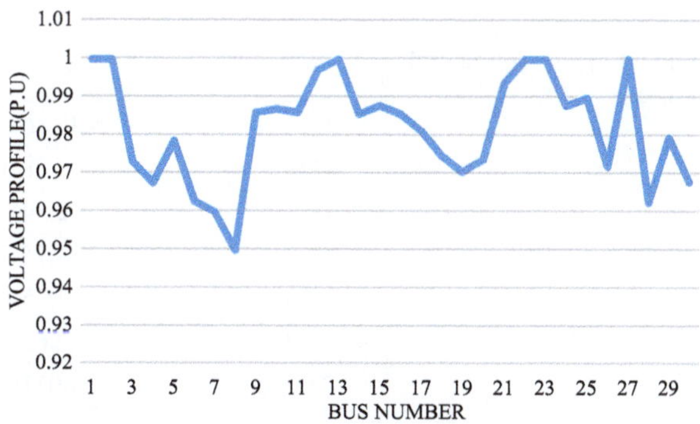

Fig. 4 Network bus voltages

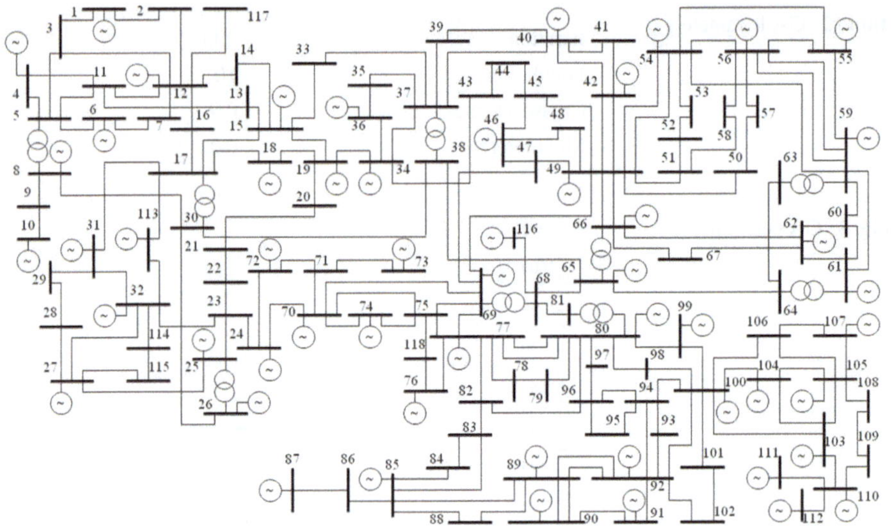

Fig. 5 One-line diagram of the IEEE 118-bus test system, by Illinois Institute of Technology, version of 2003 [34]

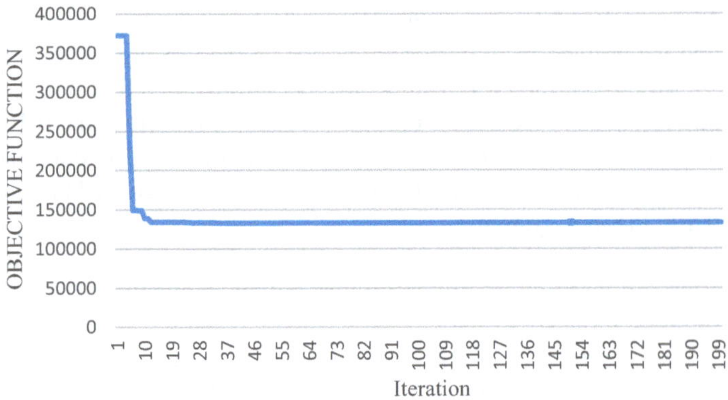

Fig. 6 GA performance for IEEE 118-bus

Table 3 The result of OPF problem

	P	Q
G1	35.8647	−1.3302
G2	48.2850	46.6214
G13	22.3454	7.3337
G22	24.7567	2.3819
G23	17.7557	5.0429
G27	42.6279	24.5683
Total cost	**5801107**	

Table 4 Parameters of the IEEE 118-Bus System including power generation restrictions and generator cost

Bus	P_{Min}	P_{Max}	Q_{Min}	Q_{Max}	V_{Min}	V_{Max}	a	b	c
1	0	1800	−300	300	106	1	0	40	0
4	0	200	−300	300	107	1	0	40	0
6	0	200	−13	50	108	1	0	40	0
8	0	200	−300	300	109	1	0	40	0
10	0	1600	−147	200	110	1	0	20	0
12	0	300	−35	120	111	1	0	20	0
15	0	200	−10	30	112	1	0	40	0
18	0	200	−16	50	113	1	0	40	0
19	0	200	−8	24	114	1	0	40	0
24	0	200	−300	300	115	1	0	40	0
25	0	800	−47	140	116	1	0	20	0
26	0	1100	−1000	1000	117	1	0	20	0
27	0	200	−300	300	118	1	0	40	0
31	0	200	−300	300	119	1	1	20	0
32	0	200	−14	42	120	1	0	40	0
34	0	200	−8	24	121	1	0	40	0
36	0	200	−8	24	122	1	0	40	0
40	0	200	−300	300	123	1	0	40	0
42	0	200	−300	300	124	1	0	40	0
46	0	200	−100	100	125	1	1	20	0
49	0	800	−85	210	1	1	0	20	0
54	0	200	−300	300	1	1	0	20	0
55	0	200	−8	23	1	1	0	40	0
56	0	200	−8	15	1	1	0	40	0
59	0	600	−60	180	1	1	0	20	0
61	0	600	−100	300	1	1	0	20	0
62	0	200	−20	20	1	1	0	40	0
65	0	1400	−67	200	1	1	0	20	0
66	0	1400	−67	200	1	1	0	20	0
69	0	200	−5	15	1	1	0	20	0
70	0	200	−10	32	1	1	0	40	0
72	0	200	−100	100	1	1	0	40	0
73	0	200	−100	100	1	1	0	40	0
74	0	200	−6	9	1	1	0	40	0
76	0	200	−8	23	1	1	0	40	0
77	0	200	−20	70	1	1	0	40	0

(continued)

Table 4 (continued)

Bus	P_{Min}	P_{Max}	Q_{Min}	Q_{Max}	V_{Min}	V_{Max}	a	b	c
80	0	1700	−165	280	1	1	0	20	0
85	0	200	−8	23	1	1	0	40	0
87	0	200	−100	1000	1	1	3	20	0
89	0	2000	−210	300	1	1	0	20	0
90	0	200	−300	300	1	1	0	40	0
91	0	200	−100	100	1	1	0	40	0
92	0	200	−3	9	1	1	0	40	0
99	0	200	−100	100	1	1	0	40	0
100	0	900	−50	155	1	1	0	20	0
103	0	200	−15	40	1	1	0	20	0
104	0	200	−8	23	1	1	0	40	0
105	0	200	−8	23	1	1	0	40	0
107	0	200	−200	200	1	1	0	40	0
110	0	200	−8	23	1	1	0	40	0
111	0	200	−100	1000	1	1	0	20	0
112	0	200	−100	1000	1	1	0	40	0
113	0	200	−100	200	1	1	0	40	0
116	0	200	−1000	1000	1	1	0	40	0
1	0	1800	−300	300	106	1	0	40	0

condition of the problem. Also, according to the Fig. 7, the OPF problem also meets the condition of the permissible voltage range. Therefore, it can be seen that the OPF problem causes the lowest fuel cost according to all the restrictions.

By comparing the two systems of 30-bus and 118-bus, we understand that the execution time of 118-bus is longer because it is a bigger system. Also, the implementation of OPF in the 118-basis system is more complicated because the number of variables is more, and therefore the genetic algorithm must perform a larger search space and may not result in an optimal response, and the parameters of the algorithm must be well adjusted, and the execution of the algorithm must be repeated several times too. We know that we have reached the optimal answer, which may take a long time due to the high execution time of the program.

5 Conclusions

This chapter explored the GA potential for solving the OPF problem. First, the OPF problem is defined and possible approaches to resolve it are discussed. Then, the advantages and disadvantages of technical and evolutionary algorithms, especially

Table 5 The result of OPF problem for IEEE 118-Bus

	P	Q
G1	658.71336	484.59115
G4	0.00000	57.01734
G6	0.04000	62.51243
G8	0.00000	38.11355
G10	401.87000	−143.20647
G12	85.79000	212.74515
G15	20.88000	44.64890
G18	13.22000	39.62051
G19	21.58000	26.83711
G24	0.00000	0.74482
G25	193.81000	49.43867
G26	279.76000	−8.96935
G27	9.92000	−13.36383
G31	7.25000	34.30773
G32	14.86000	−14.18390
G34	4.88000	142.69703
G36	10.66000	10.23696
G40	49.32000	−7.49601
G42	40.99000	−562.71647
G46	19.04000	−40.39367
G49	0.00000	−3672.60874
G54	49.54000	−748.64032
G55	32.13000	−10.26660
G56	32.56000	−860.09921
G59	149.70000	−639.67617
G61	148.41000	−169.72974
G62	0.00000	−1.17097
G65	352.24000	34.11296
G66	348.86000	−2370.10554
G69	12.40664	−78.39020
G70	0.00000	45.56103
G72	0.00000	13.21013
G73	0.00000	25.72593
G74	16.93000	−0.65608
G76	22.85000	−1.68434
G77	0.00000	−3129.14371
G80	430.84000	−2871.79753
G85	0.00000	−36.90663

(continued)

Table 5 (continued)

	P	*Q*
G87	3.63000	−5.09117
G89	501.84000	−4423.88848
G90	0.00000	−1537.50423
G91	0.00000	−97.67462
G92	0.00000	−2728.42448
G99	0.00000	−66.39718
G100	231.29000	−31.97895
G103	38.25000	12.40925
G104	0.00000	11.96693
G105	5.16000	13.21912
G107	29.03000	−12.05260
G100	7.03000	−5.06269
G111	35.24000	−2.60599
G112	36.48000	22.79406
G113	0.00000	39.18588
G116	0.00000	−106.71343
Total Cost	**133,609.602705035**	

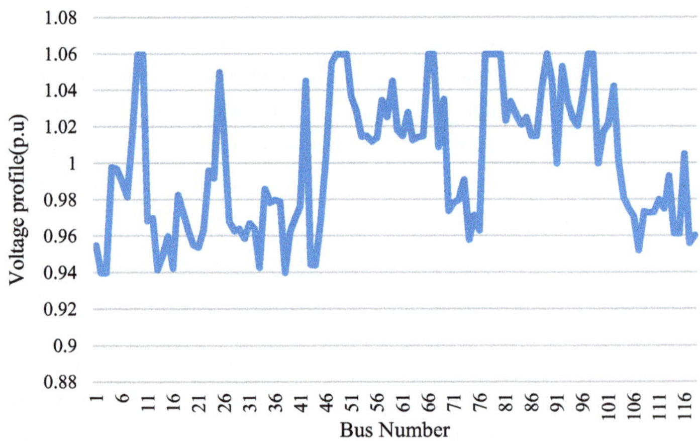

Fig. 7 Network bus voltages for IEEE 118-bus

GA, have been discussed. After defining the OPF problem, a method for applying the GA to the OPF problem is described. Afterwards, the implementation of the GA to the OPF problem with the goal of reducing fuel expenses as much as possible while still meeting all the technical and security requirements of the IEEE 33-bus and 118-bus system is provided. The simulation outputs showed that GA implementation is a viable option for dealing with such issues.

References

1. Shaheen MA, Hasanien HM, Mekhamer S, Talaat HE, editors (2020) Optimal power flow of power networks with penetration of renewable energy sources by harris hawks optimization method. In: 2020 2nd International Conference on Smart Power & Internet Energy Systems (SPIES); IEEE
2. Ayele GT, Mabrouk MT, Haurant P, Laumert B, Lacarrière B (2021) Optimal heat and electric power flows in the presence of intermittent renewable source, heat storage and variable grid electricity tariff. Energy Convers Manage 243:114430
3. Wood AJ, Wollenberg BF, Sheblé GB (2013) Power generation, operation, and control. John Wiley & Sons
4. Paranjothi S, Anburaja K (2002) Optimal power flow using refined genetic algorithm. Electr Power Compon Syst 30(10):1055–1063
5. Yuryevich J, Wong KP (1999) Evolutionary programming based optimal power flow algorithm. IEEE Trans Power Syst 14(4):1245–1250
6. Ahmed MK, Osman MH, Shehata AA, Korovkin NV, editors (2021) A solution of optimal power flow problem in power system based on multi objective particle swarm algorithm. In: 2021 IEEE Conference of Russian Young Researchers in Electrical and Electronic Engineering (ElConRus). IEEE.
7. Kahourzade S, Mahmoudi A, Mokhlis HB (2015) A comparative study of multi-objective optimal power flow based on particle swarm, evolutionary programming, and genetic algorithm. Electr Eng 97(1):1–12
8. Abdullah M, Javaid N, Chand A, Khan ZA, Waqas M, Abbas Z, editors (2019) Multi-objective optimal power flow using improved multi-objective multi-verse algorithm. Workshops of the International Conference on Advanced Information Networking and Applications; Springer
9. Chen G, Qian J, Zhang Z, Sun Z (2019) Multi-objective improved bat algorithm for optimizing fuel cost, emission and active power loss in power system. IAENG Int J Comput Sci 46(1):118–133
10. Abido M (2011) Multiobjective particle swarm optimization for optimal power flow problem. Springer, Handbook of swarm intelligence, pp 241–268
11. Abbasi M, Abbasi E, Mohammadi-Ivatloo B (2021) Single and multi-objective optimal power flow using a new differential-based harmony search algorithm. J Ambient Intell Humaniz Comput 12(1):851–871
12. Shehata A, Ahmed M, editors (2019) State estimation accuracy enhancement for optimal power system steady state modes. In: IOP Conference Series: Materials Science and Engineering; IOP Publishing.
13. Bouktir T, Slimani L, Belkacemi M (2004) A genetic algorithm for solving the optimal power flow problem. Leonardo J Sci 4:44–58
14. Fletcher R (2000) Practical methods of optimization. John Wiley & Sons, Chichester
15. Abou El Ela A, Abido M, Spea S (2010) Optimal power flow using differential evolution algorithm. Electr Power Syst Res 80(7):878–85
16. El-Fergany AA, Hasanien HM (2015) Single and multi-objective optimal power flow using grey wolf optimizer and differential evolution algorithms. Electr Power Compon Syst 43(13):1548–1559
17. Abido MA (2002) Optimal power flow using particle swarm optimization. Int J Electr Power Energy Syst 24(7):563–571

18. Haghighi R, Yektamoghadam H, Dehghani M, Nikoofard A (2021) Generation expansion planning using game theory approach to reduce carbon emission: A case study of Iran. Comput Ind Eng 162:107713
19. El-Fergany AA, Hasanien HM (2018) Tree-seed algorithm for solving optimal power flow problem in large-scale power systems incorporating validations and comparisons. Appl Soft Comput 64:307–316
20. Attia A-F, El Sehiemy RA, Hasanien HM (2018) Optimal power flow solution in power systems using a novel Sine-Cosine algorithm. Int J Electr Power Energy Syst 99:331–343
21. Shaheen MA, Hasanien HM, Mekhamer SF, Talaat HE (2019) Optimal power flow of power systems including distributed generation units using sunflower optimization algorithm. IEEE Access. 7:109289–109300
22. Qais MH, Hasanien HM, Alghuwainem S, Nouh AS (2019) Coyote optimization algorithm for parameters extraction of three-diode photovoltaic models of photovoltaic modules. Energy 187:116001
23. Dehghani M, Bagheri M, Nurmanova V, Taghipour M, Karimyan P, Gharehpetian GB, et al., editors (2020) An improved PSO-GA based fuzzy controller of STATCOM for cascaded induction motors of drilling system. In: 2020 IEEE international conference on environment and electrical engineering and 2020 IEEE industrial and commercial power systems Europe (EEEIC/I&CPS Europe); IEEE.
24. Dehghani M, Han W, Karimipour H, editors (2018) Coordinated fuzzy controller for dynamic stability improvement in multi-machine power system. In: 2018 IEEE International Conference on Smart Energy Grid Engineering (SEGE); IEEE.
25. El-Fergany AA, Hasanien HM (2020) Salp swarm optimizer to solve optimal power flow comprising voltage stability analysis. Neural Comput Appl 32(9):5267–5283
26. Jalalzad SH, Yektamoghadam H, Haghighi R, Dehghani M, Nikoofard A, Khosravy M et al (2022) A game theory approach using the TLBO algorithm for generation expansion planning by applying carbon curtailment policy. Energies 15(3):1172
27. Prioste FB. Optimal power flow using genetic algorithm. In: Roberto CJeAB-FaHVSaDDFaDWBa, editor. {Anais do 15 Congresso Brasileiro de Intelig\^encia Computacional}. Joinville, SC2021. p 1–6
28. Dashtdar M, Najafi M, Esmaeilbeig M (2020) Calculating the locational marginal price and solving optimal power flow problem based on congestion management using GA-GSF algorithm. Electr Eng 102(3):1549–1566
29. Anwer A, Almosawi AH, Alshabbani G, editors (2020) Optimal power flow for Iraqi power system. In: 2020 International congress on human–computer interaction, optimization and robotic applications (HORA); IEEE.
30. Stanković S, Söder L, editors (2018) Optimal power flow based on genetic algorithms and clustering techniques. In: 2018 Power Systems Computation Conference (PSCC); IEEE.
31. Osman M, Abo-Sinna MA, Mousa A (2004) A solution to the optimal power flow using genetic algorithm. Appl Math Comput 155(2):391–405
32. Bakirtzis AG, Biskas PN, Zoumas CE, Petridis V (2002) Optimal power flow by enhanced genetic algorithm. IEEE Trans Power Syst 17(2):229–236
33. Alsac O, Stott B (1974) Optimal load flow with steady-state security. IEEE Trans Power Appar Syst 3:745–751
34. Index of Data Illinois Institute of Technology [Online]. Available from: http://motor.ece.iit.edu/data/

Chapter 8
Multi-Objective Optimal Power Flow Considering Offshore Wind Farm

Wissem Nemouchi, Youssouf Amrane, Nour El Yakine Kouba, Lakhdar Nadjib Boucetta, and Hichem Nemouchi

1 Introduction

Electrical energy is the heart of our modern society. The steady increase in electrical energy consumption can be primarily attributed to the combined effects of industrialization and population growth [1]. To fight against climate change and air pollution, the supply of renewable energy is an attractive solution reducing fossil fuel harmful residues, among them, wind energy [2].

Technical challenges arise when integrating large wind farms into the transmission system related to the wind variability (intermittent, random, and unprediction) [3]. This concept can be considered as a disturbance of the power injected into the electrical system and can degrade the performance of the electricity grid and cause instances of instability [4].

Hence, the interest includes improving voltage stability margin in transmission network within the wind power plant with an automatic voltage regulation system supplying or assimilating the reactive power necessary to sustaining the voltage in

W. Nemouchi · Y. Amrane · N. E. Y. Kouba (✉) · L. N. Boucetta · H. Nemouchi
Laboratory of Electrical and Industrial Systems, University of Science and Technology Houari Boumediene, 16111 El Alia, Bab Ezzouar, Algiers, Algeria
e-mail: koubanour@gmail.com

W. Nemouchi
e-mail: nemouchiwissem@gmail.com

Y. Amrane
e-mail: youcef.lsei@gmail.com

L. N. Boucetta
e-mail: l_boucetta@enst.dz

H. Nemouchi
e-mail: hichusthb@gmail.com

M. Khosravy et al. (eds.), *Frontiers in Genetics Algorithm Theory and Applications*, Springer Tracts in Nature-Inspired Computing,
https://doi.org/10.1007/978-981-99-8107-6_8

a steady state within the permissible operating limits of the transmission network [5, 6].

The solution lies in total loss optimization [7, 8] but with an increase in production costs which is a contrast. This problem is presented in terms of optimal power flow (OPF), which aims to maximize consumer profit and minimize the total cost by adjusting generated active power [9, 10] in such way as both active-reactive power losses are acceptable and the constraints on the power flows in the transmission lines are satisfied and to control the active power output of the generators as well as their voltage levels [11]. The OPF can also control load and phase angle transformers and all other devices in flexible AC transmission systems named FACTS.

In addition, Flexible Alternative Current Transmission Systems (FACTS) are an effective device that employs to ensure power system flexible operation [12]; it means minimizing the total losses and consequently the production cost [13].

Meta-heuristics techniques like Particle Swarm Optimization (PSO), Differential Evolution (DE), and Genetic Algorithm (GA) are widely used to optimize large complex problems tackling a system that is multimodal, highly constrained, multi-objective, and discrete in nature [14].

The primary drawback of evolutionary algorithms lies in their significant computational expense, typically requiring a substantial number of function evaluations, often reaching several thousands.

Genetic Algorithms (GA) is a nature-inspired computation technique and a subset of Artificial Intelligence (AI), originally focused on single-objective optimization problems; these applications encompass a wide range of domains such as self-organized control [15], maintenance [16], energy management [16, 17], cyber security [18, 19], predictive analysis [20], and lightweight AI model design [21, 22]. Genetic Algorithms (GA) have emerged as a powerful force in the industrial sector, offering numerous benefits and solutions to various challenges. However, NSGA-II and NSGA-III extended the capabilities of GA to address multi-objective optimization problems.

The framework is focused on the multi-objective optimization of power flow with integration of SVC in an electrical grid in the presence of a renewable source (wind turbine) with two types: onshore and offshore; they are similar in operation but the conditions at sea involve the construction of more robust devices.

The first section introduces a brief description of FACTS devices. We presented in the second section wind turbine modeling. The third section is devoted to NSGA-II and NSGA-III algorithms; then we shed light on problem formulation; Sect. 6 provides an in-depth exploration of the case study, outlining its details and characteristics. The last part will contain the results and discussions of F to the 114 network nodes with the integration of SVC devices.

2 Facts System

FACTS devices are deployed in transmission systems for various advantages, including reducing active power transmission losses, controlling power flow, and enhancing voltage stability and profile [13]. The evolution of FACTS devices has introduced new possibilities for more efficient network operations through continuous and rapid adjustments of different parameters such as phase shift, voltage, and impedance [23]. As a result, power transfers can be better managed and voltages can be more effectively regulated, thereby increasing stability margins and enabling the utilization of thermal limits on transmission lines [24, 25].

However, maintaining a balance between production and consumption necessitates continuous monitoring of the system to ensure service quality.

Different methods of FACTS device modeling have been developed for the study of the steady state. They differ mainly in the method used to integrate FACTS in the load flow calculation. There are three modeling techniques; most often encountered are detailed below:

- **Power injection at extremities line**

One of the most common methods is to represent FACTS devices as power injections at nodes The principle consists in replacing the effect due to the FACTS device on the power transits in the line by power injections at both ends. These are calculated so that the effect produced is equivalent to that of the device in question [26].

The expressions for the active and reactive power injections at i node are provided by Eq. (1):

$$P_i^F = P_{ij} - P_{ij}^F \tag{1}$$

$$Q_i^F = Q_{ij} - Q_{ij}^F \tag{2}$$

where

P_i^F, Q_i^F: Equivalent injection of active and reactive power in i node.

P_{ij}, Q_{ij}: Transits of active and reactive power in the FACTS device.

P_{ij}^F, Q_{ij}^F: Transits of active and reactive power with FACTS in the line.

- **Creating dummy node**

This study presents a FACTS model that focuses on controlling active and reactive power transfers presented in the model of a UPFC device [27]. To preserve the power balance, the power injected at i' node is subtracted from i node [26].

- **Nodal admittance matrix modification**

The FACTS are inserted in the line according to their type; they are considered as elements that modify the nodal admittance matrix. The device can be placed in the middle or at the end of the line [26].

The equivalent line parameters are determined and substituted for those of the line without FACTS in the nodal admittance matrix. The latter is modified as follows:

$$Y\text{mod} = \begin{bmatrix} Y'_{ii} & Y'_{ij} \\ Y'_{ji} & Y'_{jj} \end{bmatrix} = \underbrace{\begin{bmatrix} Y_{ii} & Y_{ij} \\ Y_{ji} & Y_{jj} \end{bmatrix}}_{\text{Line}} + \underbrace{\begin{bmatrix} y^F_{ii} & y^F_{ij} \\ y^F_{ji} & y^F_{jj} \end{bmatrix}}_{\text{FACTS}} \tag{3}$$

Static Var Compensator (SVC) is a type of FACTS appliance designed to compensate reactive power on networks and maintain bus voltage magnitude [28]. In situations where the system voltage is below the desired threshold, the SVC develops reactive power (capacitive SVC). However, in the case of high system voltage, it absorbs reactive power (inductive SVC) [29]. The SVC is based on ordinary inductive and capacitive elements, and the control is done through power electronics-based elements (ordinary thyristors). The main functions of an SVC are Voltage support, dampening power oscillations, and improving static and transient stability margins [29].

3 Wind Turbine Modeling

The wind turbine converts the kinetic energy of the air mass into rotational motion, thereby generating mechanical power that is characterized by the rotational speed and the mechanical torque [30]. Equation (4) represents the mechanical power extracted from the wind:

$$P_u = \frac{1}{2}\rho S u^3 \tag{4}$$

The aerodynamic power appearing at the rotor is given in Eq. (2):

$$P_{aér} = C_p \cdot P_u = \frac{1}{2}C_p(\lambda)\rho S u^3 \tag{5}$$

$C_p(\lambda)$: Power coefficient.

ρ: The density of the air.

S: The swept area.

u: The wind speed.

The wind turbine is defined by its curve or speed ratio which is given by Eq. (3):

$$\lambda = \frac{\Omega_{turbine} R_t}{u} \tag{6}$$

$\Omega_{turbine}$: The mechanical turbine speed.

R_t: The turbine radius.

4 Proposed Algorithm

The NSGA-II (Elitist Non-dominated Sorting Genetic Algorithm) is a highly efficient algorithm widely used for identifying the Pareto optimal set, offering a diverse range of solutions [29]. It is characterized by three key features: the utilization of elitism, the prioritization of non-dominated solutions, and the incorporation of explicit solution diversity.

The NSGA-II begins by randomly generating an initial population, denoted as P_0, consisting of N parent individuals. At each generation (t), a population of N children (Q_t) is created from the parent population (P_t) through genetic operators such as selection, crossover, and mutation. The parent and child populations are then combined to form a new population (R_t) of size 2N. The search for non-dominated solutions involves categorizing individuals in R_t into different fronts based on their ranks. This process involves comparing each individual with all others using the concept of dominance. Non-dominated individuals are assigned to the first rank, representing the Pareto front. The algorithm iteratively eliminates these individuals to identify subsequent ranks and construct the new parent population ($P_{(t+1)}$).

In cases where there are more solutions in the last front than available spots in the new population ($P_{(t+1)}$), the individuals are sorted based on their crowding distance [30] in ascending order. This approach ensures the optimal distribution of individuals across the highest ranking front.

However, NSGA-III, introduced by Deb and Jain, represents a recent methodological advancement in the field [30]. The NSGA-III algorithm [37] starts the random initialization of the population size N.M-dim. Predefined more distributed, the plane incorporates reference points H with hyperplanes, each defined by a normal vector, providing comprehensive coverage across various R_+^M regions. Every reference point within the hyperplane is strategically positioned to intersect with each objective function, giving rise to a distinct function known as Das and Dennis's technique or $H = \binom{M+p-1}{p}$. Each range consists of $(p+1)$ points, while N is a multiple of 4, ensuring it is greater than H and satisfies the desired conditions for the algorithm.

At generation t, the entire population is transformed into non-dominated solutions using the same methodology as the sorting mechanism NSGA-II algorithm, after which P_t produces a new descendant population Q_t.

Through the utilization of mutation and recombination operators, every member of the population is associated with a specific reference point, while the selection operator enables competition among distinct reference points. A combined population $R_t = P_t \cup Q_t$ is created by merging the undominated solutions P_{t+1} until each solution cannot be included from the entire edges. If there exists an edge that cannot be associated with a selection of F_L after P_{t+1} and F_L perform a nested and normalized mechanism is employed. In this mechanism, each member of the population is linked to a specific reference point based on the shortest perpendicular distance (d()) between the member and a reference line connecting the origin to the given reference point. Following the niching mechanism, the member F_L linked with the minimum reference points is selected. The overall process aims to identify a population member for each reference point in P_{t+1} that is in proximity to the Pareto optimal front. This is achieved through the application of crossover, mutation, and recombination operators, which facilitate the development of uniform solutions [31].

Its basic framework is similar to that of NSGA-II that uses a pre-allocated reference set mechanism to choose better various solutions in the population size in free space, while NSGA-II algorithm does not require any pre-allocation method on the objective space. Thus, more time is taken to generate the first solution in spaces; NSGA-III easily generated the first solution so NSGA-III is better than NSGA-II algorithm to solve many objective problems [16]. In our study, the NSGA-III method is recommended for its relatively better ability to handle many objectives (up to 50 objectives) and that it also has a uniform diversity to obtain a Pareto optimal front in a set of undominated solutions [27] (Fig. 1).

5 Problem Formulation

The scheduling of reactive power in the network is mathematically formulated as a multi-objective optimization problem, subject to various constraints. The first consists in minimizing the loss cost and investment cost of compensation means. The second function problem aims to improve the stability and the security of voltage (sum of two objective functions).

The optimization problem is addressed using the NSGA-II and NSGA-III algorithms, both of which are employed to find solutions for the problem.

5.1 Variables

U, X are the state vector systems defined by Eqs. (7) and (8) [32–36]:

$$U = [P_{Gi}, Q_{Gi}, V_{Nb}, Q_{svci}, P_{windi}, T_i] \quad (7)$$

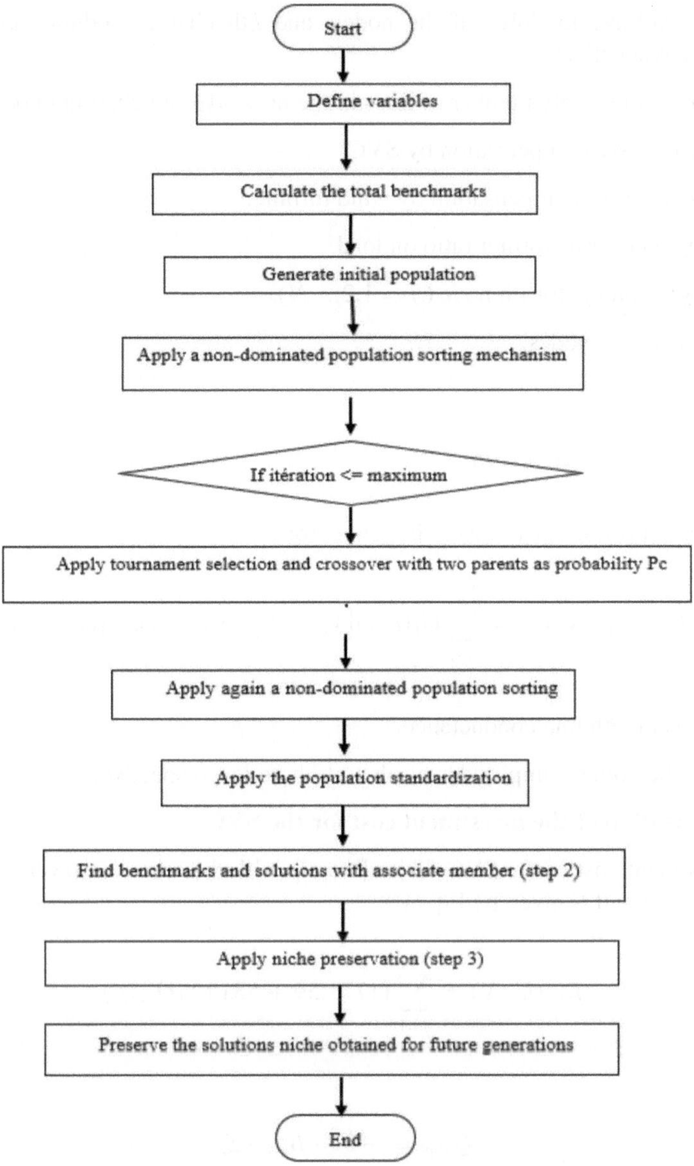

Fig. 1 NSGA-III algorithm flowchart [27]

$$X = [V_{Lk}, \theta_j] \tag{8}$$

where

P_{Gi}, Q_{Gi}: the ith generator's active and reactive power.

V_{Nb}, V_{Lk}: voltage modulus of the nodes, and kth charge modulus nodes (k = 1,2,...N_L), respectively.

N_b, N_L are respectively number of lines in the network, and charging node.

Q_{svci}: the reactive compensation by SVC.

P_{windi}: the reactive compensation by wind turbine.

T_i : the regulator transformer ratio on load.

θ_j: the angle voltage for jth node (j = 1,2,...N).

N is total network nodes.

5.2 Objective Function

- **Minimization of active losses in generators**

$$Ploss_{opt}(X, U) = \sum_{k=1}^{NL} G_i(i, j)\left[V_i^2 + V_j^2 - 2V_i V_j \cos(\theta_i - \theta_j)\right] \tag{9}$$

where G_i is the ith line conductance.

V_i, V_j are the voltage amplitudes of ith and jth nodes, respectively.

- **Minimization of the investment cost for the SVC**

The investment cost of the SVC varies linearly with the reactive power of the SVC to be installed and is given by Eq. (10):

$$f_{svc}(U, X) = \sum_{i=1}^{Nsvc} (1771.59 + 5314.8|Q_{svci}|) \tag{10}$$

with

$$Q_{svci} = -U^2 \cdot B_{svc} \cdot S_n \tag{11}$$

- **Maximization of voltage stability**

The Fast Voltage Stability Index (FVSI) is a metric used to assess and predict the voltage stability state of a system, specifically with reference to a particular transmission line. In the context of an interconnected network, Fig. 2 depicts a single line. The FVSI, as proposed by Musirin et al. [25], is derived from a single line power flow concept.

Fig. 2 Typical diagram of the power transmission line

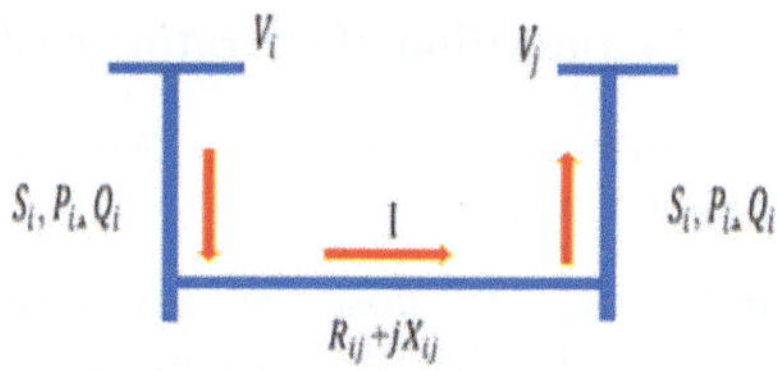

The quadratic voltage equation at the receive node is given in Eq. (12):

$$V_2^2 - \left(\frac{R}{X}\sin\delta + \cos\delta\right)V_1 V_2 + \left(X + \frac{R^2}{X}\right)Q_2 = 0 \quad (12)$$

By imposing the condition that the discriminant of the equation is greater than or equal to zero, we obtain the following inequality:

$$\left[\left(\frac{R}{X}\sin\delta + \cos\delta\right)V_1\right]^2 - 4\left(X + \frac{R^2}{X}\right)Q_2 \geq 0 \quad (13)$$

We obtain Eq. (10):

$$\frac{4Z^2 Q_2 X}{V_1(R\sin\delta + X\cos\delta)^2} \leq 1 \quad (14)$$

Given that δ is close to 0: $\delta \approx 0$, $R\sin\delta \approx 0$, and $X\sin\delta \approx X$.

We consider that '*i*' and '*j*' are the start and receive nodes, respectively. The FVSI can be mathematically expressed using Eq. (15):

$$FVSI = \frac{4Z^2 Q_2}{V_i^2 X} \quad (15)$$

In the given context, *Z* refers to the line impedance, *X* represents the line reactance, Q_j represents the reactive power at the receiving node, and V_1 denotes a specific voltage value.

- **Minimization of the voltage safety**

F_{VSta} is the voltage safety objective function given by Eq. (16):

$$F_{VSta}(X, U) = \begin{cases} if(V_i > V_{i\max})\ F_{VSta} = (V_i - V_{i\max}) \\ if(V_{i\min} < V_i)\ F_{VSta} = (V_{i\min} - V_i) \end{cases} \quad (16)$$

where $Vi_{\max}$, $Vi_{\min}$ are respectively maximum and minimum voltage at the *i*th bus.

- **Maximization of operating cost of wind turbine**

The operational cost of a wind turbine $F_{windturbine}$ can be categorized into 3 compenents:

$WPCost_{oe}$: the cost resulting from the overestimation of imbalances and the cost associated with reserving the availability of wind power. Overestimation occurs when the actual wind power generated is lower than the expected output, leading the operator to procure additional active power from an alternate source. The expected value of wind power for the *j*th turbine is denoted as $E(Y_{oe,j})$ and is determined by Eq. (19).

$WPCost_{ue}$: the cost resulting from the underestimation of imbalances and the penalty cost associated with the availability of wind power. Underestimation occurs when the actual wind power generated exceeds the expected production, resulting in the loss of the excess energy and the operator having to compensate the wind power supplier. The expected value of wind power for the underestimation of the *j*th turbine is represented as $E(Y_{ue,j})$ and is determined by Eq. (20).

$WPCost_{dir}$: the direct cost [25].

The total cost of wind energy is therefore represented in Eqs. (17) and (18):

$$F_{windturbine} = \sum_{j=1}^{Nw} \left(WPCost_{oe,j} + WPCost_{ue,j} + WPCost_{dir,j} \right) \tag{17}$$

$$F_{windturbine} = \sum_{j=1}^{Nw} \left(C_{rw,j} \times E\left(Y_{oe,j}\right) + C_{pw,j} E\left(Y_{ue,j}\right) + g_j \times W_j \right) \tag{18}$$

$C_{rw,j}$ is the overestimated cost coefficient for *j*th wind turbine.

$C_{pw,j}$ is the underestimated cost coefficient for *j*th wind turbine.

N_w is the total number of wind turbines.

$$\begin{aligned} E(Y_{oe,j}) &= W_j \times \left[1 - \exp(-(V_{in,j}/C_j)^{k_j} + \exp(-(V_{out,j}/C_j)^{k_j} \right] \\ &+ \left(W_{r,j} \cdot V_{in,j}/(V_{r,j} - V_{in,j}) + W_j \right) \times \left[\exp(-(V_{in,j}/C_j)^{k_j}) - \exp(-(V_{1,j}/C_j) \right] \\ &+ \left(W_{r,j} \cdot C_j/(V_{r,j} - V_{in,j}) \right) \times \Gamma\left[1 + 1/k_j.(V_{1,j}/C_j)^{k_j} \right] - \Gamma\left[1 + 1/k_j.(V_{in,j}/C_j)^{k_j} \right]. \end{aligned} \tag{19}$$

where

$$V_{1,j} = V_{in,j} + \left(V_{r,j} - V_{in,j} \right) W_j / W_{r,j} \tag{20}$$

$$E(Y_{ue,j}) = (W_{r,j} - W_j) \times \left[exp(-(V_{r,j}/C_j)^{k_j}) - \exp(-(V_{out,j}/C_j)^{k_j})\right]$$
$$+(W_{r,j}.V_{in,j}/(V_{r,j} - V_{in,j}) + W_j) \times \left[exp(-(V_{r,j}/C_j)^{k_j}) - \exp(-(V_{1,j}/C_j)^{k_j})\right]$$
$$+(W_{r,j}.C_j/(V_{r,j} - V_{in,j})) \times \left\{\Gamma\left[1 + 1/k_j, (V_{1,j}/C_j)^{k_j}\right] - \Gamma\left[1 + 1/k_j, (V_{r,j}/C_j)^{k_j}\right]\right\} \quad (21)$$

W_j, $W_{r,j}$ are respectively active and rated power for *j*th wind turbine. C_j, k_j are respectively scale factor and form factor in Weibull distribution for *j*th wind turbine. $V_{in,j}$ and $V_{out,j}$ are respectively cut in and cut out wind speed for jth wind turbine. $\Gamma(\cdot)$ is the Gamma function found by Matlab software:

$$\Gamma(\beta) = \int_0^\infty y^{\beta-1} e^{(-y)} dy \quad (22)$$

5.3 *Constraints*

- **Equality constraints**

These constraints present the power flow equations, which express the balance between the generation and consumption of active and reactive power for each node in the network, Eqs. (20) and (21):

$$P_{Gj} + P_{wi} - P_{Di} - V_i \sum_{j=1}^{NB} V_j \left(G_{ij} \cos\theta_{ij} + B_{ij} \sin\theta_{ij}\right) = 0 \quad i = 1, \ldots N_B \quad (23)$$

$$P_{Gj} + P_{wi} - P_{Di} - V_i \sum_{j=1}^{NB} V_j \left(G_{ij} \cos\theta_{ij} + B_{ij} \sin\theta_{ij}\right) = 0 \quad i = 1, \ldots N_B \quad (24)$$

P_{Di},, Q_{Di} are respectively active and reactive load power.

P_{Gi}, Q_{Gi} are respectively generated active and reactive power.

G_{ij}, B_{ij} are respectively conductance and susceptance between *i* and *j* node.

- **Inequality constraints**

These constraints present the physical and technical limitations of different equipment installed in elnetwork; they are summarized as follows:

$$P_{Gi}^{\min} \leq P_{Gi} \leq P_{Gi}^{\max} \quad (25)$$

$$Q_{Gi}^{\min} \le Q_{Gi} \le Q_{Gi}^{\max} \tag{26}$$

$$V_{Gi}^{\min} \le V_{Gi} \le V_{Gi}^{\max} \quad i = 1, 2 \ldots N_G \tag{27}$$

where N_G is generator number.

V_{Gi}, P_{Gi}, Q_{Gi} are respectively voltage, active, and reactive generator power.

$V_{Gi}^{\min}$ and $V_{Gi}^{\max}$ are respectively lower and upper generator voltage limitations.

$P_{Gi}^{\min}$, $P_{Gi}^{\max}$ and $Q_{Gi}^{\min}$, $Q_{Gi}^{\max}$ are respectively lower and upper active and reactive power generator limitations.

The SVC susceptance is limited by lower and upper limits in Eq. (28):

$$Q_{svci}^{\min} \le Q_{scvi} \le Q_{svci}^{\max} \quad i = 1, 2 \ldots N_{svc} \tag{28}$$

where N_{svc} and b_i are respectively number and susceptance of SVC; $Q_{svci}^{\min}$ and $Q_{svci}^{\max}$ are respectively lower and upper SVC limitations.

Wind power is limited in Eq. (29):

$$Pwind_i^{\min} \le Pwind_i \le Pwind_i^{\max} \quad i = 1, 2 \ldots N_{windturbine} \tag{29}$$

$N_{windturbine}$ is wind turbine number. $Pwind_i$ is active power generated by wind turbine.

The ratio of the on-load regulator is limited as in Eq. (30):

$$T_i^{\min} \le T_i \le T_i^{\max} \quad i = 1, 2, \ldots N_T \tag{30}$$

N_T is transformers number.

The load node voltage and the power flows represent the following constraints:

$$V_{Li}^{\min} \le V_{Li} \le V_{Li}^{\max} \quad i = 1, 2 \ldots N_{Load} \tag{31}$$

$$S_i \le S_i^{\max} \quad i = 1, 2 \ldots N_{LI} \tag{32}$$

$$S_i = \sqrt{(P_i)^2 + (Q_i)^2} \quad j = 1, 2 \ldots N_b \quad i = 1, 2 \ldots N_b \tag{33}$$

$$P_{ij} = V_i^2 G_{ii} + V_i V_j \left(G_{ij} \cos\theta_{ij} + B_{ij} \sin\theta_{ij} \right) \tag{34}$$

$$P_{ji} = V_j^2 G_{jj} + V_i V_j \left(G_{ji} \cos\theta_{ji} + B_{ji} \sin\theta_{ji} \right) \tag{35}$$

$$Q_{ij} = -V_i^2 B_{ii} + V_i V_j (G_{ij} \sin \theta_{ij} - B_{ij} \cos \theta_{ij}) \quad (36)$$

$$Q_{ji} = -V_j^2 B_{jj} + V_i V_j (G_{ji} \cos \theta_{ij} - B_{ij} \sin \theta_{ij}) \quad (37)$$

where

N_{LI} is the number network line, V_{Li} is voltage modulus of charge nodes. $V_{Li}^{\min}$ and $V_{Li}^{\max}$ are respectively upper and lower voltage modulus limits of charge nodes. S_i, P_i, and Q_i are respectively apparent, active, and reactive power transmitted by ith line. $S_i^{\max}$ is lower apparent power limit. P_{ij}, Q_{ij} are respectively the active and reactive power transmitted from i node to j node. P_{ji}, Q_{ji} are respectively the active and reactive power transmitted from j node to i node. G_{ii} and B_{ii} are respectively real and imaginary part of the diagonal elements for the admittance matrix, G_{ij} and B_{ij} are respectively the real and imaginary parts of the non-diagonal elements for the admittance matrix from i node to j node, G_{ji} and B_{ji} are respectively the real and imaginary parts of the non-diagonal elements of the admittance matrix from j node to i node.

6 Results and Discussions

6.1 *Case Study*

In this part, three sites were chosen going from west to east: Ténès, EL-Kala, and Bejaia. Table 1 presents the monthly fluctuations of wind speed and power density.

6.2 *NSGA-II and NSGA-III Test (In the Presence of SVC and Offshore Wind Turbines)*

The electrical network test is composed of 114 buses, and 15 generators at busbar Nos. 4, 5, 11, 15, 17, 19, 22, 52, 80, 83, 98, 100, 101, 109, 111, and 99 PQ nodes through 175 transmission lines, and 16 transformers taps. The permissible range for transformer load adjusters is between 0.90 and 1.10 per unit (pu). Nodes 49, 96, 80, and 12 are chosen for the location of wind turbines with 80 offshore wind turbines in the first three nodes each, and 240 wind turbines in the Adrar node, the vector of control variables for the OPRF consists of the active power of the generators. The parameters considered include the voltage magnitudes of the generators, the ratios of the 16 transformers, and the reactive power outputs of the Static Var Compensators (SVCs) and wind turbines.

The lower and upper limits on voltage modules of all PQ busbars are 0.90 pu and 1.1 pu, the active power is 3146.2 MW, and the reactive power is 1799.4 MVAR; we record 98.902 MW of active losses.

In order to resolve our problem, we will study two cases: the first case is the comparison between the three offshore sites (Ténès, Bejaïa, and El-Kala) and the site of Adrar located in Algrian Saha by using the NSGA-II method. The second case is the same comparison using the NSGA-III method (Table 2).

F1: wind turbine cost **($/h); F2**: active losses **($/h)**;

F3: security and stability voltage **(PU); F4**: SVC cost **($/h)**.

From this, we can observe that the minimum of fitness function obtained by the NSGA-III method is smaller than the one obtained by the NSGA-II method. We can also deduce that the cost of the offshore wind turbine obtained by the two methods is greater than the one of the onshore wind turbine. It can also be seen that the active

Table 1 Monthly variations of mean velocity and mean power density at 10 m

Stations	Ténès		Bejaia		El-Kala	
Parameters	V (m/s)	P (w/m^2)	V (m/s)	P (w/m^2)	V (m/s)	P (w/m^2)
January	5,40	128,85	3,62	42,46	3,74	83,01
February	4,34	85,05	3,41	35,73	5,03	222,1
March	4,73	108,51	3,22	28,69	4,03	94,05
April	4,78	98,02	3	22,78	4,2	82,8
May	4,62	88,09	2,85	17,59	3,71	65,12
June	4,05	55,05	2,94	20,17	3,6	55,61
July	4,04	53	2,97	19,88	3,06	44,12
August	3,87	46,2	2,92	19,17	3,03	30,5
September	3,97	51,44	2,88	18,42	3,31	48,3
October	3,74	46,35	3,11	34,3	3,77	90,5
November	4,06	67,84	3,41	34,56	3,7	57,23
December	4,53	92,25	3,78	48,76	4,09	103,89

It's clear from the table that monthly wind speeds decrease between the month of May and October

Table 2 [F2 F2 F3 F4] results for offshore and onshore using NSGA-II and NSGA-III

	NSGA-II		NSGA-III	
	Offshore	Onshore	Offshore	Onshore
F1 ($/h)	$9.6333*10^5$	$5.1565*10^5$	$8.1858*10^5$	$4.5695*10^5$
F2 ($/h)	$0.9571*10^3$	$1.2192.10^3$	$0.8632*10^3$	$1.2335*10^3$
F3 (pu)	0.7377	1.0470	0.6794	1.1102
Svc ($/h)	$5.3317*10^4$	$3.5994*10^5$	$1.1227*10^5$	$3.1077*10^5$

Table 3 F1 results for the two types of wind turbines with NSGA-II and NSGA-III methods for 25 years' operation

F1 ($/h)	NSGA-II		NSGA-III	
	Offshore	Onshore	Offshore	Onshore
	$9.6333*10^5$	$30.939*10^5$	$8.1858*10^5$	$27.417*10^5$

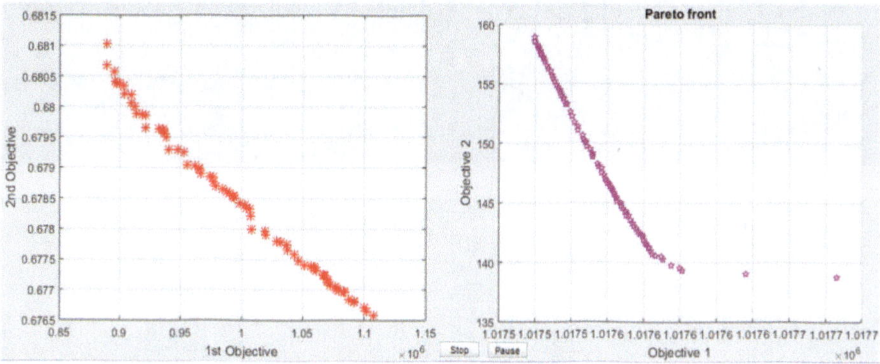

Fig. 3 Pareto cost/stability voltage curve for 144 network nodes

losses in the lines for the offshore wind turbine decrease compared to the onshore wind turbine losses.

For 25 years' operation, by multiplying the cost of onshore wind turbine by a coefficient (A = 7/12) that represents the months of operation of wind turbine and by another coefficient (B15/25) that represents the life of onshore wind turbine, the results are presented in Table 3.

These two results are obtained after some generations, which show the choice of the objective function related to the SVC installation. We can distinguish that NSGA-II provides well-distributed Pareto fronts than NSGA-III (Figs. 3 and 4).

From the optimal reactive power flow results, for both methods the minimum obtained voltage stability for the offshore wind turbine ($4.6921*10^{-5}$ pu) and ($4.6921*10^{-5}$) is lower than that of the onshore wind turbine (0.0032) and (0.0038); it should be noted that all the voltages for the offshore and onshore wind turbine are between their allowed limits (Table 4).

7 Conclusions

The work presented constitutes the improvement of the efficiency of electrical networks through the integration of renewable energy (offshore wind) as well as flexible AC transmission systems (FACTS) and the resolution of the active and reactive power flow problem by taking into account the constraints of equality and inequality.

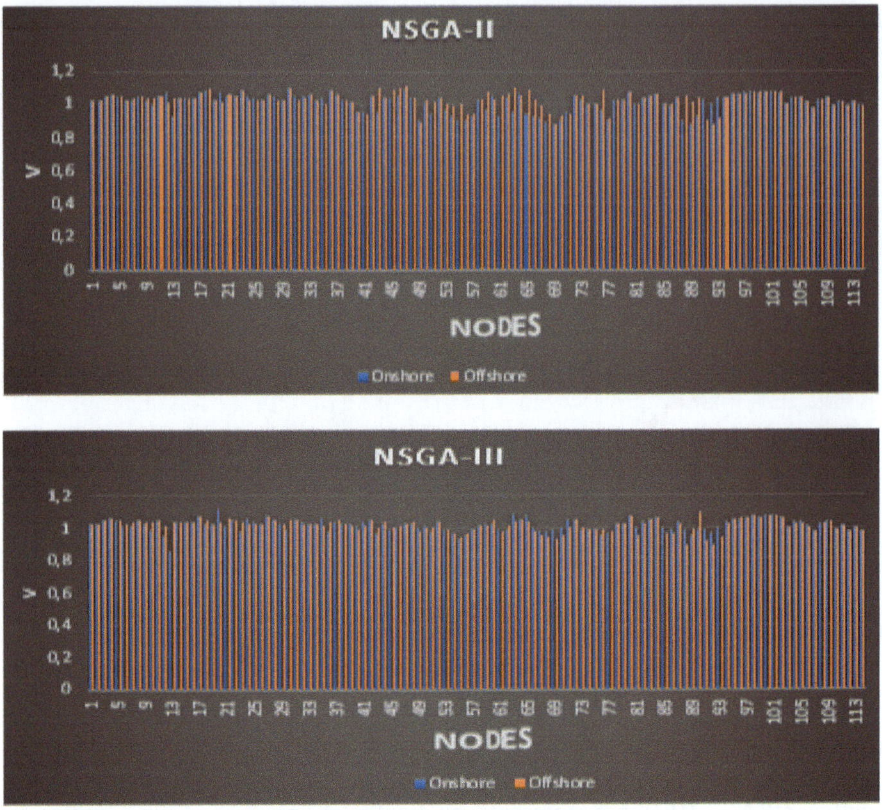

Fig. 4 Optimal power flow voltages with NSGA-II and NSGA-III

Indeed, the optimal power flow (OPF) is able to minimize both the installation cost of wind turbines and SVC, active losses in the lines as well as voltage deviations at each node.

A first phase of this work, a synthesis is carried on the state of the art that has allowed to highlight the interest on the wind; we have also presented the variable speed wind turbines that are able to participate in the regulation of voltage and frequency through the presence of power electronic converters. A brief description on the onshore and offshore wind types with their advantages and disadvantages has been presented in this work based on the comparison between these two types.

We made a description and modeling of FACTS devices, and we chose the SVC to solve our problem.

In this work, the NSGA-II and NSGA-III methods were chosen and applied to solve the optimal power flow problem in the 114-node network in the presence of static reactive power compensators (SVC) and offshore wind sources.

Table 4 Control variable results for the two types of wind turbines with the NSGA-II and NSGA-III methods

NSGA-II								
Nodes	Offshore				Onshore			
	Pg	Vg	Qsvc	Wind turbine	Pg	Vg	Qsvc	Wind tubine
4	−1.1	1.07			−0.87	1.07		
5	6.5	1.05			6.5	1.05		
11	1	1.05			1	1.05	64.38	250.13
15	1	1.04			1	1.04		
17	5	1.08			5	1.08		
19	1.5	1.03			1.5	1.03		
22	1.5	1.05			1.5	1.05		
49			0.00064	83.341				
53	2	0.999			2	0.999		
80	2.5	1.039	0.00015	83.333	2.5	1.039		
83	2	1.070			2	1.070		
96			0.00015	83.3338				
99	2	1.070			2	1.070		
101	2	1.060			2	1.060		
102	2	1.0396			2	1.039		
110	3.5	0.9928			3.5	0.992		
112	3	0.9848			3	0.9848		
NSGA-III								
Nodes	Offshore				Onshore			
	Pg	Vg	Qsvc	Wind turbine	Pg	Vg	Qsvc	Wind tubine
4	−1.18	1.07			−0.8	1.07		
5	6.5	1.05			6.5	1.05		
11	1	1.05			1	1.05	64.380	250.1
15	1	1.04			1	1.04		
17	5	1.08			5	1.08		
19	1.5	1.03			1.5	1.03		
22	1.5	1.05			1.5	1.05		
49			0.00064	83.341				
53	2	0.9992			2	0.99		
80	2.5	1.039	0.00015	83.333	2.5	1.03		
83	2	1.0702			2	1.07		
96			0.00015	83.3338				
99	2	1.0702			2	1.07		

(continued)

Table 4 (continued)

101	2	1.0600			2	1.06		
102	2	1.0396			2	1.03		
110	3.5	0.9928			3.5	0.99		
112	3	0.9848			3	0.98		

We can conclude that the NSGA-III method is better than NSGA-II in terms of execution time, being faster with a minimum of iteration (50 iterations)

The obtained results show that the proposed approach is effective in solving the multi-objective reactive compensation problem. The obtained non-dominated solutions are well distributed and have satisfactory characteristics with the obtaining of a well-distributed Pareto front by NSGA-III which certainly is the fastest.

References

1. Ahmad AS, Hassan MY, Abdullah MP, Rahman HA, Hussin F, Abdullah H, Saidur R (2014) A review on applications of ANN and SVM for building electrical energy consumption forecasting. Renew Sustain Energy Rev 33:102–109. https://doi.org/10.1016/j.rser.2014.01.069
2. Mutezo G, Mulopo J (2021) A review of Africa's transition from fossil fuels to renewable energy using circular economy principles. Renew Sustain Energy Rev 137:110609. https://doi.org/10.1016/j.rser.2020.110609
3. Tang X, Yin M, Shen C, Xu Y, Dong ZY, Zou Y (2018) Active power control of wind turbine generators via coordinated rotor speed and pitch angle regulation. IEEE Trans Sustain Energy 1–1. https://doi.org/10.1109/tste.2018.2848923
4. Mararakanye N, Bekker B (2019) Renewable energy integration impacts within the context of generator type, penetration level and grid characteristics. Renew Sustain Energy Rev 108:441–451. https://doi.org/10.1016/j.rser.2019.03.045
5. Bubshait AS, Mortezaei A, Simoes MG, Busarello TDC (2017) Power quality enhancement for a grid connected wind turbine energy system. IEEE Trans Ind Appl 53(3):2495–2505. https://doi.org/10.1109/tia.2017.2657482
6. Galvani S, Bagheri AA, Farhadi Kangarlu M, Nikdel N (2022) A multi-objective probabilistic approach for smart voltage control in wind-energy integrated networks considering correlated parameters
7. Wang P, Qiuwei W, Sheng H, Bin Z, Canbing L (2022) Distributed optimal voltage control strategy for AC grid with DC connection and offshore wind farms based on ADMM
8. Bibhu PG, Subrat KB, Byamakesh N (2022) Genetic algorithm optimized and Type-I fuzzy logic controlled power smoothing of mathematical modeled Type-III DFIG based wind turbine system
9. Shuijia L, Wenyi G, Ling W, Qiong G (2022) Multi-objective optimal power flow with stochastic wind and solar power
10. Zamzam AS, Baker K (2020) Learning optimal solutions for extremely fast AC optimal power flow. In: 2020 IEEE international conference on communications, control, and computing technologies for smart grids (SmartGridComm)
11. Ferdinando F, Terrence WKM, Pascal Van H (2020) Predicting AC optimal power flows: combining deep learning and Lagrangian dual methods
12. Gaur D, Mathew L (2018) Optimal placement of FACTS devices using optimization techniques: a review. IOP Conf Ser Mater Sci Eng 331

13. Naderi E, Pourakbari-Kasmaei M, Abdi H (2019) An efficient particle swarm optimization algorithm to solve optimal power flow problem integrated with FACTS devices. Appl Soft Comput 80:243–262
14. Ahmad AAL, Sirjani R (2020) Optimal placement and sizing of multi-type FACTS devices in power systems using metaheuristic optimisation techniques: an updated review. Ain Shams Eng J 11(3):611–628
15. Variengien A, Pontes-Filho S, Glover TE, Nichele S (2021) Towards self-organized control: using neural cellular automata to robustly control a cart-pole agent. Innov Mach Intell (IMI) 1:1–14. https://doi.org/10.54854/imi2021.01
16. Takano H, Iwase N, Nakayama N, Asano H (2021) Decision support in maintenance for electrical components of distribution networks. Innov Mach Intell (IMI) 2:1–11. https://doi.org/10.54854/imi2022.001
17. Incremental conductance based particle swarm optimization algorithm for global maximum power tracking of solar-PV under nonuniform operating conditions
18. Khosravy M, Nakamura K, Hirose Y, Nitta N, Babaguchi N (2022) Model inversion attack by integration of deep generative models: privacy-sensitive face generation from a face recognition system. IEEE Trans Inf Forensics Secur 17:357–372
19. Khosravy M, Nakamura K, Nitta N, Babaguchi N (2020) Deep face recognizer privacy attack: model inversion initialization by a deep generative adversarial data space discriminator. In: 2020 Asia-Pacific signal and information processing association annual summit and conference (APSIPA ASC). IEEE, pp 1400–1405
20. Joshi A, Khosravy M, Gupta N (2020) Machine learning for predictive analysis. In: Proceedings of ICTIS
21. Gupta N, Khosravy M, Gupta S, Dey N, Crespo RG (2022) Lightweight artificial intelligence technology for health diagnosis of agriculture vehicles: parallel evolving artificial neural networks by genetic algorithm. Int J Parallel Program 1–26
22. Neeraj G, Khosravy M, Patel N, Dey N, Gupta S, Darbari H, Crespo RG (2020) Economic data analytic AI technique on IoT edge devices for health monitoring of agriculture machines. Appl Intell 50:3990–4016
23. Maouedj R, Bousalem S, Benyoucef B (2008) Algeria wind energy resources, vol 6. Scientific Technical Centre, TATA
24. Bhesdadiya RH, Trivedi IN, Jangir P, Jangir N, Kumar A (2016) An NSGA-III algorithm for solving multi-objective economic/environmental dispatch problem
25. Amrane Y, Boudour M, Ladjici AA, Elmaouhab A, Belazzoug M, Lamari M (2018) Optimal reactive power flow in the presence of wind power loss minimization. In: International conference on electrical sciences and technologies in Maghreb (CISTEM), October 2018. IEEE, pp 1–6
26. Segundo-Ramirez J, Medina A (2009) Modeling of FACTS devices based on SPWM VSCs. IEEE Trans Power Deliv 24(4):1815–1823
27. Gu Z-M, Wang G-G (2020) Improving NSGA-III algorithms with information feedback models for large-scale many-objective optimization. Futur Gener Comput Syst 107:49–69
28. Chopade P, Bikdash M, Kateeb I, Kelkar AD (2011) Reactive power management and voltage control of large transmission system using SVC (Static VAR Compensator). In: 2011 Proceedings of IEEE Southeastcon
29. Bimali B, Uprety S, Pandey RP (2021) VAR compensation on load side using Thyristor switched capacitor and Thyristor controlled reactor. J Inst Eng 16(1)
30. Valenciaga F, Puleston PF (2005) Supervisor control for a standaloneHybrid generation system using wind and photovoltaic energy. IEEE Trans Energy Convers 20(2):389405. (Modélisation simulation et gestion d'énergie d'un système Hybride avec stockage d'énergie)
31. Ethmane IA, Yahfdhou A, Mahmoud AK, Maaroufi M (2019) Statcom integration in a power grid to enhance voltage stability
32. Ransikarbum K, Scott J Mason (2020) A bi-objective optimisation of post-disaster relief distribution and short-term network restoration using hybrid NSGA-II algorithm

33. Cerf R (1994) An asymptotic theory of genetic algorithms. PhD thesis, University of Montpelier II
34. Kalyanmoy D, Himanshu J (2013) An evolutionary many-objective optimization algorithm using reference-point based non-dominated sorting approach, part I: solving problems with box constraints
35. Guesmi T, Alshammari BM, Almalaq Y, Alateeq A, Alqunun KH (2021) New coordinated tuning of SVC and PSSs in multimachine power system using coyote optimization algorithm
36. Yahiaoui M, Bouanane A, Boumediene L (2020) Distribution network reconfiguration for loss reduction using PSO method
37. Deb K, Jain H (2014) An evolutionary many-objective optimization algorithm using reference-point-based nondominated sorting approach, Part I: Solving problems with box constraints, in IEEE Transactions on evolutionary computation, 18(4):577–601. https://doi.org/10.1109/TEVC.2013.2281535

Chapter 9
Genetic Algorithm Applications to Power Management

Asal Rudbari, Mahdi Khosravy, and Olaf Witkowski

1 Introduction

Electricity is a vital part of human life and as the population increases, the demand for reliable energy supply is also increasing. Electricity generation and demand are two critical factors that need efficient management to ensure a balance between them. In traditional centralized systems, electricity is generated at a large scale from remote generating stations and is transmitted over long distances to reach consumers. However, this system has many limitations and is not efficient enough to meet the growing energy demands. The result has been the development of modernized Decentralized Energy Grids (DEG), which are an alternative to the traditional centralized networks. Unlike traditional centralized networks, DEG systems require less infrastructure, making them less capital-intensive. These systems can provide off-grid electricity access to remote and rural areas that lack electricity connection, leading to considerable economic, social, and environmental benefits [1]. Some sources of renewable energy that can be used to generate power in a DEG include:

- Solar power
- Wind power
- Hydropower
- Geothermal power
- Biomass power
- Fuel cells power
- Battery storage power

A. Rudbari (✉)
Sharif University of Technology, Tehran, Iran
e-mail: asal.roudbari@ee.sharif.edu

M. Khosravy · O. Witkowski
Cross Labs, Cross Compass Ltd., Tokyo, Japan

M. Khosravy et al. (eds.), *Frontiers in Genetics Algorithm Theory and Applications*, Springer Tracts in Nature-Inspired Computing,
https://doi.org/10.1007/978-981-99-8107-6_9

Electricity cannot be stored, and it must be generated according to demand. Thus, demand prediction is essential in the field of energy management. In today's world, energy management has become increasingly important due to a variety of factors, including rising energy costs, concerns over climate change, and the need for energy security. Energy management refers to the process of controlling and optimizing energy usage, to improve efficiency, reduce costs, minimize environmental impact, and ensure reliable supply. Many limitations exist in energy management methods at a large scale. One of the biggest challenges associated with energy management is minimizing costs. Energy is often a significant expense for businesses and households alike, and it can be difficult to balance the need for energy with the need to keep expenses in check. This requires careful planning and optimization, such as reducing energy waste, using energy-efficient equipment and technologies, and optimizing energy consumption based on usage patterns and demand. Another challenge is reducing carbon emissions, which is a key part of fighting climate change.

Incorporating AI and smart technologies in the system is an emerging trend that can greatly enhance the performance of DEG systems. Smart grids are an example of implementing intelligent functions into the electricity grids, resulting in improved reliability and sustainability of energy systems. The use of AI in smart grids has enabled better forecasting and prediction of energy demand, allowing for dynamic load management and efficient energy use without the need for human intervention. Furthermore, AI-powered sensors and monitoring systems can provide insights into the condition of the system, leading to real-time monitoring of energy usage and enabling predictive maintenance. This helps to detect faults and malfunctions in the system, reducing downtime and the need for repairs. Demand response is a strategy that plays a significant role in balancing electricity supply and demand [2]. It involves adjusting the demand for electricity in response to changes in the availability of supply, particularly during peak demand periods. The goal of demand response programs is to lower the need for additional power generation by reducing electricity consumption during periods of high demand. In addition to that, the application of the demand response concept in decentralized energy systems significantly reduces energy costs (Fig. 1).

Artificial Intelligence (AI) has made great strides in the energy domain recently and is developing computer programs to go beyond human capabilities and it is gaining growing attention nowadays. AI-based methods can be used to collect datasets from energy grids, and these methods can help overcome the issues in demand prediction. Using AI in energy grids can also help lower energy costs while balancing supply and demand. AI can be divided into branches such as Machine Learning (ML), deep learning, natural language processing, robotics, expert systems, and fuzzy logic. They can be used widely for improving the performance of machines and networks and applied in different areas such as health care [3], robotics [4], smart city [5], Internet of things [6], power quality [7], power events recognition [8] power planning [9], smart house [10], morphological filter [11], autonomous vehicle [12], fake data generation and detection [13], gaming [14] cyber-security [15], etc.

Many algorithms exist in these fields and are getting developed every day. One of these ML-based algorithms, called Genetic algorithm (GA), is a computational

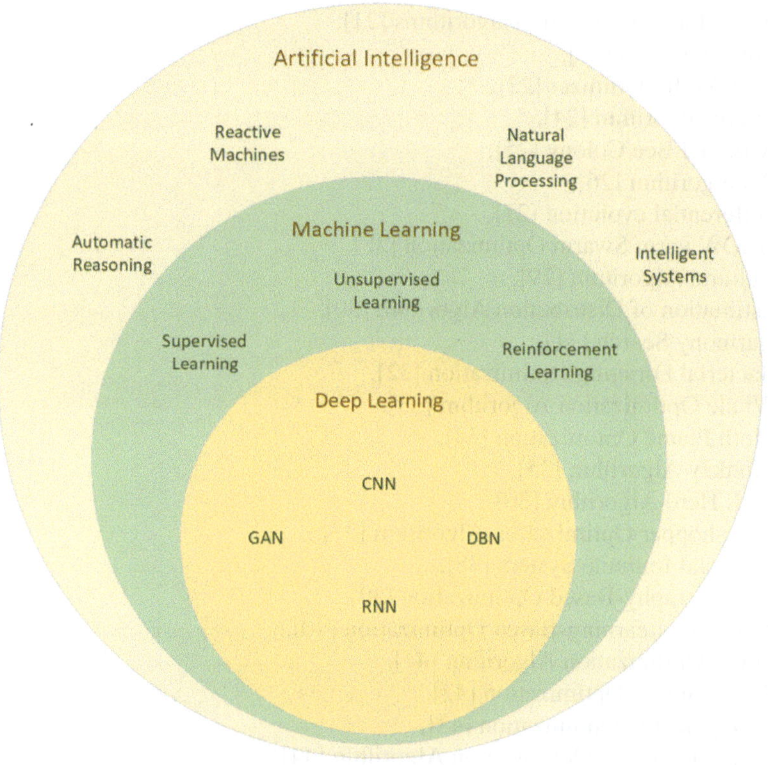

Fig. 1 Comparison between AI, ML, and DL

model of biological evolution introduced and developed by Professor Holland in 1975 [16, 17]. GA offers a wide range of flexibility in application as it can be useful as a search method for solving complex problems or for modeling systems [13]. GA is mostly applied in stochastic optimization that uses natural selection to approximate solutions for a given problem. The search starts from a randomly chosen set of individuals and uses a population of solutions to make a generation. A performance index called the fitness function (also called the cost function) is getting used to calculate the fitness of the solutions over iterations. Parents for the next generation get chosen from their fitness score after applying the fitness function to the present generation. The best possible solution will be found through iterations by applying the leading operators of GA, such as the crossover function and mutation function. GA has been applied in different studies in energy or power management. Some examples of these applications will be discussed in the following sections. Other optimization techniques are used for energy management systems including other nature-inspired algorithms as well, as the main ones are listed below:

1. particle swarm optimization (PSO) [18, 19],
2. simulated annealing (SA) [20],

3. Ant colony optimization algorithms [21],
4. Cuckoo search [22],
5. Grey Wolf Optimizer [23],
6. Firefly algorithm [24],
7. Artificial Bee Colony [25],
8. Bat algorithm [26],
9. Differential evolution [27],
10. GLOWworm Swarm Optimization [28],
11. Cultural Algorithm [29],
12. Estimation of Distribution Algorithm [30],
13. Harmony Search [31],
14. Bacterial Foraging Optimization [32],
15. Whale Optimization Algorithm [33],
16. Moth Flame Optimization [34],
17. Monkey Algorithm [35],
18. Krill Herd Algorithm [36],
19. Grasshopper Optimization Algorithm [37],
20. Artificial Immune System [38],
21. Biogeography-Based Optimization [39],
22. Teaching–Learning-Based Optimization [40],
23. Spiral Optimization Algorithm [41],
24. Social Spider Optimization [42],
25. Group Search Optimization [43],
26. Electromagnetic Optimization Algorithm [44],
27. Moth-Flame-inspired Optimization Algorithm [45],
28. Spiral Dynamic Algorithm [46],
29. Brain Storm Optimization Algorithm [47],
30. Self-Organizing Migrating Algorithm [48],
31. Adaptive Particle Swarm Optimization [49],
32. Artificial Physics Optimization Algorithm [50],
33. Grey Relational Analysis-Based Genetic Algorithm [51],
34. Crow Search Algorithm [52],
35. Social Group Optimization Algorithm [53],
36. Cheetah Algorithm [54],
37. Dolphin Echolocation Algorithm [55], etc.

In addition to the above list of evolutionary algorithms, there is great potential for the recently invented techniques such as Mendelian Evolutionary Theory Optimizer (METO) [56]. These methods have their own strengths and weaknesses and are applied differently depending on the specific requirements of the EMS. Besides power management, GA along with AI has a key role in a variety of industrial applications in areas like self-organized control [57], maintenance [58], power quality and control [7, 59], predictive analysis [60], lightweight AI model design [61, 62], etc.

2 GA in Smart Grids (SGs)

An electric grid is a complex network of generators, transmission lines, and distribution systems that generate, transmit, and distribute electricity to different regions of a large area. The increasing demand for reliable and efficient energy delivery has resulted in the transformation of traditional power grids into smart grids. Smart grids have been envisioned as the future of energy distribution, as they employ automation and power electronics in a two-way advanced power distribution system to provide reliable, efficient, and sustainable energy. One of the key features of smart grids is their ability to control power flow and perform self-healing. Self-healing is a novel feature of smart grids that enables them to perform self-repair and control power flow in the event of network breakdowns. Self-healing smart grids are characterized by their ability to detect and correct faults automatically. This feature leads to enhanced grid reliability and takes security against network breakdowns while it is against network breakdowns. The integration of advanced technologies such as sensors, smart meters, and communication networks is a key feature of smart grids, as it enables real-time monitoring and management of energy usage. This real-time information allows grid operators to optimize the distribution of energy resources effectively, leading to significant reductions in energy wastage and greenhouse gas emissions. Energy management systems play a critical role in the efficient utilization of energy resources in smart grids. As the world moves toward a clean energy future, energy management in smart grids will continue to play a crucial role in promoting sustainability and combating climate change [63] (Fig. 2).

In smart grid technology, GAs have been used for optimization problems, such as the optimization of power flow, voltage control, and energy management. One of the key applications of GAs in smart grid technology is the optimal placement of renewable energy sources (RES), including solar panels, wind turbines, and hydroelectric turbines. The primary goal of using GAs in this context is to maximize the penetration of renewable energy sources in the existing energy infrastructure. The distribution of RES throughout a smart grid network can be a significant challenge that GAs can help solve. By using GAs, the optimal placement of RES can be determined, and the network can be optimized to ensure a more reliable and efficient power supply. Energy storage system plays a critical role in ensuring the stability and reliability of the electricity grid while enabling the integration of renewable energy sources [64].

In the concept of renewable energy generation and storage, using historical hourly data for wind and PV generation, and cooling load, then normalizing them, scaling parameters can be obtained by minimizing the cost function [65]. This also can give us storage capacity. According to that, the objective function is defined as a relationship between the storage capacity, installation costs, and capacities. This:

$$f(x) = min(C_{pv}.\alpha_{pv} + C_W.\alpha_W + C_s.S + C_p.P) \tag{1}$$

where C_{pv} is the storage capacity, C_W is the installation cost of wind, C_S is the installation cost of the storage system, and C_P is the cost of power rating of the

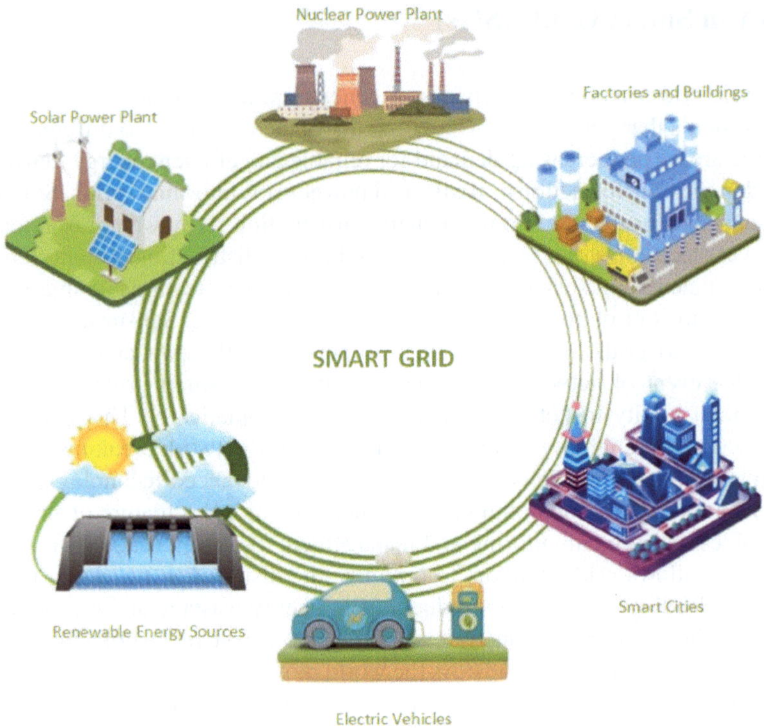

Fig. 2 Smart Grid

storage system. α_{PV}, α_W, S, and P are the capacities. According to the formula, an energy balance equation has to be obtained to guarantee that the combination of wind and PV generation meets the required HV or cooling load over the optimization period. Also, a balance between generation and storage has to be met. The balance of the storage system takes different values whether the storage is getting charged or discharged. The storage also has limitations as its maximum and minimum capacity govern it. Last but not least, the residual energy at any time should be equal to the capacity of the storage system [66].

Another area where GAs have been applied to smart grid technology is in predicting the load demand. Prediction of load demand is essential in energy management and can be difficult to predict accurately. These models use historical data on energy consumption and analyze weather conditions, day of the week or time of day, and other factors and apply GA to find optimal combinations of variables that can accurately predict future load demand. The predictions generated by these models help grid operators plan and manage the distribution of energy resources, reduce the risk of blackouts, and optimize energy use. Furthermore, voltage control optimization in distribution grids is another concept where GAs are applied. Voltage control involves regulating the voltage levels to within an acceptable range, in order to prevent damage to equipment and ensure a reliable supply of electricity. Some of the applications of GAs in voltage control include optimizing the placement and settings of voltage regulators, determining the optimal control strategies for reactive power compensation, and optimizing the allocation of distributed energy resources. By optimizing voltage levels, it is possible to minimize the power demand, losses, and overall network cost. Lastly, GAs have been employed in the optimization of power flow through the smart grid. Power flow optimization involves balancing the amount of power demand and supply through the grid. In this regard, GA can aid in determining the best routes and the amount of power to be transmitted through each.

3 GA in Microgrids (MGs)

A Microgrid (MG) is a self-contained and flexible energy system consisting of a combination of distributed energy resources (DERs), such as renewable energy sources and conventional generators, energy storage systems, and local loads [67]. The energy system can operate in either a grid-connected or islanded mode. The grid-connected operation relies on the host grid for balancing power exchange, and regulation of frequency and voltage. In contrast, the islanded operation requires internal frequency and voltage regulation.

To manage the MG's operation in a grid-connected mode, a fuzzy expert system is used to control the output power. The fuzzy expert system applies analytical techniques and sets of rules to integrate the various variables and evaluate the production and consumption levels of the MG in real time, ensuring proper energy management and optimal performance. To optimize the MG's operation, a cost function with predefined parameters should be defined, representing the problem we want to solve. By minimizing the cost function, the MG's error will be minimized, guaranteeing enhanced performance.

In the operation of MGs, it is crucial to define clear boundaries and limitations while ensuring specific conditions are met. The consumption pattern is one of such critical conditions that must be checked as it pertains to load energy consumption patterns. This consideration helps identify instances of excess energy provision and deficit within the MG. To ensure service continuity, it is vital to check if the battery should be charged or discharged, depending on the energy demand. Advanced control techniques, including Model Predictive Control (MPC) and Model-Based Predictive Control (MBPC), can be used to regulate power flow and maintain the balance between supply and demand [68].

3.1 Power Flow in MGs

MGs are increasingly using renewable sources of energy, such as solar PV and wind, to power their communities. However, they also face challenges in managing fluctuations in supply and demand, which require the use of energy storage systems. MGs rely on energy storage systems such as batteries to manage fluctuations in supply and demand. These systems help manage fluctuations in supply and demand by absorbing excess energy and releasing it when demand exceeds supply. They also help maintain a stable power flow and avoid voltage and frequency fluctuations, which can lead to outages, equipment damage, and increased costs. To optimize the use of renewable energy within MGs, most of the wind power is consumed locally. In this regard, a relationship between energy components should be found. These energy components are the variables in the process of optimization. The objective function is defined as follows [69]:

$$f(x) = min\left(\sum_{k=0}^{n} p_{grid}(k)\right) \tag{2}$$

where n is the number of variables. The total wind power fed into the grid will be used in three main ways. The first and most important one is to supply the electrical demand of the targeted household application. The second way is to be stored in the form of hydrogen and oxygen. The rest of the total power will be fed into a smart grid that we name p_{grid}. The goal is to minimize p_{grid}. The process of optimization needs some boundaries to be defined and should be done step by step as described in the flowchart Fig. 2. Time patterns can be divided into m groups for a more accurate result. Then the GA should be applied in every pattern to reach the best solution.

3.2 Memory-Based GA (MBGA)

Memory-Based Genetic Algorithm (MBGA) is a type of GA that incorporates a memory mechanism that can store and reuse previously evaluated solutions during the search process. This approach makes MBGA more efficient and effective in finding optimal solutions for complex optimization problems. MBGA makes use of memory to improve the amplification capability of GA. Researchers have shown that MBGA outperforms traditional GAs in terms of convergence speed and solution quality. The considered MG in this part includes wind plants, photovoltaic (PV) plants, etc. Sharing the requested power among DERs can be challenging in energy management. The generation cost of the required power should be minimized in an MG for several reasons:

1. Reduced Operational Costs: Minimizing the generation cost of power in a microgrid can reduce the operational costs associated with running the microgrid. This is because the cost of the fuel source used to generate power is often the most significant variable cost associated with operating a microgrid.
2. Improved Energy Efficiency: By minimizing the generation cost of power, a microgrid can maximize its energy efficiency. This means that it can generate

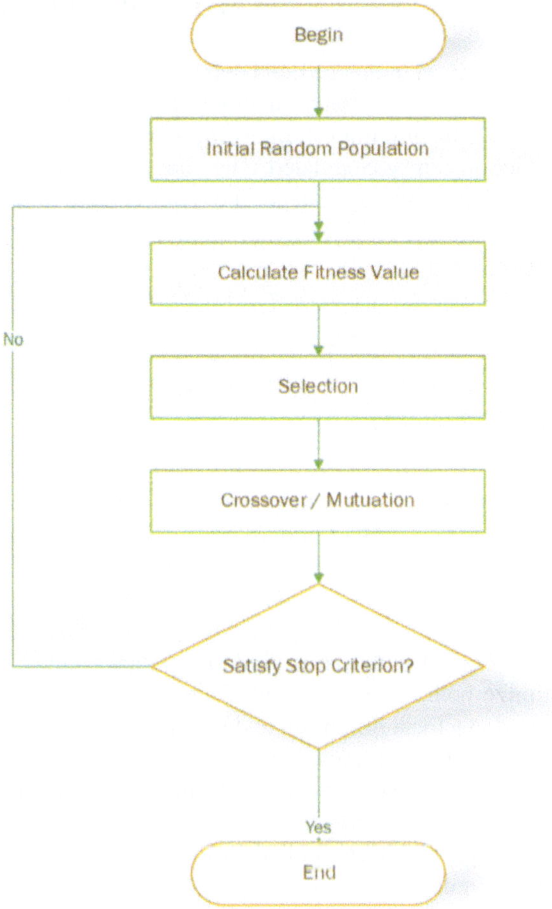

Fig. 3 Flowchart of the standard genetic algorithm

the same amount of power with less fuel, or generate more power with the same amount of fuel, which ultimately reduces costs.

3. Increased Reliability: Microgrids that are designed to minimize generation costs are typically more reliable. By using a variety of power sources, such as solar, wind, and batteries, a microgrid can ensure that it always has an available source of power, even if one source is not working properly.

It's also worth noting that microgrids can be powered by a combination of these energy sources, which can help to further reduce costs and increase efficiency.

In this regard, the generation cost of the required power should be minimized. The cost function for DERs in microgrids ($) is considered a quadratic function of generated power by DER (MW) in i-th DER. The hourly cost can be formulated as below:

$$C_i(P_i) = aP_i^2 + bP_i + c \tag{3}$$

Coefficients in this formula depend on the DER technology (fuel cost, efficiency, etc.). The objective function should be defined based on the cost function defined in Eq. 3. By summing the generation cost of each DER, the total cost of energy production can be calculated. The constraint in this work can be the total generated energy that should be equal to the requested amount of energy by DERs.

$$f(x) = min\left(\sum_{i=0}^{N_{DER}} C_i(P_i)\right) \tag{4}$$

3.3 Binary GA (BGA)

The Binary Genetic Algorithm (BGA) is an optimization technique commonly used in various fields. BGA works by encoding solutions as a string of binary digits instead of decimal values. The algorithm begins by randomly generating an initial population of these binary strings, assessing their fitness using a defined objective function, and then evolving the population over many generations through selection, crossover, and mutation to find an optimal solution. BGA is unique in that it can optimize both continuous and discrete variables while being less prone to getting trapped in a local minimum and thus more likely to search for the global minimum. One of its shortcomings, however, is the high probability of information loss when dealing with continuous variables due to the discreteness of the binary encoding.

In energy management, BGA-GPR (Binary Genetic Algorithm-Gaussian Process Regression) has been applied to improve the short-term forecasting of electricity demand in decentralized energy systems. GPR models are widely used in machine learning applications because of their flexibility and inherent uncertainty measures over predictions. A feature selection approach based on BGA-GPR is developed to identify the most influential variables in electricity demand datasets for four building types. The starting point is to generate an initial population of binary strings of features that are chosen or not chosen based on the values of "1" and "0", respectively. The fitness of each string is evaluated using the objective function, and the parents are then selected for crossover to produce offspring. Through multiple generations, the algorithm evolves towards an optimal solution by exchanging genes between individuals.

In conclusion, BGA-GPR is a promising approach to addressing optimization and prediction problems in energy management. The use of binary encoding allows for more stable and efficient optimization, while GPR models offer a flexible and robust prediction framework. The feature selection approach based on BGA-GPR has shown great potential in identifying the most influential variables in electricity demand datasets for building types, which is crucial for energy efficiency and sustainability.

4 GA in Electric Vehicles (EVs)

The automotive industry has evolved in recent years, leading to a paradigm shift towards the use of electric vehicles (EVs). EVs are becoming increasingly popular due to their numerous advantages over fuel vehicles in terms of efficiency, operating costs, and environmental impact. In addition, EVs offer the potential to integrate renewable energy sources into the power grid. Through Vehicle-to-Grid (V2G) technology, EVs can exchange power with the grid by using their battery storage. This functionality allows for more efficient and environmentally friendly use of energy in the transportation system. Furthermore, with a smart grid recharging infrastructure in place, EVs can help reduce the strain on the smart grid by taking power from the grid when they are charging and giving back power when demand is high.

However, there are many challenges to implementing V2G technology, such as determining the most efficient use of the EVs' batteries and optimizing the cost of production. The use of EVs has significant potential to address energy management problems and integrate renewable energy sources into the power grid. By using optimization algorithms such as GAs, the challenges of implementing V2G technology can be addressed, making the integration of EVs into transportation and energy systems more efficient and cost-effective. GAs can be used as a strategy to address these challenges by minimizing costs, particularly in terms of battery aging costs, electricity prices, and total production costs.

4.1 Vehicle To Grid Technology (V2G)

Vehicle-to-grid (V2G) technology aims to enable electric vehicles (EVs) to not only draw power from the grid but also feed power back into the grid. This is achieved by employing two main types of charging systems: AC and DC charging systems. AC charging systems, also known as onboard chargers, use an alternating current (AC) to charge the EV's battery. These chargers are typically slow charging and have a power output range of 3.7–22 kW. They require the use of a charging cable and a domestic power outlet or a dedicated charging station. AC charging systems are commonly used in residential and commercial settings where the EV is parked for several hours. On the other hand, DC charging systems, also known as fast chargers or rapid chargers, directly charge the EV's battery using a direct current (DC) via an AC–DC converter on the charger side. These chargers can offer power output ranges of 50–350 kW, enabling rapid charging times. They are commonly used in public charging stations that aim to provide fast charging to multiple EVs (Fig. 4).

The communication between the grid and EVs in V2G systems is crucial to effectively transfer data related to travel, battery, and EV conditions, among others. This communication enables intelligent decision-making regarding charging mode, both on the part of the grid and the EVs. It results in a complex communication structure that is dynamic and depends on the requirements of the grid and the EVs. The

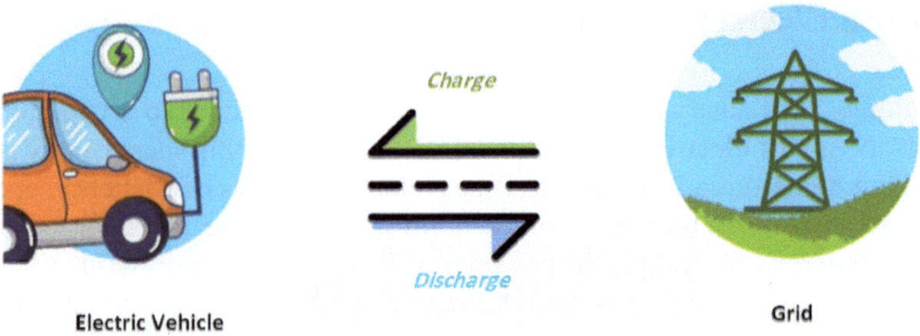

Fig. 4 Vehicle to grid

communication structure also needs to be flexible, adaptable, and secure to ensure reliable and fast information flow.

In this concept, GA is used to optimize the charging and discharging of EVs in a V2G system, taking into account factors such as the availability of renewable energy sources and the state of the electrical grid. GA can help to increase the efficiency of V2G systems, reduce carbon emissions, and lower the cost of electricity. As EVs can act as energy storage units, GA is used to optimize the scheduling of EVs for charging and discharging so that the energy demand of consumers is met without exceeding the capacity of the electrical grid.

4.2 Hybrid Electric Vehicles (HEVs)

Hybrid Electric Vehicles (HEVs) combine two power sources, an electric motor and an internal combustion engine that runs on gasoline, to offer several benefits over traditional gasoline-powered vehicles. Studies have found that HEVs can significantly improve fuel efficiency, reducing fuel consumption by up to 40% compared to conventional vehicles, and save drivers money at the pump. HEVs also offer reduced emissions, as the electric motor helps to lower greenhouse gas emissions and other harmful pollutants. Additionally, HEVs have a greater driving range than pure electric vehicles, as the gasoline engine can provide additional power when needed.

Advancements in technology and increasing consumer demand for eco-friendly vehicles have driven the development of HEVs over the last decade. Battery technology has improved dramatically, resulting in higher energy densities, longer life spans, and reduced costs. Furthermore, the development of more powerful and efficient engines, along with improved aerodynamics, has resulted in even greater fuel efficiency in HEVs. One of the ways to optimize HEVs for fuel economy and reduction in pollution is to apply GA optimization techniques, which involves finding an optimal set of control parameters for the HEV's power train based on a set of constraints and objectives. This method has been widely studied by researchers and has been proven effective in reducing fuel consumption and emissions in HEVs.

HEVs offer several benefits over traditional gasoline-powered vehicles, including improved fuel efficiency, reduced emissions, and greater driving range. As technology continues to advance and consumer demand for eco-friendly vehicles grows, HEVs are expected to become even more popular in the coming years. The application of GA optimization techniques can further improve fuel economy and reduce the pollution emissions of HEVs. Control variables to minimize NO and CO_2, and brake-specific fuel consumption (BSFC) should be set. These variables include ignition angle, EGR rate, AFR, etc.

4.3 Fuel-Cell-Based Hybrid Electric Vehicles (FCHEVs)

Fuel-Cell-based Hybrid Electric Vehicles (FCHEVs) and Hybrid Electric Vehicles (HEVs) are both types of hybrid vehicles, but they differ in the way they generate and store energy and operate slightly differently. FCHEVs use a fuel cell stack to generate electricity from hydrogen and oxygen, which is then used to power an electric motor. FCHEVs also typically have a battery to store excess energy generated by the fuel cell and by regenerative braking. The main advantage of FCHEVs is that they produce water as a byproduct and emit no harmful pollutants, making them an environmentally friendly option for transportation. HEVs, while still more efficient and sustainable than traditional gasoline-powered cars, do emit some greenhouse gases and pollutants.

However, FCHEVs also face some challenges, including the lack of hydrogen refueling stations and the high cost of producing and storing hydrogen. HEVs, on the other hand, have more established infrastructure and are generally more affordable. FCHEVs usually do not have rechargeable batteries and use fuel cells as the main source of power. In FCHEVs, the energy is stored as hydrogen and is converted to electricity by the fuel cell. Energy management in these vehicles can be seen in the reduction of hydrogen consumption by varying the parameters while respecting the constraint. The battery's state of charge (SoC) should remain the same if the FCHEV is not a plug-in. The SoC varies in its optimal SoC which means absorbing the peaks of power from the battery while the fuel cell is running within its optimal power zone. GA can be applied based on the battery SoC level and the consumption of hydrogen can be managed (Fig. 5).

5 GA in Household Energy Demand Management (HEDMS)

Household Energy Demand Management Strategy (HEDMS) [70] is a promising approach to reducing energy consumption, cutting down costs, and improving the efficiency of energy systems. In this strategy, peak load reduction is a vital aspect that

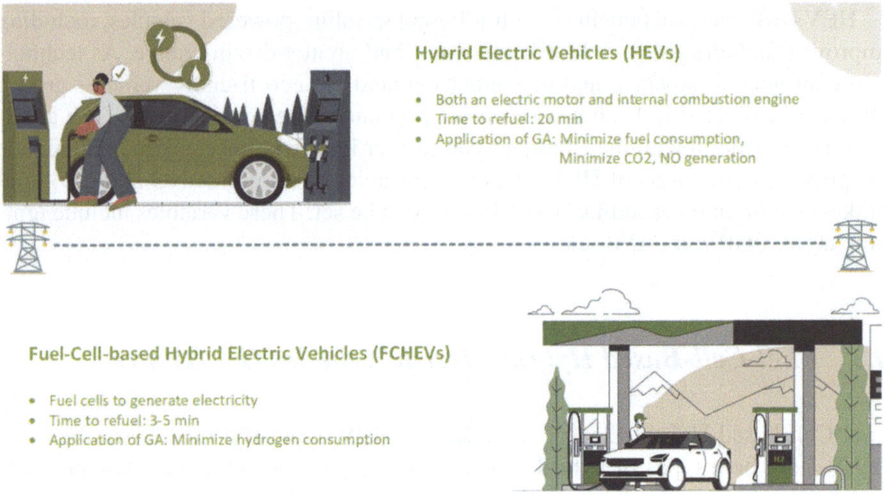

Fig. 5 HEVs versus FCHEVs

helps to keep the stress on the energy grid to a minimum during times of high demand. GAs have been used to optimize the control of devices and appliances during peak times, enabling households to consume less energy.

Many studies investigated the effectiveness of using a GA-based HEDMS to reduce electricity cost and peak load. The study used multiple linear regression models to predict household energy consumption, and a GA algorithm to optimize the scheduling of appliances. The results showed that the GA-based strategy could reduce peak load which is leading to a reduction in the electricity cost.

The specific process of the HEDMS involves the user defining operating parameters, such as comfort settings and rated power, and simulating a daily power profile for all household equipment. The total daily power consumption is then estimated by combining the simulated power profiles with electricity prices. The HEDMS identifies equipment that is expected to participate in the demand management program based on user preferences and then extracts previous power data in 1-minute intervals. Operating parameters for participating devices are optimized using GA to reduce peak demand while maintaining user comfort. Several studies have demonstrated the effectiveness of GA-based demand management strategies in reducing peak load and energy costs while improving overall energy efficiency. One effective approach to achieving this goal is by using GAs that can help to find the optimal combination of devices to use during peak times. By determining the most efficient use of appliances and devices, households can reduce their overall energy usage and contribute to a more sustainable energy system. By adjusting load shifting and utilizing dynamic pricing mechanisms, GA-based energy management systems can distribute energy consumption more evenly and reduce the peak demand during high energy demand periods. Studies have shown that GA-based approaches have achieved significant reductions in peak load during periods of high energy demand while improving overall energy efficiency and reducing energy cost for households. The overall load can be calculated from the summation of the hourly load which is a function of the number of appliances and the average power consumption of each appliance. This relationship is mathematically described below:

$$Load = \sum_{hours=1}^{24} P_{App} \times App \tag{5}$$

The total daily power can be obtained from the summation of equipment powers which are simulated based on the established models and parameters. Combining the total daily power and electricity price, EC can be estimated. Then, the user selects the equipment that is expected to participate in the HEDMS. The data unit extracts the previous power data in a time step of 1 min. Operating parameters of the equipment involved in the control and the power data of the equipment are obtained using GA. The time range set by the user, and the preferred temperature according to the user's settings are the two constraints in this application.

$$f(x) = min\left(\sum_{j=1}^{N}\sum_{i=1}^{M_j}\sum_{t_{f,i}}^{t_{f,i}+t_{di}} \frac{\lambda(t) \times P_j^i(t)}{60}\right)$$

Note that $\lambda(t)$ is the electricity price at time t ($\frac{\$}{kW.h}$), P_j^i is the power change with time of the jth equipment in the ith action cycle. Also, t_{di} is the working time in the ith action cycle, and $t_{f,i}$ is the optimal opening time of the equipment in the ith action cycle.

6 Conclusion

The purpose of this study is to conduct a comprehensive review of the use of GA as an optimization technique for energy management. Compared with standard power grids, smart grids benefit from superior efficiency in multiple aspects owing to their integration with AI methods. GA, an efficient algorithm that utilizes natural selection and learning capabilities, has demonstrated effectiveness in energy management optimization. Moreover, it has the potential to promote the implementation of sustainable energy practices in various energy management systems such as renewable energy, load forecasting, and power distribution. GA applications have been shown to effectively reduce costs, decrease CO_2 emissions that contribute to global warming, as well as enhancing energy efficiency. The practical implementation of GA may still have limitations, such as complexity, parameter tuning, and computational resources. Yet, the benefits it provides to energy management systems far outweigh the limits.

With continual technological improvements, GA's limitations are anticipated to be resolved in the future.

References

1. Kahraman C, Kayakutlu G (2018) Energy management–collective and computational intelligence with theory and applications, vol 149. Springer, Berlin
2. Aslam S, Iqbal Z, Javaid N, Khan ZA, Aurangzeb K, Haider SI (2017) Towards efficient energy management of smart buildings exploiting heuristic optimization with real time and critical peak pricing schemes. Energies 10(12):2065
3. Panesar A (2019) Machine learning and AI for healthcare. Springer, Berlin
4. Murphy RR (2019) Introduction to AI robotics. MIT Press
5. Navarathna PJ, Malagi VP (2018) Artificial intelligence in smart city analysis. In: International conference on smart systems and inventive technology (ICSSIT). IEEE, pp 44–47
6. Khosravy M, Nakamura K, Nitta N, Dey N, Crespo RG, Herrera-Viedma E, Babaguchi N (2022) Social iot approach to cyber defense of a deep-learning-based recognition system in front of media clones generated by model inversion attack. IEEE Trans Syst, Man, Cybern: Syst 53(5):2694–2704
7. Senjyu T, Khosravy M (2022) Power system planning and quality control
8. Mahela OP, Shaik AG, Gupta N (2015) A critical review of detection and classification of power quality events. Renew Sustain Energy Rev 41:495–505
9. Dehghani M, Taghipour M, Sadeghi Gougheri S, Nikoofard A, Gharehpetian GB, Khosravy M (2021) A deep learning-based approach for generation expansion planning considering power plants lifetime. Energies 14(23):8035
10. Bregman D (2010) Smart home intelligence-the ehome that learns. Int J Smart Home 4(4):35–46
11. Khosravy M, Gupta N, Witkowski O, Pasquali A (2021) Neighborhood base matched morphological filters: Cross-fertilization with linear lowpass filtering. In: 2021 international conference on computational science and computational intelligence (CSCI). IEEE, pp 1623–1628
12. Ma Y, Wang Z, Yang H, Yang L (2020) Artificial intelligence applications in the development of autonomous vehicles: a survey. IEEE/CAA J Automatica Sinica 7(2):315–329
13. Khosravy M, Echizen I, Babaguchi N (2022) Frontiers in fake media generation and detection. Springer, Berlin
14. Riedl MO, Zook A (2013) Ai for game production. In: 2013 IEEE conference on computational inteligence in games (CIG). IEEE, pp 1–8
15. Khosravy M, Nakamura K, Hirose Y, Nitta N, Babaguchi N (2022) Model inversion attack by integration of deep generative models: Privacy-sensitive face generation from a face recognition system. IEEE Trans Inf Forensics Secur 17:357–372
16. Holland JH (1992) Genetic algorithms. Sci Am 267(1):66–73
17. Singh G, Gupta N, Khosravy M (2015) New crossover operators for real coded genetic algorithm (rcga). In: 2015 international conference on intelligent informatics and biomedical sciences (ICIIBMS). IEEE, pp 135–140
18. Badi M, Mahapatra S, Raj S (2023) Hybrid boa-gwo-pso algorithm for mitigation of congestion by optimal reactive power management. Optim Control Appl Methods 44(2):935–966
19. Khosravy M, Gupta N, Patel N, Senjyu T, Duque CA (2020) Particle swarm optimization of morphological filters for electrocardiogram baseline drift estimation. In: Applied nature-inspired computing: algorithms and case studies, pp 1–21
20. Xiong R, Wang S, Yu C, Fernandez C, Xiao W, Jia J (2023) A novel nonlinear decreasing step-bacterial foraging optimization algorithm and simulated annealing-back propagation model for long-term battery state of health estimation. J Energy Storage 59:106484

21. Khan FA, Ullah K, ur Rahman A, Anwar S (2023) Energy optimization in smart urban buildings using bio-inspired ant colony optimization. Soft Comput 27(2):973–989
22. Balasubramanyam P, Sood VK (2023) A novel hybrid swarm intelligence and cuckoo search based microgrid ems for optimal energy scheduling. Distrib Gener Altern Energy J 1119–1148
23. Vijaya Kumar N, Charles Raja S, Arun Mozhi S, Jeslin Drusila Nesamalar J (2023) Effective power loss management in the distribution system by the hybrid cuckoo search grey wolf optimizer. IETE J Res 1–13
24. Balaji K, Sai Kiran P, Sunil Kumar M (2023) Power aware virtual machine placement in iaas cloud using discrete firefly algorithm. Appl Nanosci 13(3):2003–2011
25. Qin D, He Z, Zhao X, Liu J, Zhang F, Xiao L (2023) Area and power optimization for fixed polarity reed-muller logic circuits based on multi-strategy multi-objective artificial bee colony algorithm. Eng Appl Artif Intell 121:105906
26. Abdelqawee IM, Emam AW, ElBages MS, Ebrahim MA (2023) An improved energy management strategy for fuel cell/battery/supercapacitor system using a novel hybrid jellyfish/particle swarm/bat optimizers. J Energy Storage 57:106276
27. Kar MK, Kumar S, Singh AK, Panigrahi S (2023) Reactive power management by using a modified differential evolution algorithm. Optim Control Appl Methods 44(2):967–986
28. Nanda B, Jena RK, Muni BP (2023) Enhancing power quality in microgrids with hybrid tuna-glowworm swarm optimization strategy for renewable energy sources. Authorea Preprints
29. Haldar V, Chakraborty N (2015) Power loss minimization by optimal capacitor placement in radial distribution system using modified cultural algorithm. Int Trans Electr Energy Syst 25(1):54–71
30. Li Y, He H, Wang Y, Xu X, Jiao L (2015) An improved multiobjective estimation of distribution algorithm for environmental economic dispatch of hydrothermal power systems. Appl Soft Comput 28:559–568
31. dos Santos Coelho L, Mariani VC (2009) An improved harmony search algorithm for power economic load dispatch. Energy Convers Manag 50(10):2522–2526
32. Dubuisson F, Chandra A, Rezkallah M, Ibrahim H (2020) A bacterial foraging optimization technique and predictive control approach for power management in a standalone microgrid. In: IEEE electric power and energy conference (EPEC). IEEE, pp 1–7
33. Qais MH, Hasanien HM, Alghuwainem S (2020) Enhanced whale optimization algorithm for maximum power point tracking of variable-speed wind generators. Appl Soft Comput 86:105937
34. Mei RNS, Sulaiman MH, Mustaffa Z, Daniyal H (2017) Optimal reactive power dispatch solution by loss minimization using moth-flame optimization technique. Appl Soft Comput 59:210–222
35. Li Y, Sun B, Zeng Y, Dong S, Ma S, Zhang X (2022) Active distribution network active and reactive power coordinated dispatching method based on discrete monkey algorithm. Int J Electr Power Energy Syst 143:108425
36. Roy PK, Paul C (2015) Optimal power flow using krill herd algorithm. Int Trans Electr Energy Syst 25(8):1397–1419
37. Alhejji A, Hussein ME, Kamel S, Alyami S (2020) Optimal power flow solution with an embedded center-node unified power flow controller using an adaptive grasshopper optimization algorithm. IEEE Access 8:119020–119037
38. Hatata A, Osman G, Aladl M (2018) An optimization method for sizing a solar/wind/battery hybrid power system based on the artificial immune system. Sustain Energy Technol Assess 27:83–93
39. Rarick R, Simon D, Villaseca FE, Vyakaranam B (2009) Biogeography-based optimization and the solution of the power flow problem. In: 2009 IEEE international conference on systems, man and cybernetics. IEEE, pp 1003–1008
40. Bouchekara H, Abido M, Boucherma M (2014) Optimal power flow using teaching-learning-based optimization technique. Electr Power Syst Res 114:49–59
41. Benasla L, Belmadani A, Rahli M (2014) Spiral optimization algorithm for solving combined economic and emission dispatch. Int J Electr Power Energy Syst 62:163–174

42. Nguyen TT (2019) A high performance social spider optimization algorithm for optimal power flow solution with single objective optimization. Energy 171:218–240
43. Hagh MT, Teimourzadeh S, Alipour M, Aliasghary P (2014) Improved group search optimization method for solving chped in large scale power systems. Energy Convers Manag 80:446–456
44. Bouchekara H (2020) Solution of the optimal power flow problem considering security constraints using an improved chaotic electromagnetic field optimization algorithm. Neural Comput Appl 32(7):2683–2703
45. Bandopadhyay J, Roy PK (2020) Application of hybrid multi-objective moth flame optimization technique for optimal performance of hybrid micro-grid system. Appl Soft Comput 95:106487
46. Hao MR, Ismail RMTR, Ahmad MA (2017) Using spiral dynamic algorithm for maximizing power production of wind farm. In: International conference on applied system innovation (ICASI). IEEE, pp 1706–1709
47. Jadhav H, Sharma U, Patel J, Roy R (2012) Brain storm optimization algorithm based economic dispatch considering wind power. In: 2012 ieee international conference on power and energy (PECon). IEEE, pp 588–593
48. dos Santos Coelho L, Mariani VC (2010) An efficient cultural self-organizing migrating strategy for economic dispatch optimization with valve-point effect. Energy Convers Manag 51(12):2580–2587
49. Eltamaly AM, Al-Saud M, Abokhalil AG, Farh HM (2020) Simulation and experimental validation of fast adaptive particle swarm optimization strategy for photovoltaic global peak tracker under dynamic partial shading. Renew Sustain Energy Rev 124:109719
50. Teeparthi K, Vinod Kumar D (2018) Security-constrained optimal power flow with wind and thermal power generators using fuzzy adaptive artificial physics optimization algorithm. Neural Comput Appl 29:855–871
51. Pandey AK, Gautam GD (2018) Grey relational analysis-based genetic algorithm optimization of electrical discharge drilling of nimonic-90 superalloy. J Braz Soc Mech Sci Eng 40:1–16
52. Li LL, Liu ZF, Tseng ML, Jantarakolica K, Lim MK (2021) Using enhanced crow search algorithm optimization-extreme learning machine model to forecast short-term wind power. Expert Syst Appl 184:115579
53. Madhavi G, Harika V (2018) Implementation of social group optimization to economic load dispatch problem. Int J Appl Eng Res 13(13):11195–11200
54. Ghaedi H, Kamel Tabbakh Farizani SR, Gaemi R (2022) A novel meta-heuristic framework for solving power theft detection problem: Cheetah optimization algorithm. Int J Ind Electron Control Optim 5(1):63–76
55. Lenin K, Reddy BR, Kalavathi MS (2014) Dolphin echolocation algorithm for solving optimal reactive power dispatch problem. Int J Comput (IJC) 12(1):1–15
56. Gupta N, Khosravy M, Patel N, Dey N, Mahela OP (2020) Mendelian evolutionary theory optimization algorithm. Soft Comput 24:14345–14390
57. Variengien A, Pontes-Filho S, Glover T, Nichele S (2021) Towards self-organized control: Using neural cellular automata to robustly control a cart-pole agent. Innov Mach Intell (IMI) 1:1–14. https://doi.org/10.54854/imi2021.01
58. Takano H, Iwase N, Nakayama N, Asano H (2022) Towards self-organized control: Using neural cellular automata to robustly control a cart-pole agent. Innov Mach Intell (IMI) 2:1–11. https://doi.org/10.54854/imi2022.001
59. Jalalzad SH, Yektamoghadam H, Haghighi R, Dehghani M, Nikoofard A, Khosravy M, Senjyu T (2022) A game theory approach using the tlbo algorithm for generation expansion planning by applying carbon curtailment policy. Energies 15(3):1172
60. Joshi A, Khosravy M, Gupta N (2021) Machine learning for predictive analysis: proceedings of ICTIS 2020. Springer, Berlin
61. Gupta N, Khosravy M, Patel N, Dey N, Gupta S, Darbari H, Crespo RG (2020) Economic data analytic ai technique on iot edge devices for health monitoring of agriculture machines. Appl Intell 50:3990–4016

62. Gupta N, Khosravy M, Patel N, Dey N, Crespo RG (2021) Lightweight computational intelligence for iot health monitoring of off-road vehicles: Enhanced selection log-scaled mutation ga structured ann. IEEE Trans Ind Inform 18(1):611–619
63. Alasali F, Nusair K, Alhmoud L, Zarour E (2021) Impact of the covid-19 pandemic on electricity demand and load forecasting. Sustainability 13(3):1435
64. Skrobek D, Krzywanski J, Sosnowski M, Uddin GM, Ashraf WM, Grabowska K, Zylka A, Kulakowska A, Nowak W (2023) Artificial intelligence for energy processes and systems: applications and perspectives. Energies 16(8):3441
65. Arabali A, Ghofrani M, Etezadi-Amoli M, Fadali MS, Baghzouz Y (2012) Genetic-algorithm-based optimization approach for energy management. IEEE Trans Power Deliv 28(1):162–170
66. Domínguez-Navarro JA, Tedeschi E (2016) Evaluation of the fluid model approach for the sizing of energy storage in wave-wind energy systems. Energies 9(3):162
67. Hu J (2013) Advanced control in smart microgrids. PhD thesis
68. Ali S, Khan I, Jan S, Hafeez G (2021) An optimization based power usage scheduling strategy using photovoltaic-battery system for demand-side management in smart grid. Energies 14(8):2201
69. Chen X, Cao W, Xing L (2019) Ga optimization method for a multi-vector energy system incorporating wind, hydrogen, and fuel cells for rural village applications. Appl Sci 9(17):3554
70. Jiang L, Chen X, Xue B (2019) Features, driving forces and transition of the household energy consumption in china: a review. Sustainability 11(4)

Part IV
Application to Electrical Vehicles

Chapter 10
Application of Genetic Algorithms to Electrical Vehicle Industries

Rishabh Duggal, Mahdi Khosravy, and Olaf Witkowski

1 Introduction

When we speak of the influence of genetic algorithms in electric vehicles, we aim to cover almost every aspect of the EV industry and particularly the drivetrains. Our approach is to find the variables in a particular scenario, develop an algorithm based on it, and apply the genetic operators to maximize the output. Battery and motors are the critical constituents of an EV, but we just can't consider these two as our sole variables and apply the operators, as the spectrum of possibilities would then be infinitesimal and definitely out of our control. It is the inner constituents of the battery and motor which we have to study and decide upon which characteristics would benefit us the most when the operators are applied on them, which would vary according to the different situations the EV has been put in and the outcomes desired out of it.

When compared to internal combustion (IC) engine-powered vehicles, charging an EV has always been a disadvantage, notably in terms of time taken. Additionally, increasing the range is regarded in this industry as difficult. The fact that a battery powers every component of an EV—aside from the motor(s)—must not be overlooked. In addition to the motor(s), current vehicles also have air conditioning, music systems, displays, GPS, heated/cooled seats, countless sensors, and their central processing units (CPUs). All these factors conjointly decide the sustainability of an EV. When we study the current global EV market, we can observe that

R. Duggal (✉)
Thapar Institute of Engineering and Technology, Prem Nagar, Patiala Punjab, India
e-mail: rishabhduggal249@gmail.com

M. Khosravy · O. Witkowski
Cross Labs, Cross Compass Ltd., Tokyo, Japan

M. Khosravy et al. (eds.), *Frontiers in Genetics Algorithm Theory and Applications*, Springer Tracts in Nature-Inspired Computing,
https://doi.org/10.1007/978-981-99-8107-6_10

overall, these vehicles are expensive and along with that, their fundamental features limit them to be considered as primary/sole-owned vehicles, that is, most of the customers cannot risk keeping an EV as their only vehicle, at least as of today. There are undoubtedly some examples from the leading manufacturers which provide the perfect EV for this use case, but again those EVs are quite expensive and their customers are comfortably in a position to afford an IC engine vehicle that has much more comfort or whatever their primary use case is.

This does not, however, imply that these vehicles will never replace the ones we currently drive, largely because this industry is extremely young and expanding quickly. The concerns that were briefly described above are undoubtedly there, but they are dissipating swiftly as the days go by. With the implementation of genetic algorithms to this industry, one can definitely gain confidence on the sustainability of this industry in the near future. This chapter has been divided into several sections, each presenting a firm rationale as to how this approach is lucrative enough to provide advancements in both the EV and GA industries.

2 Electric Vehicles—A Technical Brief

Although the first EV was introduced to us back in the late nineteenth or early twentieth centuries, it was not until the 1960s that the world realized that an alternative means of driving vehicles other than internal combustion engines must be introduced. Increasing demand and the depreciating arsenal of fossil fuels which power these IC engines motivate us even further to explore other means which could end up being more lucrative.

An EV consists mainly of two components—a battery and a motor, which are the soul of an EV. The battery provides the juice for the motor which in turn translates into the desired maneuvers. But it is not just these two which fulfill the embodiment of the vehicle, we need something to control them. This crucial task is handled by an ECU (electronic control unit) and a BMS (battery management system) [1]. Both these systems are a set of electronics working on a very high level. ECU handles almost every aspect of the vehicle such as the driving dynamics which cover the motor characteristics, stability control, torque vectoring, differential management, power delivery, and other passive usage tasks such as air conditioning (AC), radio, music systems, and safety components such as airbags and anti-lock braking system (ABS). Basically, it manages everything except the batteries, which are handled by the BMS. This system focuses solely on the battery, keeping it safe and always in optimum working condition. A deep dive into all these crucial aspects which humanizes the vehicle are discussed in further sections. Figure 1 depicts the fundamental parts of an EV [2].

Along with these, the EVs have a much easier working principle [3], that is, due to the versatile range of induction motors, there is no need for additional mechanical parts which essentially limits the hindrance occurring in the real-life performance of the vehicle when compared to the simulated or computed expectations. An elec-

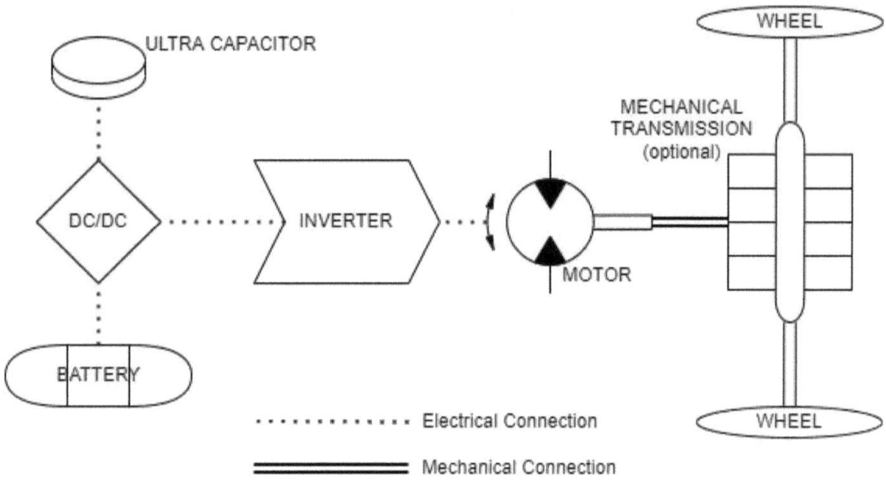

Fig. 1 Fundamental parts of an electric vehicle

tric vehicle has an initial infinite torque and the horse power remains consistent throughout the revolutions per minute (RPM) range. Apart from this, the nature of induction motors allows them to rev to very high RPMs, possibly up to 18000. This eliminates the use of a gearbox (however, sometimes a two-speed gearbox is used to achieve maximum performance, like the Porsche Taycan). Not only these, if individual motors are used to power individual wheels, the use of differentials and thus the limited slip differentials (LSDs) can also be eliminated.

With the technology improving as the days proceed, the range of these EVs is also being consistently improved (discussed in a further section) [4]. There are still many more advantages of EVs like the battery recharge can be done from anywhere as long as there is a power supply and the pollution reduction is conspicuous. But there are some downsides to the use of EVs, like even though the vehicle maneuver limits air pollution, battery production which requires lithium mining causes tremendous harm to the environment [5]. Most EVs still do not have adequate range, the passive uses in the vehicle drink up a significant amount of juice from the battery and EVs are still not affordable for an average person. But these milestones are still achievable and we can overcome obstacles much faster as we witness the growth in the technology associated with these electric vehicles. With the depreciating assets we possess which run the IC engines and the versatility and simplistic production associated with a single EV, the future is inevitably dominated by them. Figure 2 depicts the global growth in EV sale.

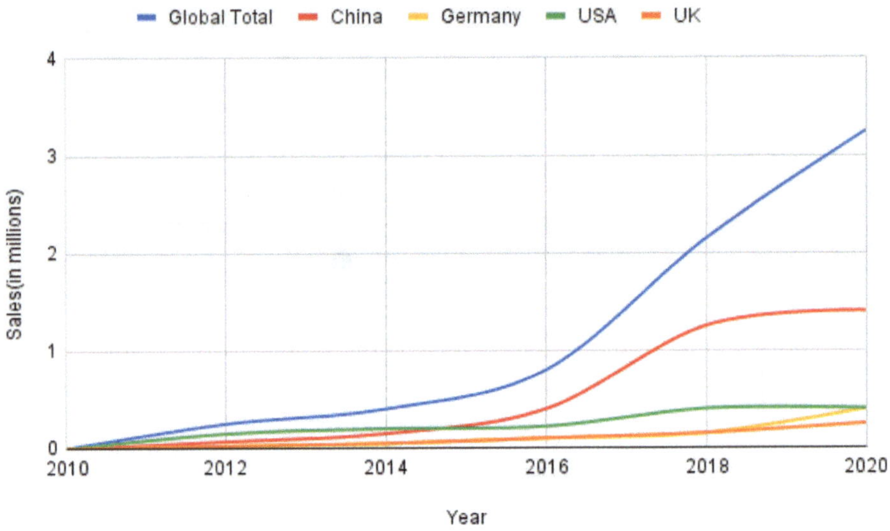

Fig. 2 Global sales of electric vehicles by years

3 Genetic Algorithms in Electrical Vehicle Industries

It has not been a long time since the first practical electric vehicle was invented, which would show up at a stretch of assembly lines, which makes this industry fairly new. The growth is very evident, which also means that this industry is still in a learning phase and is yet to explore many options to choose to proceed further. There are a lot of variables associated with the several aspects upon which an EV is constructed. These variables require a massive amount of computations considering several permutations to yield the best outcomes. This is where the use of genetic algorithms could end up being an astute choice [6]. We can use the basic ideology of GAs to make some of the most efficient algorithms which would seriously develop the driving dynamics of an EV. Not only the characteristics of the battery and motor can be chosen as parents, but their respective relationships with each of the vehicle's components can also be worked upon genetically. Apart from this, as time proceeds, the data gathered in this industry would also pile up which will further improve the driving dynamics.

Figure 3 shows the number of publications on GA (genetic algorithm), GP (genetic programming), DE (differential evolution), ES (evolution strategies), and EP (evolutionary programming) according to Web of Science (WoS) in recent years. Besides GA, currently, other evolutionary algorithms are in use in different industrial applications, and also there are some recently invented techniques highly inspired by GA like Mendelian Evolutionary Theory Optimizer (METO) [7]. Evolutionary computation along with artificial intelligence plays a key role in electrical vehicle industries and also is applied in many areas like maintenance [8], self-organized control [9],

Fig. 3 Number of publications in the WoS database

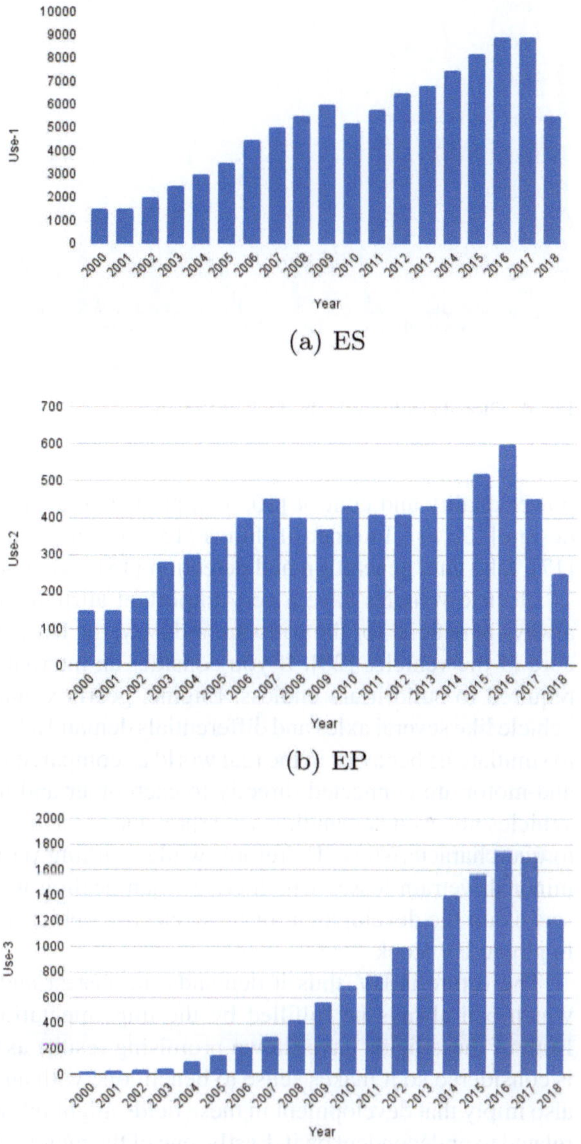

(a) ES

(b) EP

(c) GA, GP, DE, ES, EP

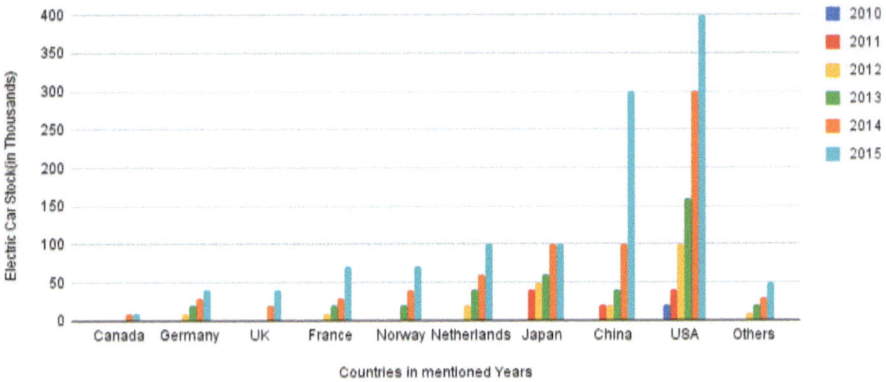

Fig. 4 Global electric vehicle stock evolution

power quality and control [10, 11], predictive analysis [12], lightweight AI model design [13, 14], Internet of things [15], power planning [16], morphological filter [17], fake data generation and detection [18], cyber-security [19], etc.

Electric vehicles have a very important attribute associated with them which is always beneficial for the software industry, the least mechanical complexities in the core of the vehicles [20]. If you consider an internal combustion vehicle, the parts required to build it are endless. Engine, gearbox, and their links to the rest of the vehicle like several axles and differentials demand a lot, which makes it more difficult to simulate its behavior in the real world as compared to electric vehicles. The battery and motor are connected directly to each other and to the wheels with most of the vehicles not even demanding a gearbox due to the induction motor's horsepower and torque characteristics. Therefore, while working on an EV, one can neglect many minor drivetrain losses which arise when dealing with the normal cars which also speeds up the development of these electric vehicles. Figure 4 shows the evolution in global EV stock.

The future is EV, thus it demands the fastest and most productive innovations which can always be fulfilled by the implementation of genetic algorithms [21]. Both of these fields have shown promising results as far as some previous research is considered so it makes sense to benefit one with another's assistance which could also imply that development in these fields might influence the development in fields related to or dependent on it. Lastly, one of the major setbacks right now in the electric vehicle industry is that the vehicles are quite expensive. Many developed countries such as the United States have started to consider the affordability aspect and present some options to their citizens, but they are still nowhere near to a comparable practical IC car which a person might use as a daily driver, cost-wise. On the other hand, countries like India don't even have enough options to choose from right now, and the cheapest one as of now is the Nexon EV by Tata Motors which is priced at a cost of 15 Lacs INR, still an amount to reconsider when investing in a vehicle by an average Indian citizen. GAs would speed up the development in this sector, which would therefore imply better price models of these vehicles as well and 1 day we might just overcome this crucial barrier.

4 Optimizing the Charging Aspect

Charging an EV has always been a drawback as compared to the respective engine-powered vehicles, especially the time aspect [22]. When we consider battery recharge, we focus on how fast the battery can charge, the optimum charge range, its effect on the long-term health of the battery, and the number of charges throughout the battery life. To optimize this using genetic algorithms, what we can do is we can set a particular charge cycle as a generation for writing that GA. Depending on what the user demands, set the nature of that charge cycle, that is, different charging modes. Faster charging implies time-saving but it could put a toll on the battery life in the long run and slower charging (in an efficient manner as far as battery health is concerned) would not damage the battery like other methods would. This model is beneficial for the user when he gets back home and puts his vehicle for, let's say, an overnight charge.

Apart from this, the EVs can also charge from regenerative braking [23]. This is when the vehicle decelerates during a braking maneuver, the motor present in the vehicle acts as a generator and starts charging the battery. This is also mildly done while descending through a hill. We can set up this charging style in a mild format or in an aggressive format. A mild setup doesn't hinder the deceleration much apart from the brakes applied but won't recharge the battery significantly and an aggressive setup would significantly assist in the deceleration (sometimes result in a jerky ride quality) but would recharge the battery more than what other setups could perform.

Therefore, depending on the scenario, we can design particular charge cycles and analyze the outputs (charge speed and battery health) [24]. According to the manufacturer's demand, we can provide an efficient charge cycle that also charges the battery and doesn't decelerate the motor much, or we can provide several regenerative braking modes for the user to choose from. Assume these cycles as parents and generate offspring accordingly with the help of the three operators. In this manner, we can design a charge pattern that could meet our desired results using genetic algorithms. Figure 5 demonstrates the EV charging module [25].

5 Optimizing the Range Aspect

Maximizing the range is also considered challenging in this industry, especially in the younger days of EV innovations [26]. Previously, achieving a range of 100 km was considered a massive procurement. Keep in mind that these were the times when combustion engine-powered cars could easily travel 400–600 km on one full tank of

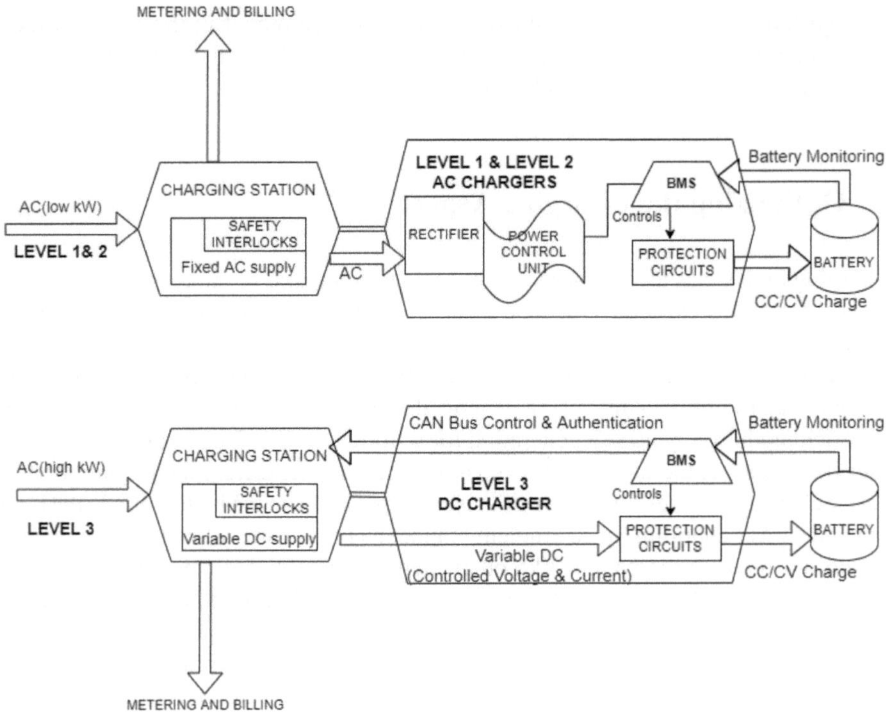

Fig. 5 EV charging module

fuel. As years went by, slowly we optimized the range and now we have EVs whose range could compete with that of IC engine vehicles. But the problem is that these EVs are a handful and very expensive [27]. Figure 6 presents a comparison of price and range for the EVs, as we have obtained it using the data on the Website of Inside EVs (https://insideevs.com) [28].

An average commuter's main demand for an EV would be to have an excellent range, otherwise many won't even consider them to be their primary vehicle. We also have to keep in mind that range depends a lot on the driving style [29], so we have two steps to work around this, firstly consider that the driver is driving in the optimum throttle range and for that, maximizing that optimum range of maneuver. Secondly, augment the range on that efficient driving maneuver. Again, we have to analyze the speed at which the vehicle is traveling and the required juice from the battery for the motor to keep up that speed. Also, we have to consider efficient ways to accelerate and decelerate, regenerative braking, etc. In this theme, the *intelligent driving feedback* is also of importance [30].

Although clubbing every aspect and then making the parents for a particular generation would not be feasible and the result might come out blunt, we can analyze different scenarios individually and make the parents out of them. According to

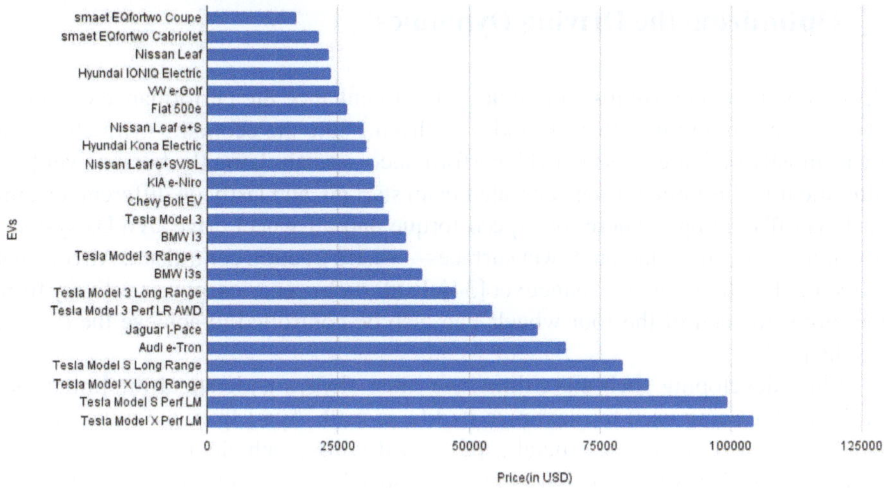

(a) Price

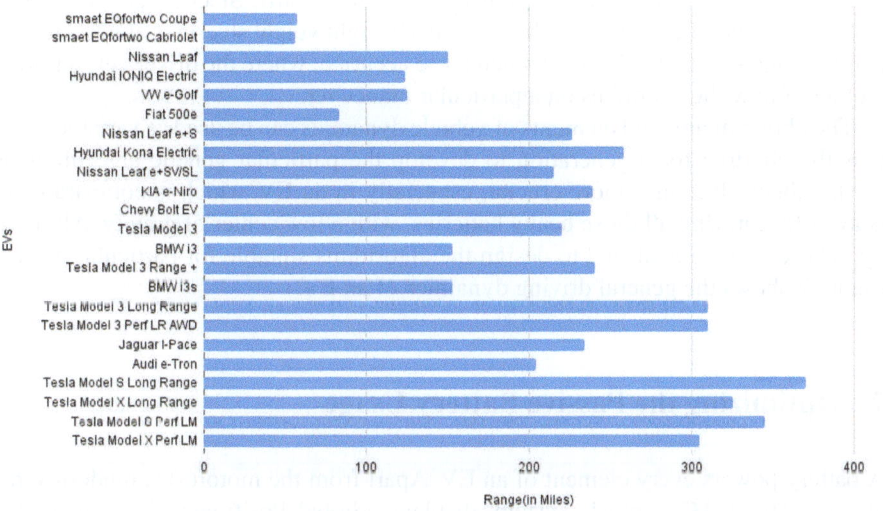

(b) Range

Fig. 6 Comparison of price and range of EVs (2019)

them, we can then make their respective genetic algorithms, use them according to the situation they are thrown into (based on different terrains, highways, and city maneuvers), and also consider the different performance modes which are being used as well. Following this workflow will definitely help us maximize the range of a particular electric vehicle [21].

6 Optimizing the Driving Dynamics

Specific genetic algorithms can be designed to enhance the performance of motor, even while running in different modes of driving like the economy—which would focus mostly on battery usage [31], performance—maximizing the horsepower [32] after the initial infinite torque generated from start go, and lastly on different terrains such as off-roading—where low-speed torque and all-wheel drive (AWD) systems dominate [33], snow and other wet surfaces—where traction control and ABS would work together to master the maneuver [34]. In all such scenarios, power delivery from the motor to each of the four wheels can also be controlled to achieve the desired results [35].

While developing such algorithms, we must always understand the limitations of the motor being used to maximize its life as well. Apart from the motor aspect, handling is also one of the crucial aspects of driving a vehicle. One needs to fully understand how a vehicle behaves in a low-speed corner and a high-speed corner [36]. A comfort-oriented ride, like city driving, demands a softer, much damper suspension, and when one needs to throw the vehicle in corners or simply conduct some high-speed driving, the suspension needs to be stiff. Steering characteristics also hamper the ride quality in which, similarly, lightweight steerings aim to provide comfort, and stiffer steerings are much more accurate, where the driver gets an exact feeling of how the car drives on a particular maneuver.

The above-mentioned elements of vehicle dynamics can be used as variables, and thus the children for a generation to develop the particular genetic algorithms as all this theory becomes more crucial, especially in an EV, which is comparatively heavier by carrying all those heavy batteries, with a low center of gravity. All these hypotheses can be analyzed to design the algorithms suitable for particular events. Figure 7 shows the general driving dynamics of an EV.

7 Optimizing the Passive Battery Usage

A battery powers every element of an EV. Apart from the motor(s), a modern vehicle consists of AC, a music system, displays, Global Positioning System (GPS), heated/cooled seats, countless sensors, and their CPUs. A simple power supply to these units would be a significant waste of the battery's juice. All these features don't always need the maximum electricity given by the battery at every stage of the vehicle's maneuver. We can apply dedicated genetic algorithms to make this system work at its optimum competence and maximum productivity. Each feature must be individually studied to understand how it is being used during every time interval of the vehicle's journey to make sure it is being used efficiently. Next, we can combine the usage of these features to create a system that delivers a structured solution to their usage [37].

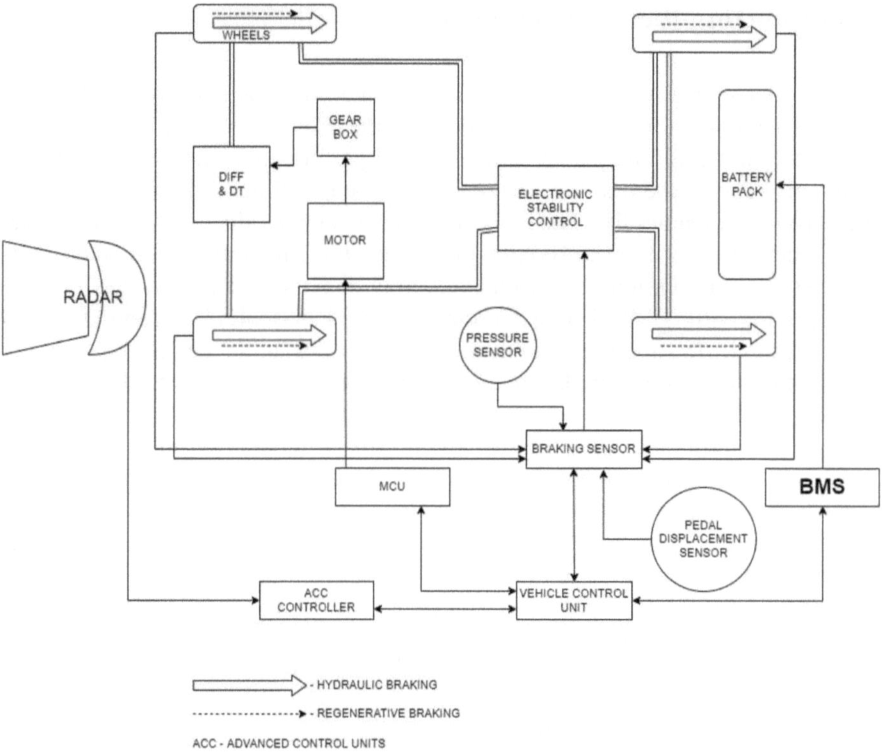

Fig. 7 General driving dynamics of an EV

For example, when the vehicle is running at a very high speed, the driver has to focus only on the road and can't afford to divert his/her mind toward the interior. In this scenario, the ECU can put the infotainment system on low priority to compute and display crucial information such as speed, tire pressure, and revolutions per minute (if the motor has a gearbox attached to it) [38]. The ECU can conserve this power to focus more on further improving the performance of the vehicle at that instance of time. AC should be given a higher priority over performance during a traffic scenario on a hot summer afternoon in a city, and on the contrary, it can even be switched off during high power runs of a performance-oriented EV. In this way, EVs could prioritize certain features on demand in particular scenarios such as performance over comfort or vice versa, or which features to prioritize when the battery is low.

8 Optimizing the Thermodynamics

Electric vehicles are quite sensitive as far as battery care is concerned, especially when the heating is considered, as compared to the IC engines. Overheated batteries or rather even a cool one puts a serious toll on performance and battery life [39]. Therefore, keeping them at the optimum temperature all the time while driving is crucial [40]. Companies like Tesla Inc. and Toyota put individual sensors in each cell of the battery pack to study and control the temperature minutely and accurately [41–43]. These sensors communicate with a very important unit of an EV, which is always a high priority—the BMS. Bosch, Continental AG, Panasonic, Intel, and Denso are one of the top BMS suppliers as of now. Sometimes managing the battery is not that difficult if it is in a more complacent environment or driven by ECO modes, but there are scenarios that demand some serious computations, such as high-performance driving, and driving the EV in regions that are extremely hot or cold. Any lack of conscience could end up having a battery seizure or even worse, a fire! As this is the most important unit of the EV alongside the battery and motor(s), the algorithms should also be designed meticulously to deliver the most optimum and safest results. Figure 8 shows the structure of the BMS based on CAN-bus.

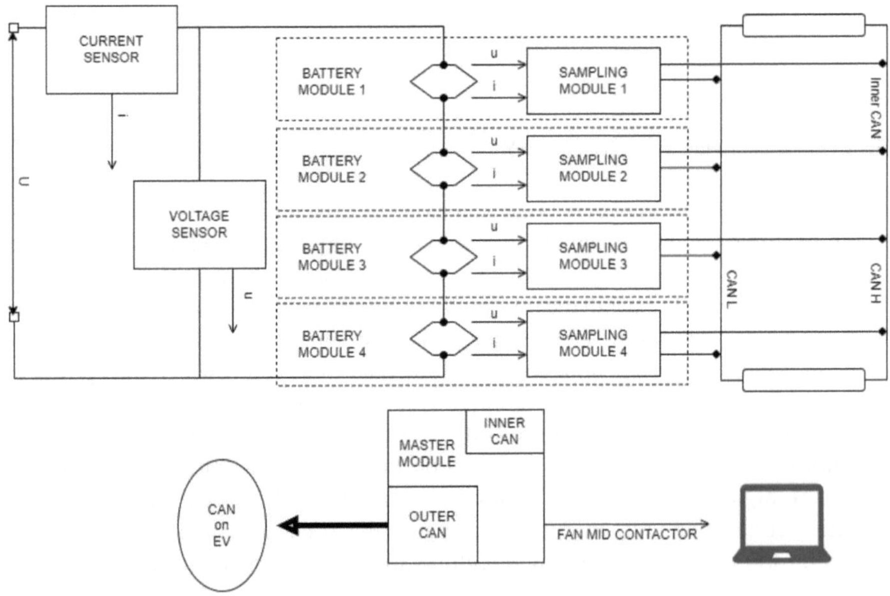

Fig. 8 Structure of the BMS based on CAN-bus

9 Limitations

There is no doubt that all the aspects discussed above would improve the development of EVs and enhance innovations in this sector. But ultimately, it is the driver who drives the vehicle and almost everything depends on the way the driver is driving. While designing such algorithms, we assume a lot of scenarios in which we ourselves decide how the car would be driven, which is not always the case in the real world. Maybe all the computations and algorithms generated would be perfect when a computer drives it, but the systems we design will always not be the best possible solution to a situation when a human drives [44].

To overcome this, a broad solution would be to divide every event into the smallest possible time frame, but as we reduce that, the number of computations would increase [45], increasing the demand for more power, thus more battery, especially when the GAs are concerned, which already implement recurring powerful computations. If we use the batteries which we currently use, we could drain them much faster, and implementing the batteries which could handle these computations, with ease is either not yet discovered or would be very expensive, which is a totally opposite path as EVs are already very expensive. As of today's technology, this challenge is yet to be overcome and there's a lot of development which is required to make this transition seamless and the communication between the driver and car much more predictable (Fig. 9).

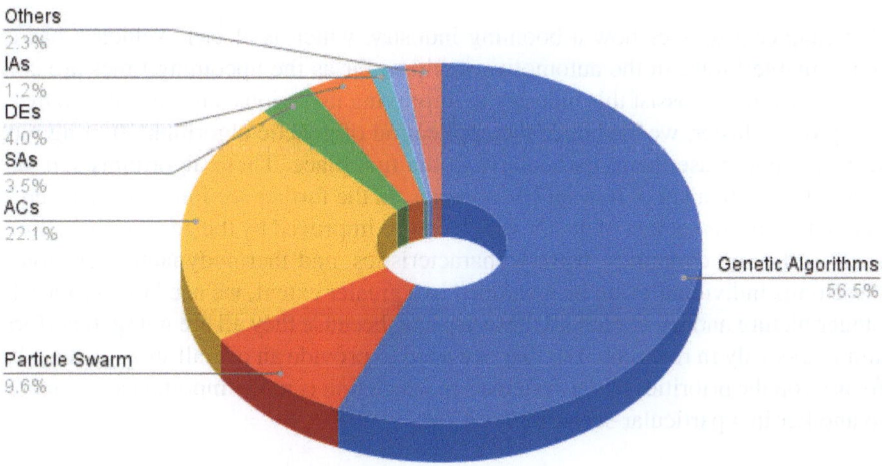

Fig. 9 Share of different evolutionary algorithms being used

10 How GAs Benefit from EV Industry's Advancement

Although we have now understood how crucial GAs are for the development of electric vehicles, we also have to consider the fact that EVs are a great platform to train genetic algorithms as well. An electric vehicle has numerous units associated with it which can be developed with the help of GAs as discussed in the previous sections, the fact that this industry is fairly new and has a lot of scope for innovation would definitely carve the path for GAs success as well. Along with that, the simplistic nature of an EV which doesn't demand many mechanical parts can be considered to be drivable by a computer with much more ease as compared to an IC engine. This holds another strong ground on which one can justify the use of genetic algorithms, thus experimenting with them in many scenarios, ultimately developing the GA industry as well. Most of the scenarios which are possible in an EV maneuver can be simulated, and thus algorithms can be designed to master that particular maneuver, this makes an electric vehicle a perfect tool for developing genetic algorithms in a certain systematic manner. It was only after we started implementing such algorithms to supplement this industry, did we realize that the help could also be supplied the other way around.

11 Conclusion

The chapter discusses how a booming industry, which is electric vehicles, has an impregnable future in the automotive world as far as the upcoming times are concerned. To further assist this industry in improving its various units and also to make this process faster, we introduce the application of genetic algorithms starting with why we should use them, particularly in the first place. These algorithms suit perfectly the mechanism of how an EV operates. In the further sections, we analyzed in depth the various sectors of an EV which can be improved by the genetic algorithms, such as driving dynamics, battery characteristics, and thermodynamics. Although these units individually can be revamped to a greater extent, we need to consider the bigger picture and try to club all these aspects, because they all are going to perform simultaneously in real time. For that, we need to provide an overall strategy that also focuses on the priorities of the system as to which unit is more important as compared to another in a particular scenario.

References

1. Brandl M, Gall H, Wenger M, Lorentz V, Giegerich M, Baronti F, Fantechi G, Fanucci L, Roncella R, Saletti, R (2012) Batteries and battery management systems for electric vehicles. In: Design, automation and test in europe conference and exhibition (DATE). IEEE, pp 971–976
2. Husain I (2021) Electric and hybrid vehicles: design fundamentals. CRC Press

3. Sarode R, Eadke A (2020) Electric vehicle (construction and working principle). Int Res J Eng Technol 7(10):140–142
4. Xiao B, Ruan J, Yang W, Walker P, Zhang N (2021) A review of pivotal energy management strategies for extended range electric vehicles. Renew Sustain Energy Rev 149:111194
5. Aijaz I, Ahmad A (2022) Electric vehicles for environmental sustainability. Smart Technol Energy Environ Sustain 131–145
6. Slowik A, Kwasnicka H (2020) Evolutionary algorithms and their applications to engineering problems. Neural Comput Appl 32:12363–12379
7. Gupta N, Khosravy M, Patel N, Dey N, Mahela OP (2020) Mendelian evolutionary theory optimization algorithm. Soft Comput 24:14345–14390
8. Takano H, Iwase N, Nakayama N, Asano H (2022) Towards self-organized control: using neural cellular automata to robustly control a cart-pole agent. Innov Mach Intell (IMI) 2:1–11. https://doi.org/10.54854/imi2022.001
9. Variengien A, Pontes-Filho S, Glover T, Nichele S (2021) Towards self-organized control: using neural cellular automata to robustly control a cart-pole agent. Innov Mach Intell (IMI) 1:1–14. https://doi.org/10.54854/imi2021.01
10. Jalalzad SH, Yektamoghadam H, Haghighi R, Dehghani M, Nikoofard A, Khosravy M, Senjyu T (2022) A game theory approach using the tlbo algorithm for generation expansion planning by applying carbon curtailment policy. Energies 15(3):1172
11. Senjyu T, Khosravy M (2022) Power system planning and quality control
12. Joshi A, Khosravy M, Gupta N (2021) Machine learning for predictive analysis: proceedings of ICTIS 2020. Springer, Berlin (2021)
13. Gupta N, Khosravy M, Patel N, Dey N, Gupta S, Darbari H, Crespo RG (2020) Economic data analytic ai technique on iot edge devices for health monitoring of agriculture machines. Appl Intell 50:3990–4016
14. Gupta N, Khosravy M, Patel N, Dey N, Crespo RG (2021) Lightweight computational intelligence for iot health monitoring of off-road vehicles: Enhanced selection log-scaled mutation ga structured ann. IEEE Trans Ind Inform 18(1):611–619
15. Khosravy M, Nakamura K, Nitta N, Dey N, Crespo RG, Herrera-Viedma E, Babaguchi N (2022) Social iot approach to cyber defense of a deep-learning-based recognition system in front of media clones generated by model inversion attack. IEEE Trans Syst Man Cybern: Syst 53(5):2694–2704
16. Dehghani M, Taghipour M, Sadeghi Gougheri S, Nikoofard A, Gharehpetian GB, Khosravy M (2021) A deep learning-based approach for generation expansion planning considering power plants lifetime. Energies 14(23):8035
17. Khosravy M, Gupta N, Witkowski O, Pasquali A (2021) Neighborhood base matched morphological filters: cross-fertilization with linear lowpass filtering. In: 2021 international conference on computational science and computational intelligence (CSCI). IEEE, pp 1623–1628
18. Khosravy M, Gupta N, Patel N (2022) Frontiers in nature-inspired industrial optimization. Springer, Berlin (2022)
19. Khosravy M, Nakamura K, Hirose Y, Nitta N, Babaguchi N (2022) Model inversion attack by integration of deep generative models: privacy-sensitive face generation from a face recognition system. IEEE Trans Inf Forensics Secur 17:357–372
20. Gelmanova Z, Zhabalova G, Sivyakova G, Lelikova O, Onishchenko O, Smailova A, Kamarova S (2018) Electric cars advantages and disadvantages. J Phys: Conf Ser 1015:052029. IOP Publishing
21. Montazeri-Gh M, Poursamad A, Ghalichi B (2006) Application of genetic algorithm for optimization of control strategy in parallel hybrid electric vehicles. J Frankl Inst 343(4–5):420–435
22. Froger A, Jabali O, Mendoza JE, Laporte G (2022) The electric vehicle routing problem with capacitated charging stations. Transp Sci 56(2):460–482
23. Yoong M, Gan Y, Gan G, Leong C, Phuan Z, Cheah B, Chew K (2010) Studies of regenerative braking in electric vehicle. In: IEEE conference on sustainable utilization and development in engineering and technology. IEEE, pp 40–45

24. Cho DH, Jung HK, Lee CG (2001) Induction motor design for electric vehicle using a niching genetic algorithm. IEEE Trans Ind Appl 37(4):994–999
25. Electropaedia: Battery and energy technologies. https://www.mpoweruk.com/infrastructure.htm. Accessed 27 July 2023
26. Wager G, Whale J, Braunl T (2016) Driving electric vehicles at highway speeds: the effect of higher driving speeds on energy consumption and driving range for electric vehicles in australia. Renew Sustain Energy Rev 63:158–165
27. Ajanovic A (2015) The future of electric vehicles: prospects and impediments. Wiley Interdiscip Rev: Energy Environ 4(6):521–536
28. EVs I, Pricing. https://insideevs.com/. Accessed 27 July 2023
29. Wang R, Lukic SM (2011) Review of driving conditions prediction and driving style recognition based control algorithms for hybrid electric vehicles. In: IEEE Vehicle power and propulsion conference. IEEE, pp 1–7
30. Gonder J, Earleywine M, Sparks W (2012) Analyzing vehicle fuel saving opportunities through intelligent driver feedback. SAE Int J Passeng Cars-Electron Electr Syst 5:450–461
31. Li L, Liu Q (2019) Acceleration curve optimization for electric vehicle based on energy consumption and battery life. Energy 169:1039–1053
32. Panday A, Bansal HO (2016) Energy management strategy for hybrid electric vehicles using genetic algorithm. J Renew Sustain Energy 8(1)
33. Chen H, Yan W, Gu JJ, Sun M (2018) Multiobjective optimization design of a switched reluctance motor for low-speed electric vehicles with a taguchi-cso algorithm. IEEE/ASME Trans Mechatron 23(4):1762–1774
34. Ahmadi M, Raesian N, Zarif M, Goharimanesh M (2015) Optimized regenerative brake system using genetic algorithm. In: International congress on technology, communication and knowledge (ICTCK). IEEE, pp 177–180
35. Sakai SI, Sado H, Hori Y (1999) Motion control in an electric vehicle with four independently driven in-wheel motors. IEEE/ASME Trans Mechatron 4(1):9–16
36. Welchko BA, Nagashima JM (2003) The influence of topology selection on the design of ev/hev propulsion systems. IEEE Power Electron Lett 1(2):36–40
37. Uzair M, Abbas G, Hosain S (2021) Characteristics of battery management systems of electric vehicles with consideration of the active and passive cell balancing process. World Electr Veh J 12(3):120
38. Buechel M, Frtunikj J, Becker K, Sommer S, Buckl C, Armbruster M, Marek A, Zirkler A, Klein C, Knoll A (2015) An automated electric vehicle prototype showing new trends in automotive architectures. In: 2015 IEEE 18th international conference on intelligent transportation systems. IEEE, pp 1274–1279
39. Katoch SS, Eswaramoorthy M (2020) A detailed review on electric vehicles battery thermal management system. In: IOP conference series: materials science and engineering, vol 912. IOP Publishing, p 042005
40. Lu M, Zhang X, Ji J, Xu X, Zhang Y (2020) Research progress on power battery cooling technology for electric vehicles. J Energy Storage 27:101155
41. Xu B, Arjmandzadeh Z (2023) Parametric study on thermal management system for the range of full (tesla model s)/compact-size (tesla model 3) electric vehicles. Energy Convers Manag 278:116753
42. Berdichevsky G, Kelty K, Straubel J, Toomre E (2006) The tesla roadster battery system. Tesla Motors 1(5):1–5
43. Zhu C, Lu F, Zhang H, Sun J, Mi CC (2018) A real-time battery thermal management strategy for connected and automated hybrid electric vehicles (cahevs) based on iterative dynamic programming. IEEE Trans Veh Technol 67(9):8077–8084
44. Hong J, Wang Z, Chen W, Wang L, Lin P, Qu C (2021) Online accurate state of health estimation for battery systems on real-world electric vehicles with variable driving conditions considered. J Clean Prod 294:125814
45. Kalakanti AK, Rao S (2023) Computational challenges and approaches for electric vehicles. ACM Comput Surv

Chapter 11
Genetic Algorithm in Traffic Control and Autonomous Driving

Seyed Mohamad Ali Tousi, Shadi Samizadeh, and Amirhossein Nikoofard

1 Prologue

Genetic algorithms and optimization algorithms play a crucial role in revolutionizing traffic control and enhancing the capabilities of autonomous vehicles. With the ever-increasing complexity of transportation systems, these algorithms offer powerful solutions to improve efficiency, safety, and overall performance. In this chapter, we will briefly go over some of the applications of the genetic algorithm in these areas (as shown in Fig. 1).

2 Traffic Control

2.1 Introduction

Urban transportation is one of the most critical challenges and characteristics of large cities in developed and developing countries. As the population of the cities grows, the demand for a fast, safe, and reliable transportation system will also increase. Avoiding traffic congestion in populated cities can cause reducing delays and car accidents, prevent over-emissions of hazardous gases into the atmosphere, and alleviate the users' stress. So finding a way to limit traffic congestion by optimizing the traffic

S. M. A. Tousi · S. Samizadeh · A. Nikoofard (✉)
K. N. Toosi University of Technology, Tehran, Iran
e-mail: a.nikoofard@kntu.ac.ir

S. M. A. Tousi
e-mail: s.m.ali.tousi@email.kntu.ac.ir

S. Samizadeh
e-mail: shadi.samizadeh@email.kntu.ac.ir

M. Khosravy et al. (eds.), *Frontiers in Genetics Algorithm Theory and Applications*, Springer Tracts in Nature-Inspired Computing,
https://doi.org/10.1007/978-981-99-8107-6_11

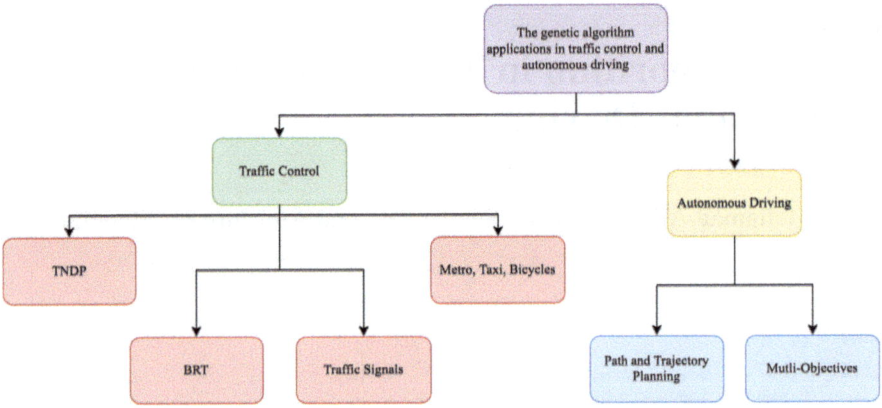

Fig. 1 The overview of the genetic algorithm's applications is to be discussed in this chapter

control parameters or optimally using other urban transportation tools is an exciting field of research for researchers. In this part, we will discover some of the primary optimization problems in the traffic control context and how the genetic algorithm helps to tackle those problems.

2.2 Transit Network Design Problem

The transit network design problem or TNDP is the problem of optimally designing bus trajectories and their frequencies to reduce the cost of passengers. In this problem, the bus network within a city is modeled in terms of some graphs with their nodes as bus stops, intersections, or zone centroids and their weighted arcs as the connection routes between nodes. The goal is to minimize the passengers' cost of using this transit network. The passengers' cost is defined as the traveling time and the waiting time that the passengers will experience if they use this transportation system.

The genetic algorithm has been employed in [1] to find an optimal way to design urban bus routes. This selection has been made in two phases. In the first phase, a set of candidate routes for the bus network have been generated by the Candidate Route Set Generation Algorithm (CRGA). As all the possible routes in the network graph will be enormous, the CRGA algorithm has been used to select and generate a set of possible ways between nodes based on the demand matrix, the designer's expertise, and the constraints of the problem.

In the second phase, the optimum route was found by employing the genetic algorithm. To represent the problem as a GA problem, any individual path generated by the CRGA has been considered as a variable, and the variables have been sorted by their performance index. In [2], three main steps have been proposed to reach the optimal network routes. The first step, similar to the previous paper, is to generate a set of possible candidate routes as the initial route set. The paper proposes an initial candidate route set generation procedure (ICRSGP). This procedure will generate all the potential and suitable routes within the network considering the practical bus transit industry guidelines. The second step is to analyze the generated ways. So a network analysis procedure (NAP) is proposed to decide the transit demand matrix, assign transit trips, determine the frequencies related to the network, and compute the performance measures. The final step combines two previous actions and employs a genetic algorithm procedure (GAP) to guide the candidate solution generation process and choose the optimal routes from the search space. The block diagram of the algorithm mentioned above is shown in Fig. 3.

Owais and Osman [3] have also employed a complete hierarchical GA to tackle the multi-objective problem of TNDP in which the genetic algorithm will generate, evaluate and guide the candidate solutions to the optimal network configurations. Many research projects that tried to propose an optimal set of solutions for the TNDP problem did not pay the necessary attention to the passengers' route choice behavior. They did not consider the "common line problem". This problem is based on the fact that the best and fastest way to get to a particular destination is to take the first vehicle arriving among an attractive set of lines. Shimamoto et al. [4] Proposes a bi-level optimization approach that tries to solve the TNDP problem with the genetic algorithm considering the common line problem.

The suitability of the genetic algorithm for finding solutions for the TNDP problem has been shown in many research projects. However, the high computational cost of implementing GA has limited the application of this optimization algorithm to some broad and complex problems. To address this problem, many researchers have proposed methods to reduce the complexity and computational costs of the algorithm. For instance, [5] offers a parallel approach to optimize the TNDP using Simultaneous Route and Frequency Coded (SRFC), proposed by [6], which is a genetic algorithm-based model. See Fig. 2. This parallel computing approach can reduce the computational cost of the SRFC and make it easier to implement the method for more complex optimization problems.

2.3 *Bus-Rapid-Transit (BRT) Scheduling Problem*

Many large cities have developed Bus-Rapid-Transit (BRT) systems to address the growing demand for public transportation. BRT is a unique corridor assigned to a fleet of buses that can operate in a dedicated bus lane specially reserved for them on a highway or an arterial road. The BRT transportation systems combine the benefits of rail transit with the flexibility of buses. Optimally scheduling the BRTs is critical for a better transportation experience. It can be represented as an optimization problem to minimize the system's total cost (e.g., the passengers' cost such as waiting time, and the operating system's cost such as the necessary size of the fleet). Liang et al. [7] Proposes a genetic algorithm-based method to solve this optimization problem by optimally scheduling the pass-by stops of the BRT and their frequency. The algorithm

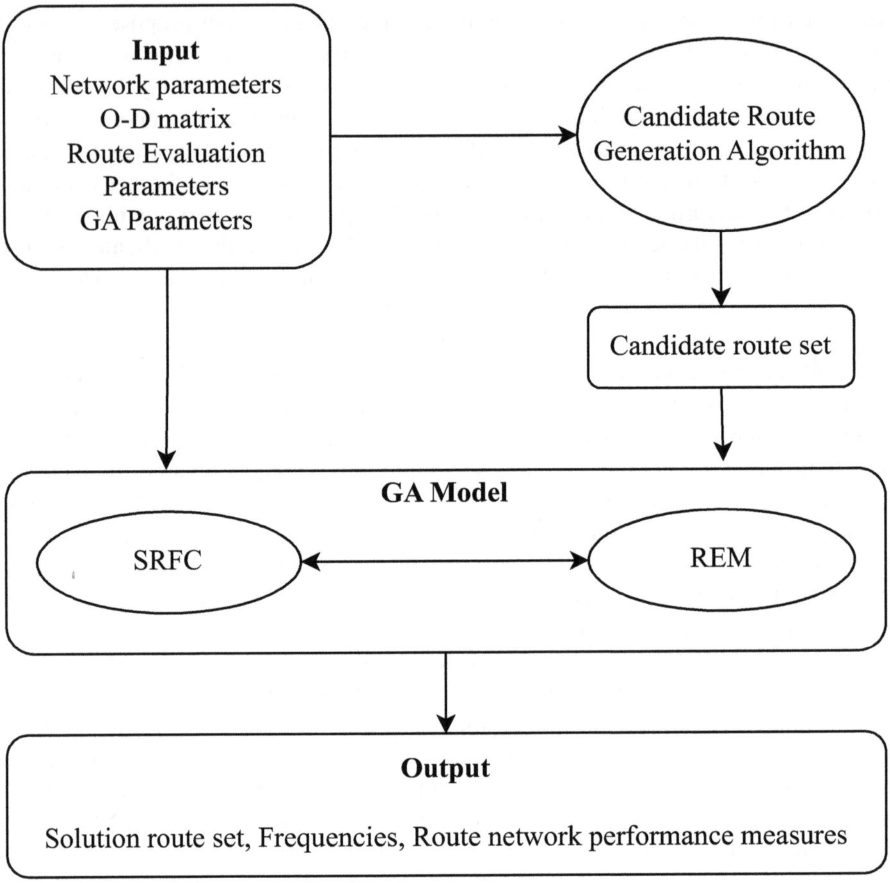

Fig. 2 The block diagram of the SFRC-based approach proposed by [5]

tries to find the optimal departure intervals (the frequency of the network) and stops to pass as its variables concerning its objective function and performance measures.

References [8, 9] Mix the BRT scheduling problem with feeder and regular buses and use genetic algorithm-based optimization approaches to tackle the issues. In [9], the researchers combine the genetic algorithm with the Tabu search method to achieve the optimal solution for the problem. Figure 4 shows the algorithm flowchart of the GA-TS optimization algorithm proposed by [9].

2.4 *Traffic Signals*

The timing cycle of the traffic lights in junctions has been a challenging optimization problem for many years. Many research projects have tried finding the optimum green

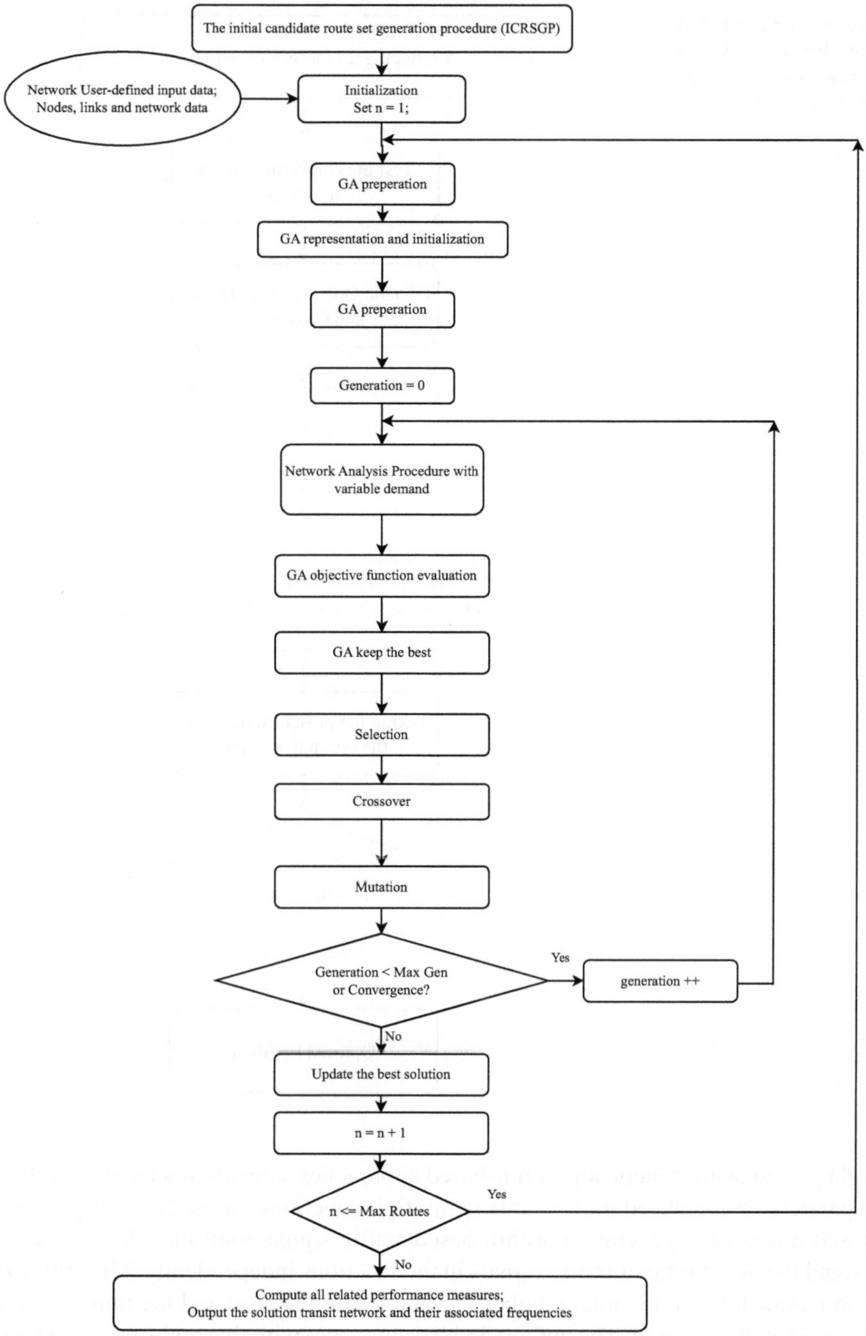

Fig. 3 The conceptual framework of the genetic optimization implementation proposed by [2]

Fig. 4 The algorithm flowchart of the GA-TS optimization algorithm proposed by [9]

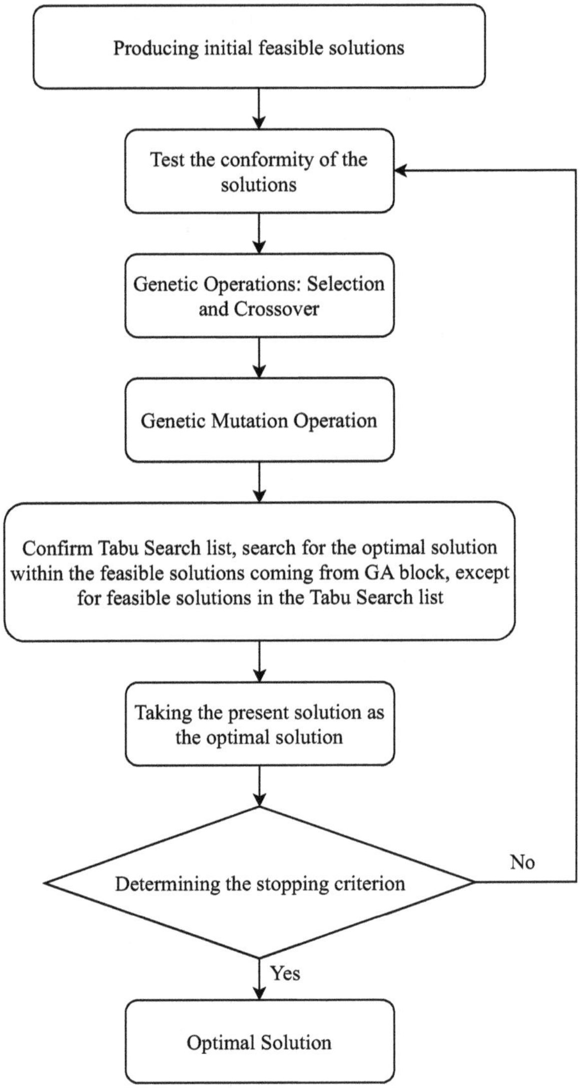

light period using genetic algorithm-based approaches, considering the intersections separately or connected traffic lights in multiple junctions. Singh et al. [10] has proposed a real-time genetic algorithm-based traffic signal controller that decides to extend the green time of traffic signals in intersections independently. The optimization parameters are the total number of vehicles on the road and the importance of the road at the junction. The authors in [10] show that with this real-time adjustment system, the traffic signal performance will be improved compared to the fixed-time traffic light systems. In [11], a hybrid fuzzy genetic algorithm-based controller is proposed for the traffic light control procedure in four connected intersections. The optimization parameter is the total number of vehicles that leave throughout the four considered intersections. The genetic algorithm has six steps to go:

1. Forming the chromosome: a chromosome or a candidate solution is a vector containing sixteen binary bits. Each bit corresponds to a traffic signal in the four considered intersections system. In each traffic signal bit, 1 represents the green light, and 0 illustrates the red light. Only one traffic signal can be 1 in an intersection at a particular time to avoid accidents.
2. Creating the initial pool of candidate solutions: concerning the conditions of the intersections, the initial population will be generated.
3. Evaluation of a candidate solution: considering the total number of vehicles that leave the intersections system, a fitness value will be assigned to each candidate solution.
4. Creating pool selection: The mating pool will be created with the Baker roulette method [12].
5. Crossover operation
6. Mutation operation

When the demand for transportation with vehicles exceeds the capability of the roads, over-saturated conditions may be accrued. These over-saturated conditions will result in heavy congestion and long queues. So many researchers have been trying to optimize traffic signal control in over-saturated conditions using genetic algorithm-based approaches. In [13], authors have tried to use the genetic algorithm to optimize four traffic signal parameters simultaneously (i.e., cycle length, green split, offset, and phase sequence). Like other GA-based algorithms, in [13], four traffic signal parameters have been rendered by a GA block and validated with a Mesoscopic Simulator to obtain the fitness values corresponding to each set of parameters. The fitness values were average delays and vehicle throughputs. The conceptual framework of the algorithm is shown in Fig. 5.

Tan et al. [14] has proposed a decentralized genetic algorithm for traffic signal control in over-saturated conditions. The authors argued that learning-based approaches such as neural networks and reinforcement learning methods suffer from high computational costs. So they proposed using a decentralized way in which every intersection will have its controller and optimization based on an adaptive objective function. The usage of adaptive objective functions is for exploiting the benefits of learning-based methods without their relatively high computational cost.

Some other researchers have focused on optimizing the traffic signal controls in a broader area than some few intersections so that they can solve some global problems such as rerouting traffics in large cities [14]. They included a network equilibrium model as a constraint in the genetic algorithm optimization process, in order to accomplish some optimum points considering a large-scale traffic area.

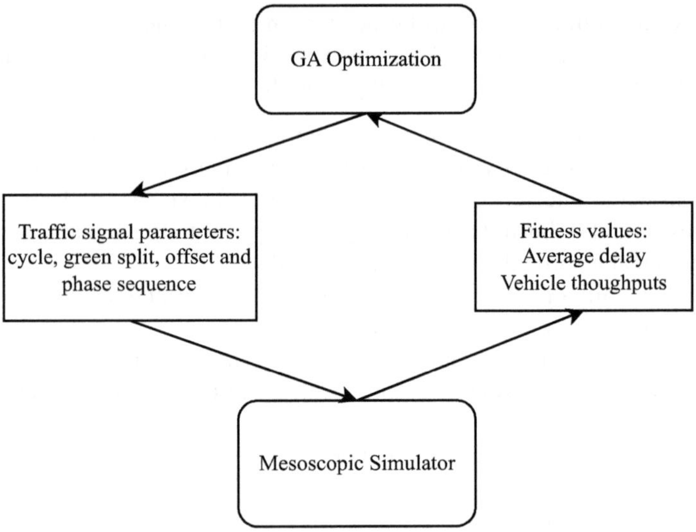

Fig. 5 The conceptual framework of the OVERSAT algorithm [13]

2.5 Metro, Taxis, and Bicycles

In addition to the above-mentioned fields of optimization, route and station design for subways, metros, and taxis can be of enormous significance in the context of traffic control and passengers' convenience. Qiu and Xu [15] has employed two main optimization algorithms (genetic algorithm and ant colony optimization algorithm) to maximize the taxi driver's profit and the passengers' cost. References [16, 17] are trying to optimize the problem of taxi-passenger matching. Gong et al. [16] is employing a fuzzy controller to assign the necessary priority score to each passenger-taxi pair. Then it uses an offline optimization algorithm to ensure that the controller is working in an optimized way. However, [17] defines a fitness function based on the pick-up rate and the passengers' waiting time and tries to optimize this fitness function as a global optimization problem with an advanced genetic algorithm. References [18, 19] are also proposing the genetic algorithm to be used in pricing and route design optimization, respectively.

The other area of optimization related to taxi transportation is to design and optimize community shuttles for metro stations [20–22]. Community shuttles are essential in developing an excellent shuttle-metro network for passengers within large cities. Xiong et al. [22] uses headways and routes as its optimization variables and then tries to utilize the genetic algorithm to optimize the cost, including passengers' demand and the suppliers' financial constraints. Yu and Lu [20] modifies the crossover and mutation operators to fit the route and frequency design problem requirements for the community shuttle bus. In their definition of crossover operation, they exchange the sequences of nodes (stops) in the routes for two or more solutions with the same destination. They also modified the mutation operator as it randomly removes an intermediate node in a solution.

The design of metro networks is subject to many constraints due to its highly complicated nature and the vast budget it requires. However, some research projects have tried to employ a genetic algorithm to optimize the design problem of the metro network, considering the vast amount of parameters needed [23].

Another transportation method that is helpful for large cities to combat the problem of the massive demand for transportation is using bicycles. A genetic algorithm has been used to optimize cyclists' safety, travel time, travel distance, and convenience [24]. They used a new mutation operator to improve the algorithm's performance. Gao and Lee [25] is also combining fuzzy C-means algorithm, genetic algorithm, and backpropagation to optimize the problem of demand and supply in bicycle-sharing transportation systems.

Besides the above-mentioned areas, various other transportation methods are being explored to be optimized by the genetic algorithm. Jordán et al. [26] tries to optimize the station placements for electric cars within the city. [27] also tries to combine different transportation systems and optimize the multi-modal route design problem.

3 Autonomous Driving

3.1 Introduction

Autonomous vehicles have been an attractive research topic for many years. With the rise of the need for unmanned vehicles to achieve some automation goals, several challenges occurred to be tackled by the research society such as car depth estimation [28, 29], trajectory tracking [30], trajectory planning and decision making Yu and Lu [31] and optimization. The demand for safe and reliable transportation tools, including vehicles, buses, and industrial facilities, has led the research communities to define some optimization problems in the area of control and maintenance of autonomous vehicles. So evolutionary optimization algorithms, the genetic algorithm above all, have found a broad field of application in unmanned cars.

The application of the genetic algorithm in the field of autonomous vehicles might fall into two main categories. The first one, the most prominent application, is to use the genetic algorithm-based optimization approaches to optimize the autonomous car's path and trajectory planning. In other words, the first application area is to better control the vehicle by having some critical factors optimized by the genetic algorithm. The second field of application of evolutionary algorithms in the research of autonomous vehicles is improving vehicle conditions using optimization methods.

The rest of the section will be dedicated to exploring some valuable research works in the above-mentioned genetic algorithm application areas.

3.2 Path Planning and Trajectory Planning

Path planning is a crucial task for auto-guided vehicles. The task is defined as finding the best path to reach a certain point in the environment that might have some obstacles. As achieving the best path requires, an optimization problem arises here to optimize the number of steps needed and the time of reaching the goal point.

The genetic algorithm has been employed in [30] to plan an optimized path for an auto-guided vehicle. In this work, each vehicle step is considered a constant movement in various directions determined by the genetic algorithm. Each direction is coded into one digit between 0 to 9 to indicate the direction of the movement. Then, a string containing a fixed number of steps in a specific direction (digits) has been rendered as a chromosome of the genetic optimization algorithm.

In [32], a similar approach has been proposed for using the genetic algorithm to optimize the path of a mobile autonomous robot. In this work, an origin-destination approach has been presented. Each origin-destination pair form a candidate solution for the solution population. They have also proposed a modified crossover operator to produce more feasible solution candidates in each epoch. This modified crossover operator could achieve faster convergence of the genetic algorithm compared to other crossover methods.

3.3 Multi-Objective

As its name indicates multi-objective optimization problem includes a group of Pareto-optimal solutions. Hence, the aim is to find the maximum number of various Pareto-optimal (or near Pareto-optimal) solutions. GA can obtain a group of Pareto-optimal solutions in a single run of multi-objective GA, instead, classic optimization methods capture a single solution in each iteration. So, to find Pareto-optimal solutions we have to apply classic optimization methods more than once. This aspect makes GA a good choice to solve multi-objective problems.

Two main points should be accomplished in solving multi-objective problems. First, the search should be guided toward the global Pareto-optimal region. Second, In the current non-dominated front, population diversity should be maintained [33].

The Non-dominated Sorting Genetic Algorithm (NSGA) was first proposed in [34] by Srinivas et al. It had three main issues; high computational complexity of non-dominated sorting, lack of elitism, and need for specifying the sharing parameter. On the other hand, NSGA-II addresses all of these issues and proposes a progressed form of NSGA. NSGA-II outperforms two other multi-objective evolutionary algorithms called PAES and SPEA, with the explicit goal of maintaining spread on the non-dominated front [35]. Non-dominated Sorting Genetic Algorithm (NSGA-II) is one of the popular multi-objective genetic algorithms.

Wang et al. [36] focuses on bus vehicle scheduling in traffic congestion. They propose a multi-objective genetic algorithm based on a dynamic scheduling approach (GA-DBVSA). Their approach consists of three steps.

First of all, a group of candidate vehicle blocks is made once traffic congestion occurs based on a timeline. Then, in the second step, a non-dominated sorting genetic algorithm (NSGA-II) is used to pick a subgroup of vehicle blocks from the group of candidate blocks to create a group of non-dominated solutions.

In the end, in order to achieve final solutions, a departure time adjustment is utilized for the non-dominated solutions. They also applied their approach to real-world problems and the results indicated that GA-DBVSA was effective and can generate scheduling schemes dynamically.

4 Future Challenges

As genetic-based optimization algorithms are widely used in various contexts, even in traffic control and autonomous driving, future challenges also lied in different areas, such as combining the optimization algorithm with learning-based algorithms or focusing on operator improvements concerning the problems' specifications.

On the other hand, most of the focus will be on the combination of Genetic Algorithms with other methodologies. Hence, this will enhance the method and underline the strengths.

- Schockenhoff et al. [37] suggests that the focus of future research should be on adapting the methodological approach to other future products.
- Sainath et al. [38] suggests that Genetic Algorithm speeds up the initial phase of generalizing features several folds compared with traditional methods.
- Also, once the network attains the basic cognitive abilities, it can be further improved upon through reinforcement learning techniques such as Deep Q-learning. So, reinforcement learning is a serious rival for Genetic Algorithms.
- It should be noted that there should be a better way to choose crossover and mutation rates. A self-organizing GA would choose these rates according to the problem. It also will boost the runtime.
- It is suggested that new methods of crossover and mutation techniques are required for convergence problems. Hence researchers in this area are working in this field [39].
- The other interesting area of research is to use some advances in the optimization algorithm developments itself to further investigate their applications in the field of unmanned vehicles and traffic control. Some recent advances in the genetic algorithm such as what is proposed in [40, 41], or [42] can be employed to improve the performance of existing optimization techniques in this field.

5 Conclusion

In this chapter, we explored some applications of the genetic optimization algorithm in traffic control, transportation, and autonomous vehicles. In each chapter section, we presented valuable research projects and papers proposing solutions for various optimization problems using genetic algorithms. Firstly, we introduced the applications of the genetic algorithm in the context of traffic control. Transit network design problems (TNDP), Bus-Rapid-Transit (BRT), traffic light and intersection design, and Metro, taxis, and bicycle route design problems have been explored in the first part of the chapter. In the second part, we explored some beneficial genetic algorithm-based solutions in the context of autonomous vehicles. Path and trajectory planning solutions based on the genetic algorithm have been explored among other multi-objective optimization problems. Finally, some future challenges in genetic-based optimization algorithms that researchers might find interesting have been presented [25].

References

1. Pattnaik SB, Mohan S, Tom VM (1998) Urban bus transit route network design using genetic algorithm. J Transp Eng 124(4):368–375
2. Fan W, Machemehl RB (2006) Optimal transit route network design problem with variable transit demand: genetic algorithm approach. J Transp Eng 132(1):40–51
3. Owais M, Osman MK (2018) Complete hierarchical multi-objective genetic algorithm for transit network design problem. Expert Syst Appl 114:143–154
4. Shimamoto H, Kurauchi F et al (2012) Optimisation of a bus network configuration and frequency considering the common lines problem. J Transp Technol 2(03):220
5. Agrawal J, Mathew TV (2004) Transit route network design using parallel genetic algorithm. J Comput Civil Eng 18(3):248–256
6. Tom VM, Mohan S (2003) Transit route network design using frequency coded genetic algorithm. J Transp Eng 129(2):186–195
7. Liang S, He Z, Sha Z (2011) Bus rapid transit scheduling optimal model based on genetic algorithm. In: ICCTP 2011: towards sustainable transportation systems, pp 1296–1305
8. Jiaqing W, Rui S, Wangtu X, Liu Y (2013) Transit network optimization for feeder bus of brt based on genetic algorithm. In: 2013 fourth international conference on digital manufacturing and automation. IEEE, pp 1627–1630
9. Chang CZ, Chen XM, Wang M (2015) Study on combinational scheduling optimization of bus transit rapid based on tabu search and genetic algorithm. Appl Mech Mater 744:1827–1831. Trans Tech Publ
10. Singh L, Tripathi S, Arora H (2009) Time optimization for traffic signal control using genetic algorithm. Int J Recent Trends Eng 2(2):4
11. Odeh SM, Mora AM, Moreno MN, Merelo JJ (2015) A hybrid fuzzy genetic algorithm for an adaptive traffic signal system. Adv Fuzzy Syst 11–11
12. Nedeljkovic I (2004) Image classification based on fuzzy logic. Int Arch Photogramm, Remote Sens Spat Inf Sci 34(30):3–7
13. Park B, Messer CJ, Urbanik T (1999) Traffic signal optimization program for oversaturated conditions: genetic algorithm approach. Transp Res Record 1683(1):133–142

14. Tan MK, Ee Chuo HS, Yin Chin RK, Yeo KB, Kin Teo KT (2017) Optimization of traffic network signal timing using decentralized genetic algorithm. In: 2017 IEEE 2nd international conference on automatic control and intelligent systems (I2CACIS). IEEE, pp 62–67
15. Qiu Y, Xu X (2018) Rpsbpt: a route planning scheme with best profit for taxi. In: 2018 14th international conference on mobile ad-hoc and sensor networks (MSN), pp 121–126. IEEE
16. Gong Y-J, Liu Y-W, Lin Y, Chen W-N, Zhang J (2019) Real-time taxi-passenger matching using a differential evolutionary fuzzy controller. IEEE Trans Syst, Man, Cybern: Syst 51(5):2712–2725
17. Liu Y-W, Zhang X-Y, Gong Y-J, Chen W-N, Zhang J (2017) A parallel genetic algorithm with region division strategy to solve taxi-passenger matching problem. In: 2017 IEEE symposium series on computational intelligence (SSCI). IEEE, pp 1–7
18. Zhang W, He R, Chen Y, Gao M, Ma C (2019) Research on taxi pricing model and optimization for carpooling detour problem. J Adv Transp
19. Nunes J, Matos L, Trigo A (2011) Taxi pick-ups route optimization using genetic algorithms. In: Adaptive and natural computing algorithms: 10th international conference, ICANNGA 2011. Ljubljana, Slovenia, Proceedings, Part I 10. Springer, Berlin, pp 410–419
20. Yu H, Lu F (2012) A multi-modal route planning approach with an improved genetic algorithm. Adv Geo-Spat Inf Sci 193
21. Xiong J, Chen B, He Z, Guan W, Chen Y (2021) Optimal design of community shuttles with an adaptive-operator-selection-based genetic algorithm. Transp Res Part C: Emerg Technol 126:103109
22. Xiong J, Guan W, Song L, Huang A, Shao C (2013) Optimal routing design of a community shuttle for metro stations. J Transp Eng 139(12):1211–1223
23. Król A, Król M (2019) The design of a metro network using a genetic algorithm. Appl Sci 9(3):433
24. Nunes P, Moura A, Santos J (2023) Solving the multi-objective bike routing problem by meta-heuristic algorithms. Int Trans Oper Res 30(2):717–741
25. Gao X, Lee GM (2019) Moment-based rental prediction for bicycle-sharing transportation systems using a hybrid genetic algorithm and machine learning. Comput Indus Eng 128:60–69
26. Jordán J, Palanca J, Martí P, Julian V (2022) Electric vehicle charging stations emplacement using genetic algorithms and agent-based simulation. Expert Syst Appl 197:116739
27. Yu H, Lu F (2012) A multi-modal route planning approach with an improved genetic algorithm. Adv Geo-Spat Inf Sci 193
28. Tighkhorshid A, Ali Tousi SM, Nikoofard A (2023) Car depth estimation within a monocular image using a light cnn. J Supercomput 1–18
29. Ali Tousi SM, Mostafanasab A, Teshnehlab M (2020) Design of self tuning pid controller based on competitional pso. In: 2020 4th conference on swarm intelligence and evolutionary computation (CSIEC). IEEE, pp 022–026
30. Griffiths IJ, Mehdi QH, Wang T, Gough NE (1997) A genetic algorithm for path planning. IFAC Proc Vol 30(7):485–490
31. Samizadeh S, Nikoofard A, Yektamoghadam H (2022) Decision making for autonomous vehicles' strategy in triple-lane roundabout intersections. In: 2022 8th international conference on control, instrumentation and automation (ICCIA). IEEE, pp 1–6
32. Lamini C, Benhlima S, Elbekri A (2018) Genetic algorithm based approach for autonomous mobile robot path planning. Procedia Comput Sci 127:180–189
33. Deb K (1999) Multi-objective genetic algorithms: problem difficulties and construction of test problems. Evol Comput 7(3):205–230
34. Srinivas N, Deb K (1994) Muiltiobjective optimization using nondominated sorting in genetic algorithms. Evol Comput 2(3):221–248
35. Deb K, Agrawal S, Pratap A, Meyarivan T (2000) A fast elitist non-dominated sorting genetic algorithm for multi-objective optimization: Nsga-ii. In: Parallel problem solving from nature PPSN VI: 6th international conference. Paris, France, September 18–20, 2000 Proceedings 6. Springer, Berlin, pp 849–858

36. Wang C, Shi H, Zuo X (2020) A multi-objective genetic algorithm based approach for dynamical bus vehicles scheduling under traffic congestion. Swarm Evol Comput 54:100667
37. Schockenhoff F, Zähringer M, Brönner M, Lienkamp M (2021) Combining a genetic algorithm and a fuzzy system to optimize user centricity in autonomous vehicle concept development. Systems 9(2):25
38. Sainath G, Vignesh S, Siddarth S, Suganya G (2021) Application of neuroevolution in autonomous cars. In: International virtual conference on Industry 4.0: select proceedings of IVCI4. 0 2020. Springer, Berlin, pp 301–311
39. Katoch S, Chauhan SS, Kumar V (2021) A review on genetic algorithm: past, present, and future. Multimed Tools Appl 80:8091–8126
40. Gupta N, Patel N, Tiwari BN, Khosravy M (2019) Genetic algorithm based on enhanced selection and log-scaled mutation technique. In: Proceedings of the future technologies conference (FTC), vol 1. Springer, Berlin, pp 730–748
41. Variengien A, Nichele S, Glover T, Pontes-Filho S (2021) Towards self-organized control: using neural cellular automata to robustly control a cart-pole agent. arXiv:2106.15240
42. Gupta N, Khosravy M, Patel N, Dey N, Mahela OP (2020) Mendelian evolutionary theory optimization algorithm. Soft Comput 24:14345–14390

Part V
Signal and Image Processing, and Robotics

Chapter 12
Nature-Inspired Computing for Wireless Sensor Network

Mojtaba Farmani, Mohammad Javad Khani, and Zahra Shirmohammadi

1 Introduction

Wireless sensor networks (WSNs) have seen a rapid increase in their applications over the past several decades [1–6]. It is made up of a base station (BS) and many sensor nodes as shown in Fig. 1. BS is a major coordinator system with a high processing and energy capacity [7, 8].

Wireless sensor networks are devices with low processing power, small size, and cheap price that have the ability to measure and communicate wirelessly. Sensor nodes configure themselves when they are placed in a network, and after collecting data, they connect to each other and transmit the information results to the base station. Then they connect to other networks such as the Internet through gateway nodes and transmit information to the end user.

The sensor node consists of main portions including a sensor unit, processing unit, transmitter–receiver, and power supply. WSNs are often randomly distributed in the environment. Therefore, their exact position is not predetermined. Such a feature allows the deployment of nodes in dangerous or inaccessible places for humans. In order to distinguish the exact position of the nodes, if necessary, the locator systems can be used in the hardware of the nodes.

Wireless sensor network is a network with high stability against sensor failures. It is very robust and when several sensors fail, the network tries to continue its activity with the remaining nodes. Also, network connectivity is constantly changing.

M. Farmani · M. J. Khani · Z. Shirmohammadi (✉)
Department of Computer Engineering, Shahid Rajaee Teacher Training University, Tehran, Iran
e-mail: shirmohammadi@sru.ac.ir

M. Farmani
e-mail: mojtabafarmani@sru.ac.ir

M. J. Khani
e-mail: m.j.khani@sru.ac.ir

M. Khosravy et al. (eds.), *Frontiers in Genetics Algorithm Theory and Applications*,
Springer Tracts in Nature-Inspired Computing,
https://doi.org/10.1007/978-981-99-8107-6_12

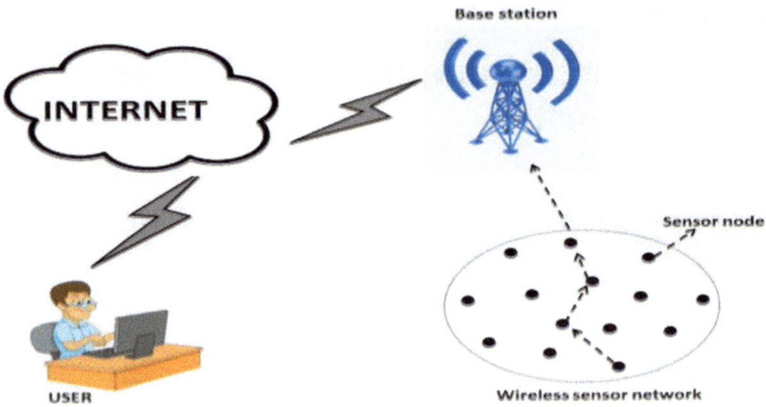

Fig. 1 Wireless sensor network [9]

Therefore, the protocols and algorithms of wireless sensor networks must have self-organizing capabilities.

Wireless sensor networks (WSN) are gaining increasing popularity for a wide range of applications in different fields likewise animal monitoring, agriculture, military, health care, smart buildings, environment, and automotive, a summary explanation of which is as follows:

- Animal monitoring: In this application, a sensor is attached to the animal's body, which can be used to identify the place of transportation and the position of the animals.
- Agriculture monitoring: In this application, using sensors, important parameters such as temperature, humidity, carbon dioxide gas level, soil minerals, etc., are monitored in order to create a suitable environment for growing crops and increasing crop productivity [10].
- Healthcare: The use of WSNs in the medical field is inevitable. In this application, vital signs of the body, such as body temperature, heart rate, blood oxygen, etc., are monitored using sensors that measure different physiological parameters and sent to doctors and nurses for further analysis [10].
- Smart buildings: A smart building is able to monitor and control its performance based on the structure of the building and the internal and external environment. In this application, various sensors are used to monitor and control smart home appliances, authentication systems for home security, and energy management [11].
- Environment: The use of WSNs in the environment can be used in applications such as coal mining, flood detection, forecasting forest fires, mountain volcanoes, water and air quality, etc. For example, people who work in coal mines face risks such as gas explosions, mine collapses, chemical spills, electric shocks, and fires. By using monitoring sensors, it is possible to detect hazards, the position of fire or electric shock, and prevent hazards to other areas. In the field of water quality,

various parameters such as pH, ammonia, dissolved oxygen in water, dissolved in water, etc., are checked using sensors [10].
- Traffic monitoring: Another application of wireless sensor networks is in smart transportation. In this application, sensors can be used to monitor traffic flow, traffic violations, identify illegal activities, car security, etc. [10].

Therefore, WSNs are used in a wide range of applications and these applications can be implemented in the actual universe. To accomplishment these networks, efficient protocols and methods are needed to solve the challenges.

The challenges of this network include:

- Limitation of physical resources: One of the most important limitations of WSNs is the limited battery power of sensor nodes. WSNs are left in an unattended environment where it is not possible to charge and replace the battery. Other limitations include limited computing power, low memory, and limited bandwidth.
- Deployment of sensor nodes: Node deployment in WSNs is based on the applications used, which affects the performance of topology control algorithms. Node deployment can be deterministic and random. The difference between these two methods is that in deterministic deployment, WSNs are placed manually and the information transmission path is also known, but in random deployment, WSNs are distributed randomly.
- Fault Tolerance: WSNs may fail due to battery depletion, physical damage, and environmental interference. Failure of WSNs should not affect the entire network and disrupt its operation.
- Scalability: The number of WSNs distributed in an application can reach hundreds or thousands of sensors. Any protocol design must be scalable and accommodate a major number of WSNs.
- Security: In some applications, communication between nodes must be secure to maintain information confidentiality.

Various methods have been proposed to solve these challenges in wireless networks. One of the effective methods in this field is the use of algorithms inspired by nature.

Nature-Inspired Computing is a powerful method for resolving difficult problems in real life that are entirely dependent on natural events [12–14]. There are three sorts of optimizations: minimization, maximization, and on-target. These are some fundamental details about optimization techniques that apply to nature-inspired computing.

The many types of optimizations include (1) single variable, (2) multi-variable, (3) single objective, (4) multi-objective, (5) constraint, (6) unconstraint, (7) linear, and (8) nonlinear.

The three steps of the process—problem, model, and solution are completed by the sorts of optimizations and variants stated above.

An iterative method in which "problem" and "model" are linked via "formulation", and "model" and "solution" are connected through "iterative process". Planning, design, execution, and optimization are the first four steps of a cycle process that generates the best solution and its circulation.

The application of biological systems that may serve as the fundamental foundation for a sensor node's choice to be globally optimized is emphasized by nature-inspired techniques in WSN, which offers the following benefits:

a. Susceptibility to probe local solutions.
b. The capacity to manage objective cost.
c. Simplicity of accomplishment.
d. Robustness, clarity, and adaptability.
e. High degree of adaptability.
f. Complex functions require a large application.
g. Solution population.

Next, in the second part, we deal with the problem detection in WSNs using the algorithm of the body's immune system. In the third part, an intelligent routing method based on the ABO algorithm is investigated. In the fourth section, a method called modified crowd intelligence for energy optimization is investigated. In the fifth section, the review of a pseudo-opposite harmony search method is discussed. In the sixth section, several multi-objective optimization methods inspired by nature have been introduced. In the seventh section, the TLBO method is reviewed for the optimization of WSNs. In the eighth section, nature-inspired methods for health care applications in wireless networks are discussed. In the ninth section, a review of a quasi-opposite harmony search algorithm is discussed, in the tenth section, several nature-inspired multi-objective optimization algorithms are introduced, in the eleventh section, the TLBO algorithm for the optimization of WSNs is reviewed, in the twelfth section deals with nature-inspired methods for healthcare applications in WSNs.

2 A GA-Based Fault-Aware Routing Algorithm for WSNs

Wireless sensor networks (WSN) [1] are the primary environmental monitoring backbone for IoT systems. WSNs are placed in the monitoring area and employed in the monitoring environment to detect metrological and hydrological information within its sensing range, like heat, light, and humidity. Since each sensor has a finite detection range, the cooperative operation of a vast number of sensors enables WSN to monitor a huge target area and transmit the monitoring environment's ambient state to the remote station (RS) for disaster risk management [2–5].

Setting up a dependable infrastructure in a dangerous environment while taking into account various WSN restrictions is the primary difficulty for IoT systems used for environmental monitoring. The primary limitation of WSN is the restricted number of deployed battery-powered WSNs, which cannot be recharged or changed [6–9].

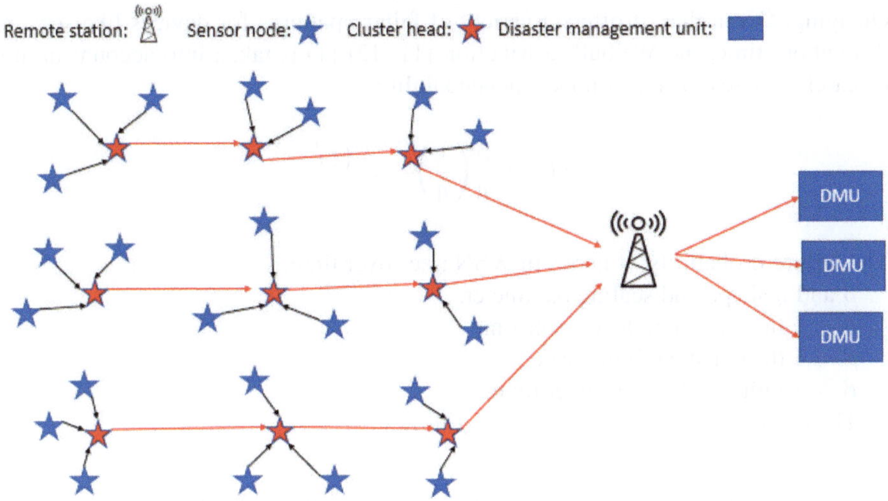

Fig. 2 Cluster Model [10]

Depending on its connectivity to the RS, the cluster head (CH) of a cluster gathers information packets from the cluster members (CMs) and forwards them to the remote station (RS) either directly or via another CHs. Clustering in WSN provides a number of benefits, including:

- The WSN uses less energy as a result.
- There is bandwidth conservation.
- Increases the network's capacity to scale.

There are k^N potential routing paths in the network for a network with N number of CHs and an average of k number of candidates for next-hop routing CH for each CH. Fig. 2 indicates the computational expense of building a routing tree in a large-scale WSN using a brute-force approach. In order to identify a nearly perfect routing alternative, it is much desired to use an optimization approach like the genetic algorithm (GA) [10].

2.1 *Terms and the System Model*

An IoT system with a heterogeneous sensor network concept for monitoring the environment. It is composed of advanced nodes with vast resources and regular sensor nodes with limited resources. The network is divided into two levels, with the upper layer consisting of advanced nodes and the lower layer composed of standard sensor nodes.

Additionally, a fault model is used to assess the likelihood of a node failure. When WSNs, particularly CHs, fail, the network splits while the topology is dynamically changing. Given that, it offers a variety of failure patterns for devices like sensors throughout time, the Weibull distribution [11, 12] (1) is taken into account in this research to assess a sensor node's dependability.

$$f(\mathrm{t}) = \frac{\beta}{\eta}\left(\frac{\mathrm{t}}{\eta}\right)^{\beta-1} \mathrm{e}^{-\left(\frac{\mathrm{t}}{\eta}\right)^{\beta}} \tag{1}$$

The failure probability of nodes in WSN rises over time.

β and η shape and scaling parameter.

$\beta < 1$ failure rate reduces over time.

$\beta = 1$ the constant failure rate.

$\beta > 1$ failure rate raises over time.

The terms and meanings used are as follows:

- $S = \{S_1, S_2, \ldots, S_n\}$ is a set of WSNs.
- $CH = \{CH_1, CH_2, \ldots, CH_k\}$ is a collection of cluster heads.
- The distance between nodes i and j is shown $Dist(i, j)$.
- The distance of node i from the RS is shown as $Dist_{RS}(i)$.
- $Energy_{res}(i)$ is the remaining energy of node i.
- R_{com} Stands for a node i's communication range.
- $CM(i)$ Stand for the cluster members of CHI.
- The collection of candidates CHs that are located within the sensor node S_i is denoted by the expression $\_CH(S_i) = \{G_j : Dist(S_i, G_j) \leq R_{com}\}$.

The following explanations:

Definition 1: The CHs in a CH's transmission range toward a distant station are known as its downstream CH set (DS_CH).

Definition 2: The CHs in a CH's transmission range that are further away from the remote station and contain at least one downstream CH other than the CH I are referred to as the CH's upstream CH collection (US_CH).

Definition 3: If a CH's transmission range does not include an upstream or downstream CH, it is referred to as an Unconnected CH (UnCon_CH).

2.2 *Proposed Algorithm (Information Sharing)*

Each node exchanges information with the RS, such as its ID, location, residual energy, etc. In the transportation range of common WSNs, a CH broadcasts HELLO message. An ordinary sensor node's Cand CH I set is modified to read "$\mathrm{Cand}_{\mathrm{CH}}$ (i) = {$\mathrm{Cand}_{\mathrm{CH}}$ (i)∪ j}" if it takes this message from a CH j. Each regular node and CH will now communicate with RS local information such as location, residual energy, $\mathrm{Cand}_{\mathrm{CH}}$ set, etc. After network setup, there is a steady phase where the CHs allocate each of their CMs a time slot for them to relay their sensed information to the appropriate CH utilizing time division multiple access (TDMA).

2.3 Network Setup

It is followed by a steady phase when the CHs give each of its CMs a time slot for them to communicate their sensed information to the appropriate CH utilizing time division multiple access (TDMA).

2.4 Steady Phase

It runs for a specified number of rounds, after which the network setup and data exchange phases are repeated to create a fresh routing tree. In order to prevent collisions, each CH uses TDMA to allocate its cluster members a time slot, during which they must communicate any detected data to the corresponding CH. Following the routing tree created during the network setup phase, the CH then collects the packets it has received and transmits them to the RS.

2.5 Simulation Setup

Utilizing the programming languages C and MATLAB 2012b. A (100×100 m^2) two-dimensional plane with 50 advanced sensor nodes and 200 standard WSNs. Ordinary and advanced WSNs have starting energies of 0.05 J and 0.1 J, respectively. A node's failure rate follows the Weibull distribution. The energy models $\varepsilon_{\text{fs}} = 10\text{PJ}/m^2/\text{bit}$ and $\varepsilon_{\text{mf}} = 0.0013\text{PJ}/\text{bit}/\text{m}^4$ parameters the simulation outcomes were contrasted with those from MHRM and EHE-LEACH to assess the effectiveness of the suggested method.

2.6 Evaluation of Experimental Results

Compare the energy efficiency, number of living nodes, and packet transference ratio of our proposed method to those of other current algorithms. The typical CH energy each round in MHRM, EHE-LEACH, and FAR. In comparison to MHRM and EHE-LEACH, CHs have, so far, conserved more energy. The burden is distributed evenly among the CHs in the FAR fitness function. How many CHs are active in a given round? If a CH can connect to RS and has enough power to transmit cluster data, it is said to be active. All CHs are not immediately active in MHRM and EHE-LEACH since these methods do not take into account any routing method for disconnected CHs.

3 GA-Based Fault Diagnosis Method for Prolonging Network Lifetime of WSNs

Different problems arise with wireless sensors, including (a) design problems with fault tolerance, low latency, scalability, transmission medium, and coverage concerns; and (b) topology problems, including coverage topology, sensor gaps, and geographic routing.

3.1 Challenges

WSN nodes are quite promising. By identifying network limitations and technological problems, it increases potential. There are a lot of obstacles to be solved using wireless sensor nodes.

a. Energy: Nodes' lifetime is limited since they run on batteries. It is possible to replenish or replace that battery power.
b. Limited Bandwidth: In order to transmit data among two WSNs, message swapping was necessary, and synchronization would be impractical without it. WSNs used a multi-hop communication channel to run and control node performance.
c. Node Prices
d. Deployment: WSNs should be placed appropriately to avoid adding to the complexity.
e. The major objective is to create a node that is smaller in scale so that it can be carried everywhere, more inexpensive so that the expense can be reduced, and more efficient in terms of electronic components.
f. Security: Resources that need to collect sensitive data must have high security. Sensor nodes' vulnerabilities to malicious infiltration and assaults are increased by their remote and unsupervised activities.

3.2 Faults, Errors, and Failures

a. A network fault
b. A hardware fault
c. A software fault

The information received by the sensor is also used to classify faults in the sensor node. They are listed below:

I. Offset fault: adding a permanent cost to the anticipated information to locate the problem node.
II. Gain fault: assess the fault node and anticipate expected gathered data to fluctuate over time.
III. Stuck-at-fault: This kind of problem is discovered when sensed data series vary as zero.
IV. Out of bound: This form of error occurs when sensed data values are discovered to be beyond the range of typical running values.

3.3 Fault Diagnosis

A network's performance, node lifetime, service type, and data security can all be improved. Additionally, a network problem may be reduced, costs can be kept under control, and nodes' service life can be extended.

Two key measures should be taken to allow flexibility in problematic circumstances.

a. Fault detection
b. Fault recovery

3.4 Proposed Method

The major suggested work combines two already-used approaches, the genetic method and the direct diffusion method. Four different message types—interest, exploratory information, reinforcement, and information messages are used by the direct diffusion method. Interest, information, gradient, and reinforcement are its four primary characteristics.

a. Interest: a question or inquiry that reveals the user's desires.
b. Data: gathered or processed data.
c. Gradient: Each node that gets attention creates a direction state.
d. Reinforcement: This optimization method is applied to learning.
e. An attribute is a pair of value descriptions of a task.

Implementation of a genetic method as a fault node recovery mechanism. This comprises five phases and is based on a set of solutions known as crowd changed into new crowd by reproducing.

a. Verify the anticipated outcome and replace it if necessary.
b. Compute a fitness function and check the substituted node information using the predicted solution.
c. A high fitness value is used as the basis for selection.
d. A single crossing method was implemented using a crossover step.

e. In order to utilize the most direction-finding pathways and extend the WSN lifespan, a mutation step is employed to flip nodes at random.

After completing the following five steps:

a. Initialization: In the initial stage, chromosomes are generated. Each chromosome was initially established with a fitness value, either 0 or 1.
b. Evaluation: The resulting fitness value is assessed using the parameters provided—chromosome genes—as a fitness function.
c. Selection: Choosing a sensor node with a better fitness value.
d. Crossover: altering the genes one at a time.

3.5 Performance Evaluation

This performance evaluation makes use of the following models:

a. Network model: For a distributed sensor network (DSN) environment, the area is L * B square meters.
b. Channel model: Data packets of varying sizes are generated in sensor networks. For media (channel) access, the S-MAC protocol of the sensor network is employed. In a sensor network, packet transmission is considered to be discrete in nature.
c. Battery model: Each sensor has a unique, finite, non-replicable battery. A joule-based initial energy was used to pinpoint the node.
d. Energy model: Power control via radio broadcast is possible. In order to meet the fault tolerance criterion, it utilizes the least amount of energy necessary.

4 A GA-Based Smart Traffic Management Method for WBANs

A network of several sensor nodes is known as a wireless body area network (WBAN) or wireless body area sensor network (WBASN) [1, 2]. WBAN has a variety of uses, including the following:

i. Health monitoring
ii. Body Area Networks (BANs) interacting with one another
iii. Mobile device-centric
iv. Distant control
v. Application in society
vi. Video stream
vii. Wearable audio.

Machine learning algorithms often follow the following steps:

i. Feature selection and yield labeling
ii. Sample accumulation
iii. Offline training
iv. Online taxonomy

The main goal of this work is to provide a genetic method-based intelligent traffic management method for WBAN (GA).

4.1 Proposed Method

Using a domain and the qualities that connect to it, the proposed technique is demonstrated. The fundamentals of WBAN and the applications and related functionality. There are certain steps throughout the entire construction. Here are some brief descriptions:

a. Initialization: denotes the initialization of the population.
b. Selection: implies creating a workable category from the population as a whole; the population is then split into two groups: (i) good people, and (ii) bad people. Here, only good people are allowed to have more children; evil people are not allowed to have children, so after one or two generations, everyone in the country is a nice person.
c. Crossover: Indication that generation should be increased.
d. Mutation: This method produces a fresh answer.

The utilized determines the stopping condition. It may specify a maximum number of iterations of 4, in which case the procedure will end after 4 iterations. In the suggested paradigm, a green signal is represented by one and a non-green signal by zero. Chromosomes are signals found in the crowd. The zero and one, or non-green and green signals, are chosen in the suggested method based on their fitness function (F(x)). GA was utilized as an intelligent strategy.

Two operations—crossover and mutation—are carried out once the chromosomes have been chosen. Crossover refers to the act of combining two different chromosomes and performing an exchange. A genetic operator called mutation is employed to produce a new result or increase the quality.

4.2 Maxone Approach for the Suggested Technique

This Maxone approach, which is based on GA, is used to increase the demanded yield, or green signal as opposed to non-green signal. A total of 50 signals, split into five slots, are used in the suggested technique. Ten signals are present in each slot. Slot is indicated by si, where i ranges from 1 to 5. Then, choosing the chromosomes using the roulette wheel method, we know that the full hermitage of the sphere is 360. Here, 20 equals 360, so the upmost cost is $S5$. Accordingly, the full percentage of $S5$ is (6/20) *100, or 30%, etc. Assign a new name to each chromosome and arrange them all according to the fitness function of a roulette wheel.

Perform the choice procedure by choosing any two couples at random; in this model, the chosen pairs are section one ($s1' - s2'$) and section two ($s4' - s5'$), and intersecting among corporators is performed. To further improve, we apply crossover between Part 3's $S2'$ and $S3'$s $S3$, as well as between $S1'$ and $S5$, etc.

s1 = 1001000101	s1' = 1000111101	s1" = 1001000101	s1'''= 0101000001
s2 = 0101000001	s2' = 1001000101	s2" = 1000111101	s2''' = 0110010001
s3 = 0110010001	s3' = 0110010001	s3" = 0110010001	s3''' = 1000111101
s4 = 0010000011	s4' = 0101000001	s4"= 0010000011	s4''' = 0010000011
s5 = 1000111101	s5' = 0010000011	s5" = 0101000001	s5'''= 1001000101
s1'''' = 0101100101 = 5 s2'''' = 0110010001 = 4 Full fitness cost is = 5 + 4 + 6 + 5 + 5=25. s3'''' = 1000111101 = 6 s4'''' = 0010101011 = 5 Prior fitness cost is 20. s5'''' = 1001010101 = 5			

4.3 Performance Evaluation

The suggested approach is based on the meta-heuristic GA methodology. The currently employed ITM [3] approach is based on the average waiting time that is utilized at every network junction point. The current TJA technique is centered on energy efficiency and traffic avoidance. It is utilized for both network monitoring and surveillance. The currently used PTS is a routing protocol that just deals with traffic and is energy-efficient. In comparison to other current approaches and the suggested method, it performs poorly.

5 Fault Diagnosis in WSNs Using a Neural Network Constructed by Deep Learning Method

Wireless sensor networks (WSNs) are made up of small, low-cost WSNs with processing, wireless relationships, and computing capabilities. A sensor node that consists of one or more sensors, powerful CPUs, memory, a battery, a wireless communication radio, and an actuator. In adverse and unsupervised areas where sensor nodes are installed. Sensor nodes are designed to function independently, be durable, and be environment-adaptive. The development of the WSNs is based on the following hypotheses. These are listed below [1]:

a. The nature of sensor nodes is homogenous, with uniform energy.
b. Sensor nodes in WSNs have the ability to transmit and receive sensed data.
c. A sensor node is said to have a difficult fault if it cannot connect with its neighbors.
d. The network is static, meaning that throughout the fault detection period, the sensor nodes' locations and the network structure stay unchanged.
e. The MAC layer handles the problem with the communication channel.
f. Sensors use the UDP/IP protocol for communication.

5.1 *Sensor Network Model*

In a square landscape of length R, N sensor nodes are randomly placed. Every sensor node in a networks$_i$, with $i = 1, 2, 3, \ldots, N$, N, has a unique identification and location,$P_i(xc_i, yc_i)$, where$0 \leq xc_i \leq R, 0 \leq yc_i \leq R$. As part of a sensor network, s_i communicates with its neighbors and uses a one-to-many broadcast primitive as its default mode of communication. Let T_r represent the transmission range of s_i and assume that it is constant across all sensors. It is considered that a sensor node is linked if it is located within T_r of s_i at n time instants. In general, a sensor network is thought of as a graph G(S, C), where S stands for a collection of WSNs, and C is the collection of relationship links between those WSNs.

5.2 *Fault Model*

When a sensor node develops a defect, it produces optional values when the error is too great. At first, it is believed that the network's sensors are fault-free and are intended to be identified according to their fault state. Throughout its operation, each sensor in the network is exposed to an intermittent defect. Let p represent the constant fault probability for each sensor that a sensor will malfunction occasionally.

5.3 Modeling of Sensor Data

Usually inaccurate or noisy, the data is collected by sensor nodes. The mistake typically arises as a result of hardware and data connectivity issues [2]. A sensor node's result $x_k(n)$ at time moment n is the result of adding the measured physical parameter's actual value a plus random error [2] (2).

$$x_k(n) = A + \upsilon_k(n), n = 1, 2, \ldots, KK = \frac{T}{\Delta T} \; and \; k = 1, 2, \ldots., N \tag{2}$$

At each of the sensor nodes, $\upsilon_k(n)$ represents incorrect data. Over the course of the whole time period T, the sensor nodes accumulate the seen information in regular intervals of time ΔT.

5.4 Problem Formulation

There are N sensor nodes scattered around a region. As specified in the closed interval $[0, T]$, data from the k th sensor s_k is presented as $x_k(t)$, where t is the time changeable and T is the entire duration of the signal (3).

$$x_s[n] = x(nT_s) \tag{3}$$

$x_s[n]$ is the sampled signal and n is the sampling index and $n = 1, 2, \ldots, L$ (4)

$$L = \left\lfloor \frac{T}{T_s} \right\rfloor \tag{4}$$

The sensors data $x_k[n]$ (5)

$$x_k(n) = A + \omega_k(n), k = 1, 2, \ldots., N \tag{5}$$

To identify the kind of failures, the data $\{x_k[n]\}_{N=1}^{L}$ from each sensor k is utilized.

5.5 Deep Learning Neural Network Algorithms for Intermittent Sensor Node Fault Detection

In machine learning, neural networks (NNs) are statistical learning algorithms that take their cues from the characteristics of biological brain networks. When implemented, NNs take the form of a network of interconnected processing units known as nodes (as opposed to sensor nodes), which are functionally similar to biological neurons.

5.6 *Basic Neural Network Design*

The feed-forward net (FFNet), cascade forward net (CFNet), pattern net (Pnet), and fit net (FitNet) are among the most often used neural networks in computing tools like MATLAB and Python. One input layer, ten neurons in hidden levels, and one output layer make up each network's architecture.

5.7 *Performance Evaluation of Various Deep Learning Algorithms and Neural Networks*

Four aspects of the captured information from every sensor node in a sensor network with N 1024 nodes make up the network's input. The total quantity of information is L = 20. Thirty of nodes are thought to be intermittently faulty. We have selected the 10% intermittent data failure for this comparison. All networks share the same input. The detection accuracy (DA) pending the confirmation and the mean squared error (MSE) of the discrepancy between the training and confirmation collection are used to calculate the neural network's efficiency. When compared to the BFG, RP, and OSS training techniques, the detection accuracy of the FitNet and feed-forward neural networks is equivalent. The best MSE performance, as predicted, is provided by the BR and LM with somewhat less DA. The efficiency of the pattern neural network is subpar. When using GD or GA as the training algorithms, it has been discovered that the DA is 85.5 and 87.77%, respectively. However, the DA can reach more than 99% in our suggested neural network with deep learning algorithm. The chance of periodic sensor node faults is rising along with the DA.

6 Wireless Sensor Network Fault Diagnosis Inspired by the Immune System

The nodes can detect the physical events that take place around them since they are sensor nodes. A sensor node may malfunction for a number of reasons, including security, energy restrictions, hardware difficulties, and more [15]. Numerous approaches, including neural networks [16], statistical methods [17], adjacent co-ordination [18], and comparison [19], are used in the literature to identify the problematic nodes.

Due to its underlying philosophy, artificial immune system (AIS) is regarded as one of the most widely used techniques [20]. AIS draws inspiration from the HIS principle, which efficiently protects our bodies against viruses and bacteria [21]. Traditional techniques' main limitations are their high energy consumption and message complexity. Because AIS is adaptable, we have employed it to get around these shortcomings. It has the ability to identify certain flaws and remember them for potential future responses. Artificial immune systems use ideas from the human immune system and apply them to mathematical issues.

6.1 AIS Methodologies for WSN Fault Diagnosis

AIS is a branch of artificial intelligence that draws inspiration from biology. The algorithms may be used to diagnose WSN faults and demonstrate the behavioral function of HIS.

The clonal choice viewpoint of the adaptive immune system, which demonstrates affinity maturation, served as the basis for the clonal selection algorithm (CSA). These algorithms concentrate on somatic hyper mutation, reproduction, and antigen–antibody interaction.

Cells are divided into self and non-self-categories using the negative selection algorithm (NSA). According to this algorithm, the body defends against pathogens like bacteria and viruses by preventing attacks on its own pattern (known pattern).

Antigen-presenting immune cells known as dendritic cells, or DCAs, are ever assistant however only become active since an outside invader enters our corporality.

The immune network algorithm (INA) focuses on the network's structure, with antibodies or cells that produce antibodies serving as nodes. The training method consists of widening edges among nodes based on affinity.

6.2 Applications of AIS Algorithms

a. Auto hydraulic brake fault diagnostics [22].
b. To locate the errors in a sizable wireless sensor network, the artificial immune system idea is applied [23].
c. AIS's clonal selection concept is used to identify defective sensor nodes, which are then classified into the appropriate kinds using a probabilistic neural network technique [24].
d. The composers considered the use of AIS to recognize fault schemes in information on the temperature of supermarket freezer cabinets [25].
e. Based on NSA [26], the authors have suggested a technique for classifying genuine noise in spoken phrases.
f. The composers proposed an innovative technique for defect identification utilizing an artificial immune network [27].

g. Using an immune network, the authors have looked into the identification of faults in plant systems [28].

7 Intelligent Routing Method Using African Buffalo in WSNs

Swarm intelligence (SI) technology known as African buffalo optimization (ABO) belongs to the category of Meta heuristics. This method is a piece of the rapidly used modern class of nature-inspired algorithms [29]. Large herbivorous animals called African buffalos travel through savannahs, champs, and forests in probe of food, or lush green grass [30]. Thus, African buffalo have developed certain traits that enable them to flourish in spite of their harsh environment [31]. The following traits are [32]:

I. Communication skills: African buffaloes use "vocalization" to talk to the herd. There are two main categories of vocalization: "maa" and "waa".
II. A vast memory: African buffalo have the ability to remember their routes over thousands of kilometers through savannah, champs, and forests. Given that African buffalo have a vast capacity for memory, it is possible.
III. Effective herd management structures: These structures enable African buffaloes to benefit from the group's collective intelligence.

7.1 The ABO's Component View

a. Initialization: ABO initializes a herd or crowd (a mass of buffaloes) by giving each individual (a buffalo) within the probe space a random position. Every buffalo's position in the search space corresponds to a potential solution.
b. Update position: Two "vocalization" methods—"maa" (for exploration) and "waa"—are used in the updating operation.
   I. Using the vocalization "maa": Memory, cooperation, and intelligence round out the list.
   II. Making the vocalization "waa": Adjustment (ii) Adjustment level
c. Refresh personal and global best places:
   I. Personal best: the position of each buffalo's best solution.
   II. Global best: the site of the most successful buffalo-related solution.
d. The whole herd is re-initialized if the fitness of the best buffalo (i.e., the world's best) does not prosper over a predetermined period of time while maintaining adequate exploration.
e. Termination: The ABO ends when any one or all of the following circumstances occur:

I. Minimum criteria
II. Exhaust solution space
III. Maximum execution time
IV. Maximum stall limit

7.2 *African Buffalo Optimization: The Algorithm*

The global optimum is solved by this algorithm in six phases:

1. Randomly set up each buffalo's starting position in the search area.
2. Determine each buffalo's level of fitness in the herd.
3. Update the herd's overall best solution and every buffalo's individual best solution.
4. Using Eqs. (6) and (7), update each buffalo's position:

$$m_{n(t+1)} = m_{nt} + C_1\left(x^g - x_{nt}\right) + C_2\left(x_n^p - x_{nt}\right) \tag{6}$$

$$\omega_{n(t+1)} = \frac{(\omega_{nt} + m_{nt})}{\lambda} \tag{7}$$

ω_{nt} and m_{nt} define discovery and elicitation of probe region

C_1 social learning parameters.

C_2 cognitive learning parameter.

λ an accidental digit in span [0,1].

x^g and x_n^p define the position of buffalo with universal and particular prime solutions, respectively.

5. Step 1 should be taken if the universal prime solution does not prosper after, say, t repetitions and the contemporary repetition is minor than the superlative repetition (or, if the conclusion criterion is met). In any other case, move on to step 6.
6. Verify the termination criteria. Return to step 2 if the termination requirement is not met. Otherwise, declare the overall best to be optimal and cease.

7.3 *Benefits of ABO Method*

a. Robustness, adaptability, and simplicity:

   The ABO method works out complicated optimization quandaries using ordinary operators likewise emulating communication and cooperative techniques.

b. Faster convergence

   By adapting social and cognitive learning parameters, the ABO algorithm causes faster convergence, and through the "voicing" technique, by creating an equivalence among discovery and elicitation, it leads to an optimal solution.

c. A smaller parameter

The ABO algorithm uses fewer parameters for the herd to probe or elicit belonging to the improvement of the method. Because the precise setting of the parameters causes faster convergence.

7.4 Utilizing the ABO Algorithm:

a. Multi-objective Optimization: This technique is used to manage multiple objective functions where every objective's nature is in discord with the others.
b. Constrained Optimization: This technique is used to improve a single objective in light of several determinations changeable that are included in the limitations.
c. Discrete Optimization: This type of optimization too relies on minimization or maximization. Additionally, it is called combinatorial optimization.
d. Optimization for Binary Integer Programming: Every variable that replaces the decision variable is either 0 or 1.

8 Modified Swarm Intelligence in WSNs

To posit a source in a sensor network, one must estimate the direction of arrival (DOA), also known as the angle of arrival. The outputs of a group of sensors may be used to calculate the DOA of signals at the sensor array.

An effective statistical technique for estimating DOA is the maximum-likelihood (ML) approach [33]. The information model, the covariance matrix of the input information vector of the array, and the noise dispensation are used to create the cost function in machine learning algorithms. When the estimated DOAs and real angles are identical throughout the optimization phase, the universal maxima of the ML task are then found. To estimate the DOA, every sensor node in the WSNs creates an optional array with its neighbors. Every node shares to its neighbors once throughout an experiment with their relative situations to create the native maximum-likelihood value task after gathering spatially uncorrelated data. The cost function is then locally optimized by the evolutionary algorithm at each sensor node. The nodes then actively participate in the diffusion mode of operation. To accomplish the universal DOA with native optimization, the obtained parameters are disseminated for further estimation.

A common clustering strategy is employed to produce effective and scalable performance in order to address the problems in spread DOA estimation. The clustering strategy decreases network communication overhead and conserves energy [34].

8.1 Maximum-Likelihood DOA Estimation of Narrow-Band Far-Field Signal

A sensor network made up of N randomly dispersed nodes. The sensor network is impacted by $(M > N)$ narrow-band far-field wellsprings. The true position of the unidentified sources is at $\theta = [\theta 1, \theta 2, ..., \theta M]$ in coordinates. The complicated response of the sensor array to a wellspring with direction θ be $a(\theta) \in C^{N\times 1}$.

$A(\theta) =$ [a (θ_1)... a (θ_M)] is the array respond matrix of the sensor array created by the sensor network. Based on the placement of the sensor nodes.

Is the source's wavelength.

The (k, l) th element of matrix $A(k, l)$ contains the complicated increase of the lth signal at the kth node (with relevance to a reference node) as [35] (8)

$$A(k,l) = exp\left\{ j\frac{2\pi}{\lambda}[x_k sin\theta_l + y_k cos\theta_l] \right\} k = 1, 2, .., Nl = 1, 2, \ldots, M \quad (8)$$

8.2 ML-DOA Estimation Quandary Formulation

The writers have provided two ML models: a conditioned model and an unconditional scheme. In a conditional scheme, signals are taken to be deterministic, whereas in an unconditional scheme, signals are taken to be random. It has been demonstrated [36] that the unconditional model performs noticeably better.

8.3 Distributed DOA Approximation Using Native ML Tasks

Every sensor must have a native value task with the similar aim when performing distributed estimating. Each node makes an iterative optimization algorithm-based attempt to guesstimate the DOA. This uses the diffusion PSO, a distributed variation of the PSO. Following the estimation of the local DOA, every node communicates its estimated values to its neighbors. Update own estimate when a node has combined all of the received estimations.

8.4 Optimization of Diffusion Particle Swarms (DPSO)

Following are the DPSO steps including, each particle gathers the best information from its neighbors and mixes it. Each sensor updates the velocities of its particles, then updates the positions of those particles, next updates the *pbest*, and all of the particles combined determine their *gbest*, shares with its neighbors its *gbest* vector.

9 Effective Optimization Method with Quasi-Oppositional Harmony Search Algorithm

Numerous distributed, tiny, and down-energy demand systems serve as sensor nodes in a WSN [33]. When sensors are used over wide areas of land, WSN experiences network lifetime issues because batteries are hard to replace or recharge. WSN's primary objectives are to (a) espy the chosen space, (b) document when events occur, and (c) evaluate the necessary parameters [34].

The formulation of the problem's mathematical stance was necessary for optimization, and solving the supplied problem is a crucial component of the assignment. The advantages of optimization approaches include the fact that they are independent of the size and structure of the issue.

The coordination probe method has been tweaked to become the QOHS algorithm (HSA). In order to improve the harmony of the music, musicians are always looking for the best tuning techniques. HSA is the result of this search. Harmony vectors, a population vector created at random and kept in HM, are the foundation of HSA. Harmony is the name given to the HSA's achieved solution, which may be represented as an n-dimensional solution vector. The following can be used to describe the HSA process:

Phase (a) Problem initialization can be expressed as Eq. (9) for a global optimization problem.

$$\min f(x) intheintervalof x_j \epsilon \left[para_j^{min}, para_j^{max}\right] j = 1, 2, \ldots, N \tag{9}$$

The objective task that must be reduced or maximized is called $f(x)$. The collection of control variables is $X = [x_1, x_2, \ldots, x_n]$. The quantity of control variables is n. The design variable's lowest and maximum limits are indicated by the variables $para_j^{min}$ and $para_j^{max}$, respectively.

Initial solution creation in phase (b): The optimization task's first solution is a population vector that was initially produced randomly. In Eq. (10), the location of the HM matrix is displayed.

$$X_j = \left[x_1^j, x_2^j, \ldots, x_n^j\right]$$

$$j = 1, 2, \ldots, HMS \tag{10}$$

$$HM = \begin{bmatrix} x_1^1 & x_2^1 & \ldots x_n^1 \\ x_1^2 & x_2^2 & \ldots x_n^2 \\ x_1^{HMS} & x_2^{HMS} & \ldots x_n^{HMS} \end{bmatrix} \tag{11}$$

Phase (c) Improvisation of the solution: A memory regard, pitch adjustment, and accidental choice are implemented as part of the improvisation process to provide an enhanced harmony vector. Equation (12) provides the pitch adjustment rule.

$$X^{new} = \left(x_1^{new}, x_2^{new} \ldots x_n^{new}\right) \tag{12}$$

$$x_j^{new} = x_j^{new} \pm r_3 \times BW \tag{13}$$

Phase (d) Solution update: The HM may be updated if the fitter vector among the novel and arrant harmony vectors, x^{new} and x^{worst}, survives. If the compatibility cost of x^{new} is superior to that of x^{worst}, x^{new} is updated by x^{worst}.

Phase (e) Method of computation: The enhanced HSA changes the costs of PAR and BW as stated in Eqs. (14) and (15), respectively, but still applies memory deliberation, pitch correction, and accidental choice [35]:

$$PAR(gn) = PAR^{min} + \frac{PAR^{max} - PAR^{min}}{NT} \times gn \tag{14}$$

$$BW(gn) = BW^{max} \times e^{\left(\frac{ln\left(\frac{BW^{min}}{BW^{max}}\right)}{NT} \times gn\right)} \tag{15}$$

9.1 Opposition-Based Learning

By introducing the idea of opposition-based learning, the previously created answer may be enhanced (OBL). According to the OBL concept, an accidentally generated solution can be strengthened by repeatedly examination the opposing solution [36, 37].

A genuine number, $x\epsilon[ub, lb]$ should be used. Equation (16) may be used to determine the opposite number of x:

$$.\breve{x} = ub + lb - x. \tag{16}$$

whereas Eq. (17) can be used to define the opposite point:

$$\breve{x}_i = ub_i + lb_i - x_i \tag{17}$$

9.2 Opposition-Based Optimization

In two parts, the OBL idea is introduced. Opposition-based concepts are introduced to enhance the first solution and HSA is selected as the parent algorithm.

Learning through quasi-opposition: A concept. When populating the basic HSA, quasi-opposing points are used in place of opposite numbers. In the HSA, the idea of quasi-oppositional generation leaping is used.

9.3 Quasi-Opposite Number

The distance [38] (18) between the probe area's center and the opposing number.

$$QOX_0 = rand\left(\frac{Para^{min} + Para^{max}}{2}, Para^{min} + Para^{max} - X_0\right) \tag{18}$$

9.4 Quasi-Opposite Point

The position that lies between the probe area's center and the opposing number (19).

$$QOX_{0i} = rand\left(\frac{Para_i^{min} + Para_i^{max}}{2}, Para_i^{min} + Para_i^{max} - X_i\right), i = 1, 2, \ldots, d \tag{19}$$

9.5 Quasi-oppositional Population Initialization

The quasi-oppositional idea in the HSA can be used to improvise a population vector that is generated at random.

9.6 Quasi-oppositional Generation Jumping

With some mechanism, the optimization procedure became more successful at finding a new candidate solution. A novel crowd vector is created using the QOHS method using the jumping rate (Jr) notion.

9.7 Benchmark Functions QOHS

In this part, the QOHS algorithm's performance is evaluated using five benchmark functions. Following is a summary of the QOHS algorithm optimization steps: Initialization, improvisation, and updating the HM are the three steps.

ACO: The fundamental idea was derived from ant foraging behavior. This method aids in locating an approximate solution that converts a discrete to a continuous optimization issue.

PSO: fundamental notion Based on the flocking of fish or fowls, this optimization strategy. By using an iterative process, it has been used to discover the best solution to a problem by attempting to amend the volunteer solution (known as a particle) from the crowd (known as a swarm) with respect to a given evaluated amount.

Fuzzy: It is predicated on true (1) and false decision-making methods (0). With the help of many basic input–output relationships, this approach may synthesize complicated nonlinear output-input relationships.

The following scenarios have been considered in relation to optimization methods used in WSNs:

Scenario (a) WSN network lifespan optimization.

Scenario (b) Communication protocol optimization for massive Wireless Sensor Network.

Scenario (c) WSN optimization with coating and lifespan.

Scenario (d) WSN dynamic spread optimization.

9.8 Evaluation of Performance

The HSA principle, which is widely applicable for locating global search or true approximate methods, is employed by the QOHS method. ACO, PSO, and fuzzy logic are the foundations of the methods OEAM, SORM, and FMRP, respectively.

10 Algorithm Multi-objective Bio-Inspired Method for WSN

A technique based on biological inspiration called Ant Colony Optimization (ACO) mimics the actions of actual ants. A troop of ants has been seen to act together to determine the quickest path between their nest and their meal.

The Artificial Bee Colony (ABC) is a significant bio-inspired technique that was motivated by how honey bees forage as a group [39]. Additionally, it looks for the best numerical answer among a lot of data.

The echoposition abilities of microbats served as inspiration for the Bat Algorithm (BA) [40]. The only animal that can detect echoposition, which aids in finding prey, is the bat.

The term "biogeography-based optimization" (BBO) refers to the investigation of the geographical dispensation of biological organisms. It is a method that optimizes a task through the repetition improvement of nomination solutions with regard to a particular evaluation of compatibility task, and it is inspired by biological evolution.

Cat Swarm Optimization (CSO) is a swarm optimization technique that was inspired by the behaviors of searching and tracking that cats frequently engage in [41]. Using the better fitness value of chosen spots, seeking decides where to travel next. Cat traces some targets in the tracking mode.

The cuckoo fowl typically sets its seeds in the dens of alternative host species. This method states that every seed in a den symbolizes a solution, whereas a cuckoo seed indicates a novel solution.

A bio-inspired program called the Chicken Swarm Optimization Algorithm (CSOA) imitates the treatments of a flock of birds, including roosters, hens, and chicks [42].

Elephant Herding Optimization (EHO) is an algorithm based on how elephants herd their young. There are several clans that make up the elephant populations. The elephants coexist in each clan under the direction of a single matriarch. The matriarch is often regarded as the oldest or fittest member of the clan.

The social behavior of fish colonies, such as looking for food and avoiding dangers, is described by the Fish Swarm Optimization Algorithm (FSOA). The excellent accuracy, adaptability, fault tolerance, and rapid convergence of this algorithm are only a few of its numerous benefits [43–47].

Grey Wolf Optimization (GWO) is drawn after ashen wolves and imitates its normal admonition construction and chasseing strategy [48].

Glowworm Swarm Optimization (GSO) was developed after studying glowworm behavior [49]. Each glowworm in the GSO algorithm [50–52] calculates a probabilistic cost for choosing the nearby glowworm with a greater luciferin intensity than its own.

The Moth Flame Optimization (MFO) method is based on the crossover tendency mode of navigation used by moths.

Particle Swarm Optimization (PSO): By repeatedly attempting to make a nominated solution better suit the specified compatibility task, PSO seeks to optimize a problem.

The Whale Optimization Algorithm (WOA) imitates how humpback whales hunt in the wild.

Researchers have already shown that bio-inspired methods, such as the cuckoo seek method, artificial bee colony, FSO, and PSO, act a considerable role in the interpretation of medical data [53–55].

10.1 Utilization of Bio-Inspired Algorithms in Various WSN Applications

These techniques can be used for the following different parameter optimizations in WSN:

(i) To determine the best route during routing.
(ii) To reduce the complexity of issues.
(iii) By identifying the problematic nodes in a wireless network, it improves security in wireless networks.
(iv) It optimizes energy usage.
(v) It contributes to extending the network's lifespan.

10.2 Issues and Challenges in Bio-Inspired Computing

Here is a summary of some of the main concerns with bio-inspired computing:

1. Theoretical evaluation
2. Dynamic Optimization
3. Optimization for multimodality
4. Topology
5. Global optimization on a large scale
6. Adjusting parameters
7. Human-centered software.

11 Intelligent Method TLBO for Cluster-Head Selection in Wireless Sensor Network

Training-learning-based optimization (TLBO) is a population-based evolutionary optimization algorithm. The principal concept behind TLBO is to simulate a traditional learning process in school, which involves the learner learning or acquiring data from educators or the educator transmitting data to students in the class. In this approach, a team of learners (i.e., students) is considered as a crowd and topics are considered as plan changeables of the optimization quandary. The output is gained in terms of mean, best, standard deviation, error, and number of function evaluations (FEs) during tests on the methods, which chiefly depends on the modality of the educator. The best out is shown as an educator in the whole crowd [56].

1. Instructor Period: Throughout the whole phase, the teacher is essential. The student's final average and improvement of their score are entirely reliant on the teacher. The population's top student is picked to serve as the teacher: (XTeacher)

2. Learner Phase: During this phase, students engage with one another and their teacher to learn new information. The primary distinction between the educator and student phases is that whereas, in the educator phase, teachers transfer information to students, in the learner phase students build upon their existing scholarship by taking classes.

This technique is admeasured into three phases including (i) network model, (ii) parameter formulation, and (iii) TLBO formula. In the first phase, it is used to show network information, in the second phase, the basic input parameters are discussed, and in the last phase, the TLBO formula is used to increase the network life cycle by optimizing multiple WSN objectives [56].

Various sensor node kinds are available to fulfill the network's stated goal of taking part in the operation. Has five different node-level kinds, with each level depending on the fundamental network characteristics that let the network accomplish its primary objective. There are six total radio ranges and five total node-level categories. Both are unique because different sensor nodes may vary depending on the situation.

Each node is located within a particular radio range depending on Euclidian distance (20).

$$Dist(n_i, n_j) = \sqrt{(x_j - x_i)^2 + (y_j - y_i)^2} \tag{20}$$

(x_i, y_i) and (x_j, y_j) are the address of notes n_i and n_j.

The first fundamental parameter is remaining energy, while the second is distance. With the aid of beginning energy and expended energy, the node's residual energy is determined (21). Energy used to transmit and receive packets is used to determine consumed energy (22).

$$Res_{eng} = Ini_{eng} - Con_{eng} \tag{21}$$

$$Con_{eng} = Eng_t + Eng_r \tag{22}$$

The interval among two incidents—(i) when the netting is initiated for a specific function and (ii) when the majority of the knots are exhausted—is referred to as the network lifespan. The set goals for each category are as follows:

a. Throughput: Demonstrates the number of information packets successfully received in a specific amount of time at the recipient end.
b. Packet Delivery Ratio: Shows the actual coefficient of packs that have been transferred and delivered at the receiver end.
c. Good put: Indicates that a data packet was successfully captured by the goal node in a specific time unit.
d. End-to-end Delay: This issue represents the overall time it takes a packet to get to a particular node.

e. Routing Overhead: Shows how many total control packets must be sent before one data packet may be successfully received.
f. The entire digit of packets dropped during a victorious information transfer is shown in Packet Loss.

12 Nature-Inspired Algorithm in Wireless Sensor Network for Perva-Sive Healthcare Applications

Wireless sensor networks (WSNs) have become more favorite, attracting the interest of both business and academics in the study and development of promising real-world solutions to assist sectors like healthcare and transportation. Numerous nodes that are linked to one another and carry out dispersed sensing duties make up a wireless sensor network. The nodes take data from the environment, transmit it via a network, and, depending on the capabilities at their disposal, could even be able to analyze the combined data from their peers. Possibly the most limiting feature of WSN design is the power supply issue. The problem of node power usage might, in principle, be reduced during transmission, reception, idle listening, or idle overhearing.

Due to the difficulties in this area, researchers have created methods that increase the lifespan of WSN while preserving the system's portability, low weight, and unobtrusiveness. An effective technique to accomplish this is by clustering the sensors into groups and designating one node as the cluster leader. However, the cluster heads' battery life is limited, and because of the increased duty, their energy is depleted more quickly. Here, the difficulty is in selecting a novel cluster head among the residual sensing motes in order to maintain network stability even when the head is damaged.

Advanced methods that address the cluster head selection problem have been widely explored using multi-objective evolutionary algorithms (MOEA), a branch of nature-inspired calculating that is frequently abbreviated.

The principal component of the networks is a big digit of sensing modules spread out across a certain area, with wireless technology being used for communication between the nodes. These nodes should be small, light, strong, and battery-powered so they may be utilized anywhere and maintain operation under challenging circumstances. Each node is made up of four essential parts: a controller, a transceiver unit, a sensing unit made up of one or more sensors and a processing unit, or any other external power source. The following are the several ways a sensor can operate: Transmission, reception, idle listening, and finally, sleep, come first.

Real-time systems are distinguished by three main elements. Time is the most valuable resource out of all of these. A real-time application is often made up of a group of interdependent activities that are called upon periodically and completed within a predetermined time frame. A real-time function's deadline can be categorized as follows: One strict deadline and two firm deadlines 3. Flexible deadline.

The following [57] design factors for the wireless sensor network subsystem in ubiquitous healthcare applications:

1. Node power consumption, 2. power runs out while transmitting, 3. the unobtrusiveness of the sensing apparatus, 4. platform portability, 5. real-time access to emergency medical services and the availability of sensors and processors, 6. reliable inter-element communication, 7. use of several hops for data relay, and 8. confidentiality and security of private medical data.

Natural systems have certain unique characteristics that serve as an inspiration and are essential in deciding the best course of action for artificial networking systems [13, 58]. In their paper "Intelligent Optimization of Wireless Sensor Network through Bio-inspired Computing: Survey and Future Directions," Sohail Jabbar and Rabia Iram [59] discuss the optimization method of bio-inspired Wireless Sensor Networks nodes, suggesting that one may improve the efficiency of computational functions from inflection, compatibility, decentralization, and fault tolerance with bio-inspired methods, which are based on the treatment of biological devices in temperament. One of the clue characteristics of biological devices that helps in determining the optimal answer is optimization.

12.1 Particle Swarm Optimization (PSO)

Eberhart and Kennedy introduced the PSO [60], a population-based optimization technique that was motivated by the gathering behavior displayed by fowls throughout the months of their migration. As they move about the issue area, a swarm of particles displays self-organization by responding to both affirmative and negative feedback.

The particles investigate the area around a favorable place rather than traveling directly toward it, giving variety to the particle movement. Each swarm member locates a solution in a higher-dimensional area that is constrained by the following four vectors:

(1) The particle's speed
(2) The top position thus far discovered
(3) The particle's current position vector
(4) The optimal position determined by nearby particles.

12.2 Ant Colony Optimization

Ant-like creatures in the wild use their sociable behavior to find meal wellsprings for their colony. An ant leaves a volatile pheromone trail on its route behind the colony after finding the food source. The further ants discover the food source by following this pheromone trail.

12.3 Artificial Immune System

Modeled after basic mammalian immunological activity. The primary goal of AIS is to effectively identify environmental changes or minute variations from the typical system behavior. In mammals, the immune system's job is to keep the corpse free from infections.

12.4 Framework for WSN Inspired by Plant Biology

For a WSN system, the structure primarily consists of three levels. The root system of the plant and its rhizosphere coordination rules are held in the first layer, while the plant's connection protocols with the microbial organism's rhizosphere are held in the second layer. The third layer is communicated with the base stations and the microbes' current communication mechanisms. The purpose of the leaves is to continuously probe for a certain incident from the sensory surroundings, similar to how real-world plants would hunt for resources like sunshine and other natural resources. In order to manufacture carbohydrates, vegetables attract carbon dioxide and juice molecules. Plants may create clusters where they can communicate with one another, sense threats in the atmosphere, and recognize the existence of other plants in the same colony. There are two types of chemotactic interactions between soil organisms: positive and negative. Positive interaction between plants and the rhizosphere is characterized by symbiotic relationships. The adverse connection is formed by parasitic associations among the host vegetable and other flora or wildlife.

13 Conclusion

Nowadays, the applications of using WSN are increasing rapidly. Wireless sensor networks (WSN) are composed of low-cost, low-power sensor nodes with wireless computing, processing, and communication capabilities. Wireless sensor networks are used due to the ease of deployment in any type of environment and a wide range of remote sensing applications such as animal monitoring, environment, health care, etc. In addition to advantages, these wireless sensor networks have challenges and limitations. Therefore, they need methods to overcome the challenges. One of the newest methods is the use of innovative techniques inspired by nature. Therefore, in this book, we explore some innovative techniques inspired by nature to solve wireless network challenges. One of the latest techniques to solve problems is to use an algorithm inspired by nature. The role of nature-inspired algorithms in achieving quality of service in WSNs has a good performance for significant parameters like energy, security, network lifetime, and coverage. TLBO algorithm is reviewed for the optimization of wireless sensor networks and nature-inspired algorithms for health care applications in wireless networks are discussed and TLBO algorithm for the optimization of wireless sensor networks is reviewed.

Acknowledgements Thank you for your cooperation and contribution.

References

1. Dener M (2017) WiSeN: a new sensor node for smart applications with wireless sensor networks. Comput Electr Eng 64:380–394
2. Kochhar A, Kumar N (2019) Wireless sensor networks for greenhouses: an end-to-end review. Comput Electron Agric 163:104877
3. Boukerche A, Sun P (2018) Connectivity and coverage based protocols for wireless sensor networks. Ad Hoc Netw 80:54–69
4. Dey N, Ashour AS, Shi F, Fong SJ, Sherratt RS (2017) Developing residential wireless sensor networks for ECG healthcare monitoring. IEEE Trans Consum Electron 63(4):442–449
5. Binh HTT, Hanh NT, Dey N (2018) Improved cuckoo search and chaotic flower pollination optimization algorithm for maximizing area coverage in wireless sensor networks. Neural Comput Appl 30(7):2305–2317
6. Roy S, Karjee J, Rawat US, Dey N (2016) Symmetric key encryption technique: a cellular automata-based approach in wireless sensor networks. Procedia Comput Sci 78:408–414
7. Barnawi AY, Mohsen GA, Shahra EQ (2019) Performance analysis of RPL protocol for data gathering applications in wireless sensor networks. Procedia Comput Sci 151:185–193
8. Fong S, Li J, SongW TY, Wong RK, Dey N (2018) Predicting unusual energy consumption events from smart home sensor network by data stream mining with misclassified recall. J Ambient Intell Humaniz Comput 9(4):1197–1221
9. Olakanmi OO, Dada A (2020) Wireless sensor networks (WSNs): Security and privacy issues and solutions. Wirel Mesh Netw-Secur, Arch Protoc, 13
10. Jino Ramson SR, Moni DJ (2017) Applications of wireless sensor networks—A survey. In: *2017* International Conference on Innovations in Electrical, Electronics, Instrumentation and Media Technology (ICEEIMT), Coimbatore, India, pp 325–329
11. Mukherjee A, Dey N, Kausar N, Ashour AS, Taiar R, Hassanien AE (2019) A disaster management specific mobility model for flying ad-hoc network. In: Emergency And Disaster Management: Concepts, Methodologies, Tools, and Applications. IGI Global, pp 279–311
12. Dey N (ed) (2017) Advancements in applied metaheuristic computing. IGI Global
13. Dey N, Ashour AS, Bhattacharyya S (2019) Applied nature-inspired computing: algorithms and case studies
14. Dey N, Ashour AS (2016) Antenna design and direction of arrival estimation in meta-heuristic paradigm: a review. Int J Serv Sci Manag Eng Technol (IJSSMET) 7(3):1–18
15. Mukherjee A et al (2019) A disaster management specific mobility model for flying ad-hoc network. In: Emergency and disaster management: concepts, methodologies, tools, and applications. IGI Global, pp 279–311
16. Swain RR, Khilar PM (2017) Composite fault diagnosis in wireless sensor networks using neural networks. Wireless Pers Commun 95(3):2507–2548
17. Panda M, Khilar PM (2015) Distributed self fault diagnosis algorithm for large scale wireless sensor networks using modified three sigma edit test. Ad Hoc Netw 25:170–184
18. Sahoo MN, Khilar PM (2014) Diagnosis of wireless sensor networks in presence of permanent and intermittent faults. Wireless Pers Commun 78(2):1571–1591
19. Mourad E, Nayak A (2012) Comparison-based system-level fault diagnosis: a neural network approach. IEEE Trans Parallel Distrib Syst 23(6):1047–1059

20. De Castro LN, Timmis J (2002) Artificial immune systems: a new computational intelligence approach. Springer Sci & Bus Media
21. Janeway CA et al (2001) The immune system in health and disease. Immunobiology. Curr Biol LteD
22. Jegadeeshwaran R, Sugumaran V (2015) Brake fault diagnosis using Clonal Selection Classification Algorithm (CSCA)—a statistical learning approach. Eng Sci Technol Int J 18(1):14–23
23. Mohapatra S,Khilar PM(2017) Artificial immune system based fault diagnosis in large wireless sensor network topology. In: Region 10 Conference (TENCON), IEEE
24. Mohapatra S, Khilar PM, Swain RR (2019) Fault diagnosis in wireless sensor network using clonal selection principle and probabilistic neural network approach. Int JCommun Syst :e4138
25. Taylor DW, Corne DW (2003) An investigation of the negative selection algorithm for fault detection in refrigeration systems. In: International conference on artificial immune systems. Springer, Heidelberg
26. de Abreu CCE, Duarte MAQ, Villarreal F (2017) An immunological approach based on the negative selection algorithm for real noise classification in speech signals. AEU-Int J Electron Commun 72:125–133
27. Jiang WK, Chen YJ, Zhang J (2013) A fault diagnosis method based on artificial immune network. In: Applied mechanics and materials, 385. Trans Tech Publications
28. Ishiguro A, Watanabe Y, Uchikawa Y (1994) Fault diagnosis of plant systems using immune networks. In: Proceedings of IEEE international conference on MFI'94. Multisensor fusion and integration for intelligent systems, IEEE
29. Dey N, Ashour AS, Bhattacharyya S (2019) Applied nature-inspired computing: algorithms and case studies, pp 1–275
30. Odili JB, KaharMNM, (2016) African buffalo optimization. Int J Soft EngComput Syst 2(1):28–50
31. Paul JD, Roberts GG, WhiteN, (2014) TheAfrican landscape through space and time. Tectonics 33(6):898–935
32. Lorenzen ED, Heller R, Siegismund HR (2012) Comparative phylogeography of African savannah ungulates 1. Mol Ecol 21(15):3656–3670
33. Kulkarni RV, Venayagamoorthy GK (2010) Particle swarm optimization in wireless-sensor networks: a brief survey. IEEE Trans Syst Man Cybern Part C Appl Rev 41(2):262–267
34. Aggarwal R, Mittal A, Kaur R (2016) Various optimization techniques used in wireless sensor networks. Int Res J Eng Technol 3(6):2085–2090
35. Mahdavi M, Fesanghary M, Damangir E (2007) An improved harmony search algorithm for solving optimization problems. Appl Math Comput 188:1567–1579
36. Tizhoosh HR (2005) Opposition-based learning: a new scheme for machine intelligence. Proceedings of International Conference on Computational Intelligence for Modelling, Control and Automation 1:695–701
37. Tizhoosh HR (2005) Reinforcement learning based on actions and opposite actions. In: Proceedings of ICGST international conference on artificial intelligence and machine learning, Egypt
38. Shiva CK, Mukherjee V (2016) Automatic generation control of hydropower systems using a novel quasi-oppositional harmony search algorithm. Electr Power Compon Syst 44(13):1478–1491
39. Karaboga D (2005) An idea based on honey bee swarm for numerical optimization. Technical report-tr06. Erciyes university, engineering faculty, computer engineering department 200:1–10
40. Yang XS (2010) A new metaheuristic bat-inspired algorithm. Nature inspired cooperative strategies for optimization (NICSO 2010). Springer, Berlin, Heidelberg, pp 65–74
41. Chu SC, Tsai PW, Pan JS (2006) Cat swarm optimization. Pacific Rim international conference on artificial intelligence. Springer, Berlin, Heidelberg, pp 854–858
42. Meng X, Liu Y, Gao X, Zhang H (2014) A new bio-inspired algorithm: chicken swarm optimization. International Conference in Swarm Intelligence. Springer, Cham, pp 86–94

43. Neshat M, Sepidnam G, Sargolzaei M, Toosi AN (2014) Artificial fish swarm algorithm: a survey of the state-of-the-art, hybridization, combinatorial and indicative applications. Artif Intell Rev 42(4):965–997
44. Zheng ZX, Li JQ, Duan PY (2019) Optimal chiller loading by improved artificial fish swarm algorithm for energy saving. Math Comput Simul 155:227–243
45. Qin N, Xu J (2018) An adaptive fish swarm-based mobile coverage in WSNs.Wirel Commun Mob Comput
46. Li X, Keegan B, Mtenzi F (2018) Energy efficient hybrid routing protocol based on the artificial fish swarm algorithm and ant colony optimisation for WSNs. Sensors 18(10):3351
47. Yin H, Zhang Y, He X (2018) WSN nodes placement optimization based on a weighted centroid artificial fish swarm algorithm. Algorithms 11(10):147
48. Kohli M, Arora S (2018) Chaotic grey wolf optimization algorithm for constrained optimization problems J. Comput Des Eng 5(4):458–472
49. Krishnanand KN, Ghose D (2005) Detection of multiple source positions using a glowworm metaphor with applications to collective robotics. In: Proceedings 2005 IEEE Swarm Intelligence Symposium, SIS 2005, IEEE. pp 84–91
50. Krishnanand KN, Ghose D (2009) Glowworm swarm optimization for simultaneous capture of multiple local optima of multimodal functions. Swarm Intell 3(2):87–124
51. Krishnanand KN, Ghose D (2008) Theoretical foundations for rendezvous of glowworminspired agent swarms at multiple positions. Robot Auton Syst 56(7):549–569
52. Liao WH, Kao Y, Li YS (2011) A sensor deployment approach using glowworm swarm optimization algorithm in wireless sensor networks. Expert Syst Appl 38(10):12180–12188
53. Menad H, Amine A (2018) Bio-inspired algorithms for medical data analysis. In: Handbook of Research on Biomimicry in Information Retrieval and Knowledge Management. IGI Global, pp 251–275
54. Giveki D, Salimi H, Bahmanyar G, Khademian Y (2012) Automatic detection of diabetes diagnosis using feature weighted support vector machines based on mutual information and modified cuckoo search. arXiv preprint arXiv:1201.2173
55. Ashour AS, Samanta S, Dey N, Kausar N, Abdessalemkaraa WB, Hassanien AE (2015) Computed tomography image enhancement using cuckoo search: a log transform based approach. J Signal Inform Process 6(03):244
56. Malakar M (2020). TLBO based cluster-head selection for multi-objective optimization in wireless sensor networks. In Nat Inspired Comput Wirel Sens Netw (pp 303–319). Springer, Singapore
57. Boitier V, Tajan P, Dilhac JM (2016) WSN nodes: design considerations and energy management. J Phys Conf Ser 773:012043. https://doi.org/10.1088/1742-6596/773/1/01204
58. Dey N (2017) Advancements in applied metaheuristic computing
59. Shehzad S (2013) Intelligent optimization of wireless sensor networks through bio-inspired computing: survey and future directions. Int J Distrib Sens Netw 2013:13
60. Kennedy J, Eberhart R (1995) Particle swarm optimization. IEEE Int Conf Neutral Networks
61. Variengien A, Pontes-Filho S, Glover TE, Nichele S (2021) Towards Self-organized control: Using neural cellular automata to robustly control a Cart-pole agent. Innov Mach Intell (IMI) 1:1–14. https://doi.org/10.54854/imi2021.01
62. Takano H, Iwase N, Nakayama N, Asano H (2021) Decision support in maintenance for electrical components of distribution networks. Innov Mach Intell (IMI) 2:1–11. https://doi.org/10.54854/imi2022.001
63. Senjyu T, Khosravy M (2022) Power System planning and quality control. Energies 15(14):4995
64. Jalalzad SH, Yektamoghadam H, Haghighi R, Dehghani M, Nikoofard A, Khosravy M, Senjyu T (2022) A game theory approach using the TLBO algorithm for generation expansion planning by applying carbon curtailment policy. Energies 15(3):1172
65. "A Bi-level Evolutionary Optimization for Coordinated Transmission Expansion Planning" and "Incremental conductance based particle swarm optimization algorithm for global maximum power tracking of solar-pv under nonuniform operating conditions"

66. Khosravy M, Nakamura K, Hirose Y, Nitta N, Babaguchi N (2022) Model inversion attack by integration of deep generative models: Privacy-sensitive face generation from a face recognition system. IEEE Trans Inf Forensics Secur 17:357–372
67. Khosravy M, Nakamura K, Hirose Y, Nitta N, Babaguchi N (2021) Model inversion attack: analysis under gray-box scenario on deep learning based face recognition system. KSII Trans Internet Inf Syst (TIIS) 15(3):1100–1118
68. Khosravy M, Nakamura K, Nitta N, Babaguchi N (2020). Deep face recognizer privacy attack: Model inversion initialization by a deep generative adversarial data space discriminator. In 2020 Asia-Pacific Signal and Information Processing Association Annual Summit and Conference (APSIPA ASC) IEEE. (pp 1400–1405)
69. Joshi A, Khosravy M, Gupta N (2020) Machine learning for predictive analysis. In: Proceedings of ICTIS
70. Gupta N, Khosravy M, Gupta S, Dey N, Crespo RG (2022) Lightweight artificial intelligence technology for health diagnosis of agriculture vehicles: parallel evolving artificial neural networks by genetic algorithm. Int J Parallel Program, 1–26
71. Gupta N, Khosravy M, Patel N, Dey N, Crespo RG (2021) Lightweight computational intelligence for IoT health monitoring of off-road vehicles: Enhanced selection log-scaled mutation GA structured ANN. IEEE Trans Industr Inf 18(1):611–619
72. Gupta, Neeraj, Mahdi Khosravy, Nilesh Patel, Nilanjan Dey, Saurabh Gupta, Hemant Darbari, Rubén González Crespo (2020) Economic data analytic AI technique on IoT edge devices for health monitoring of agriculture machines. Appl Intell 50: 3990–4016
73. Dehghani M et al. (2020) An Improved PSO-GA based fuzzy controller of STATCOM for cascaded induction motors of drilling system. In: 2020 IEEE International Conference on Environment and Electrical Engineering and 2020 IEEE Industrial and Commercial Power Systems Europe (EEEIC/I&CPS Europe), Madrid, Spain, pp. 1–6
74. Dehghani M, Taghipour M, Sadeghi Gougheri S, Nikoofard A, Gharehpetian GB, Khosravy M (2021) A Deep Learning-Based approach for generation expansion planning considering power plants lifetime. Energies 14:8035

Chapter 13
Optimization of Robust Image Watermarking Based on DCT and Decomposition Using Genetic Algorithm

Muhammad Alief Yudha Pratama, Ledya Novamizanti, and Gelar Budiman

1 Introduction

Internet is one of the main media to exchange digital information. However, the technology also has a negative impact on the exchange of information. Information becomes more vulnerable to data distortion, such as manipulation [1], theft [2, 3], and unauthorized use [4]. To maintain the security and authenticity of the digital image, it is necessary to insert additional information into the host or original image. The process of embedding information is also called digital watermarking, and the process protects the digital image from various attacks that make the image quality decrease and even damage [1, 5, 6]. The watermarking process inserts information in binary format into the host image pixel value [7].

In watermarking techniques, the information that is usually inserted is in the form of digital data, such as text, audio, imagery, and video [8, 9], which can be an authentication of image ownership [1, 10]. Evaluation of the main watermarking techniques is imperceptibility and robustness [11]. Imperceptibility is a criterion in the embedding results based on the human visual system, and there is no change between the watermarked image and the host image [12]. Robustness is a criterion for evaluating the original and extracted watermarks with various attacks [4, 12]. The watermarking technique is separated into two domain bases: the spatial domain and the transform domain [8, 13]. Spatial domains work on pixels by modifying pixel values, while

M. A. Y. Pratama · L. Novamizanti (✉) · G. Budiman
School of Electrical Engineering, Telkom University, Bandung, Indonesia
e-mail: ledyaldn@telkomuniversity.ac.id

G. Budiman
e-mail: gelarbudiman@telkomuniversity.ac.id

M. Khosravy et al. (eds.), *Frontiers in Genetics Algorithm Theory and Applications*, Springer Tracts in Nature-Inspired Computing,
https://doi.org/10.1007/978-981-99-8107-6_13

transformation domains work in transformation coefficients by modifying the transformation coefficient [8]. Commonly used transformations in watermarking schemes are Discrete Cosine Transform (DCT) [14, 15], Discrete Wavelet Transform (DWT) [14, 16], and Discrete Fourier Transform (DFT) [8, 17]. For matrix decomposition, Hessenberg decomposition (HD) and singular value decomposition (SVD) are widely used in digital image hiding [4]. SVD is used because the image has good stability [18] and is often used in wavelet domains. SVD can be used separately or in combination with transformations on watermarking [16, 19], whereas HD is widely used because it has lower computational complexity than other matrix decompositions [5].

The digital image hiding technique using Hybrid DWT-HD-SVD proposed by J. Liu, et al. [4] proves that a watermarking method in the DWT domain is difficult to resist geometric attacks. Therefore, using the DWT method requires matrix decomposition to withstand image processing such as HD and SVD from geometric attacks such as cropping, scaling, and rotating. In addition to matrix decomposition, combining two or more transformations can also improve watermarking performance that prioritizes the robustness of geometry attacks [13].

This study proposes watermarking techniques on images in the transformation domain using Hybrid DWT, DCT, HD, and SVD. The difference in the previous study [4] is the addition of a DCT transformation method which aims to improve watermark resistance, especially in geometric attacks, which are then tested for performance. This technique produces robust watermarking from various attacks, especially geometry attacks and copyright protection.

Abdulrahman and Ozturk [14] proposed a hybrid transformation method or the merger of two or more transformations to maintain resilience from various attacks, security, and increasing embedding capacity. The incorporation of DWT-DCT transformations has also been performed on pepper [15], resulting in a watermarking system with robustness-focused performance, particularly on geometric attacks that show good NC and BER values in geometry attacks. DWT is used because this method has several advantages, such as good energy compression and good quality imperceptibility in digital image watermarking [4]. DWT also divides the image into several frequency sub-bands to determine the embedding area of the watermark [13]. DCT is used because it has good properties in concentrating image energy to a small coefficient (compaction energy) and is used to calculate the quantity of image bits where information is hidden in it [14].

This paper uses a genetic algorithm (GA) to discover the optimal parameters in the watermarking technique. GA is a computational model that mimics an evolutionary model based on natural selection and natural genetics which performs various processes such as recombination, mutation and selection which are commonly used for function optimization cases [20]. Genetics in this case means getting the best parameters or individuals from a population with a collection of individuals. Then the individual becomes the solution to the original problem. This method serves as a nature-inspired computational technique and finds applications in various industrial domains such as self-organized control [21], energy management [22], cyber security [23, 24], and predictive analysis [25].

The primary contributions of this article are: (1) watermarking schemes are carried out on multiple sizes of the watermark with results in good imperceptibility and robustness on multiple sizes of the image; (2) DCT is utilized to split chosen sub-bands of the DWT on the original image into distinct and separate blocks that do not overlap; (3) GA is used to search for the most optimal alpha value, sub-bands, and decomposition levels to obtain good quality of the watermarking system regarding imperceptibility and robustness; (4) better robustness of the suggested technique compared to some state-of-the-art works, under filter attack, noise attack, compression attack, geometric attack, and signal processing attack; (5) The proposed method exhibits greater levels of invisibility compared to certain state-of-the-art techniques.

2 Related Work

Liu et al. [4] investigated a watermarking technique using the DWT-HD-SVD method, which was optimized using fruit fly optimization. The watermark image used is in the form of a copyright logo with three image sizes: 64×64, 128×128, and 256×256 pixels. The SVD in this watermarking technique is also applied to watermark images. In numerous experimental scenarios, attacks are utilized to identify the best scaling factor using fly fruit optimization techniques. This study showed good results between robustness and invisibility with various sizes of watermark images. The average PSNR value of 38.16 dB is on all three watermark image sizes and NC 1. However, this study showed poor results in geometric strikes.

Su et al. [5] proposed blind watermarking using Hessenberg's decomposition method and Contourlet transformation to protect copyright on coloured imagery. Colour imagery goes through the process of contourlet transformation, and its low-frequency sub-bands are divided into blocks of 4×4 coefficients that do not overlap. Once separated into blocks, the Hessenberg method decomposes the selected block of coefficients. The proposed method can prove that the sub-band with low frequency used can increase the robustness of the watermark, and the maximum energy value of the matrix obtained from Hessenberg decomposition can increase imperceptibility.

Jain et al. [6] designed a robust watermarking technique using the DCT-SVD methods on textured images with Arnold Transformations. Watermark images in the form of logos go through a compression process using SVD so that the diagonal coefficient is significant enough to be embedded in the host image. SVD is used on watermark images because the algorithm also qualifies for imperceptibility. The experiment resulted in watermarking on images with high imperceptibility on all types of images tested and good watermark resistance in the event of multiple attacks such as rotation, cropping, and mean filters.

Singh and Goel [11] developed image watermarking techniques to improve the invisibility and robustness of watermarking using DWT-SVD-based hybrid transformations. The LH and HL sub-bands acquired from DWT transformation are then subjected to SVD. The embedding of the watermark image is not done at the wavelet coefficient but the singular value of the DWT sub-band on the host image. The study's results proved that the method used provides resistance in watermarking images, and the resistance is also influenced by the magnitude of the scale factor used. If the scale factor used is larger, the resilience of the watermarking scheme is stronger and vice versa. This research also proves that the method used can recover watermark images fairly of high quality.

In a previous study conducted by D. Maheshwari [26] it was proposed that watermarking techniques on multiple images using SVD were based on DWT and also examined the use of SVD and DWT separately with the same research objectives. The double image used is a watermark with two image insertion in one host image. Furthermore, the host image goes through a DWT process first to divide the image into four frequency sub-bands after which SVD is applied to each sub-band. The two watermarked images are embedded one by one. Analysis of the system with the DWT-SVD method showed that DWT had a better invisibility than SVD, which was tested with various attacks such as rotation, cropping, histogram equalization, etc. Experiments using the hybrid DWT-SVD method showed that DWT and SVD have different characteristics, so the combination of the two methods has higher invisibility and correlation values and testing values compared to the single method DWT or SVD and provides more resilience when attacks occur, such as geometric attacks and other image processing manipulations.

Watermarking schemes are intended for security and authentication so that they can protect digital images. A robust digital image watermarking was performed by L. Novamizanti et al. [19] using FDCuT-DCT-SVD with a semi-blind watermarking technique. Embedding is done by dividing the host image into 3 sub-bands using FDCuT at high frequency, which will be continued using DCT. DCT is performed simultaneously on the host image and watermark image, which will then be decomposed using SVD and modified by replacing the host image's singular value in the watermark image's singular value. The results of the watermarking scheme show results both dealing with invisibility and robustness, but in experiments with various attacks, there are watermark extraction results that can be said to be vulnerable to these various attacks, namely with geometry attacks.

An algorithm for invisible digital watermarking was suggested by Wu and colleagues in their work [27]. The decomposition of DWT level 4, DCT, and SVD serves as the method's foundation. The watermark image is used in random using Arnold transform and Logistic map. Embedding is done with 4 levels of transform on a 2-dimensional host image and performs DCT on HH sub-band DWT. Two inputs, host image and watermark are also applied on SVD to change the eigenvalue of the random watermark into a eigenvalue of the host image matrix coefficients. The test results showed that the watermarking scheme has a good quality of imperceptibility where the PSNR is above 45 dB, but an experiment to see the robustness of the watermarking scheme with JPEG compression attacks (QF 30) showed a low NC result with a value of 0.6651 which indicated that the watermarking scheme was vulnerable to such attacks.

A colour image watermarking technique with a focus on robustness and imperceptibility utilizing DCT and genetic algorithms was proposed by Agarwal and Singh [28]. DCT breaks the host image into 88 blocks for the embedding process, and genetic algorithms are employed to discover the best set of pixels to embed to provide high-quality watermarking. The test findings reveal that the suggested watermarking method, which employs a host image of 512 × 512 pixels and a watermark image of 32 × 32 pixels, produces a PSNR of more than 55 dB in each colour component, including red, green, and blue. Robustness testing on the proposed watermarking scheme showed that Gaussian attacks of 0.001 and salt & pepper noise produced NC worth 0.95, and JPEG compression QF 70 produced NC 0.94 on each colour component.

The crypto-watermarking research proposed by Al-Haj et al. [29] on using SVD in the DWT domain on medical images aimed at maintaining the confidentiality and authenticity of medical images through exchanging medical information. The medical image used is separated into two regional zones: ROI and RONI. The ROI region cannot be used because it is a region with important information for diagnosis. All sub-bands on the imagery are used for the proposed scheme. The LH sub-band is used to embed patient information, the HL and LH sub-bands are for embedding the hospital logo, and the HH sub-band is for cryptographic hash watermarks. The results acquired in this work are that the developed algorithm can withstand various attacks carried out, and watermarks in the form of patient information and 5 hospital logos can be well identified to authenticate the image's ownership and source of origin with an NC value above 0.96.

Abduldaim et al. [30] proposed watermarking techniques on digital images using Hessenberg Matrix Decomposition (HMD) and DWT. The coloured host image is changed into a grayscale image, which is then DWT transformed and will then be applied as HMD on each block. After passing the HMD process, it will be separated into a 4 × 4 H and P matrix. The watermark binary image is embedded in the H sub-matrix that has gone through the HMD process. By testing NC, which produces a value of 1, it means that the genuine and recovered watermark images are of the same quality. However, testing was also carried out using various attacks, and the suggested method gave an NC value of 1% on JPEG attack testing, and an unfavourable NC resulted in an equalization histogram attack with an NC value below 0.6.

Sejpal and Borse [31] proposed a blind watermarking technique on colour imagery using Hessenberg decomposition and Schur decomposition more aimed at watermarking security. Sejpal et al. compared watermarking techniques using such decomposition. The watermark image in the form of QR is scrambled using the Fibonacci-Lucas transformation into bits. After that, the host image matrix will be performed by Hessenberg decomposition or Schur decomposition on each block previously separated into blocks of 4 × 4 size without overlapping and performed on each watermark bit. The final results show both decompositions result in good quality requirements in image watermarking techniques, but Hessenberg decomposition has better watermark quality and resistance. Testing using various attacks was also carried out, but both decompositions produced poor quality when a blurring attack was carried out with an NC value of 0.6 in Hessenberg decomposition and 0.5 in Schur decomposition.

The remaining part of this paper is explained in Sect. 3, which describes the basic concepts of DWT, DCT, HD, SVD, and genetic algorithms. Section 4 describes the proposed watermark method using DWT-DCT-HD-SVD and GA in more detail. The results of the experiment and the analysis of watermarking performance are present in Sect. 5. Section 6 provides the conclusion of this paper.

3 Preliminaries

In this section, two methods in the transformation domain (DWT and DCT), and two decomposition-based methods (HD and SVD), and GA used in the respective proposed watermarking methods are introduced as basic concepts. DWT and DCT transform the host image into several sub-frequencies, which are then HD and SVD used in the matrix transform to improve the quality of imperceptibility. Finally, the watermarking scheme tested will be improved again regarding imperceptibility and robustness by optimizing parameters using genetic algorithms.

3.1 Discrete Wavelet Transform

DWT is a wavelet transformation that provides ease of compression, transmission, and image analysis [11]. The approximation sub-band (LL) provides the low-frequency components of the image, while the horizontal (LH), vertical (HL), and diagonal (HH) sub-bands consist of the high-frequency components in the horizontal, vertical, and diagonal orientations, respectively. The DWT is used to decay an image into four non-overlapping frequency sub-bands [1]. DWT is used in watermarking schemes because it has good frequency properties in recognizing watermark embedding areas with high imperceptibility [11]. Based on the visual system, DWT is the best transformation compared to DCT because it provides better visual quality and can model a Human Visual System (HVS) more accurately.

DWT with Block A is M × N, where the contents of M and N are the pixel intensity values in the row and column positions. M × N is defined (Fig. 1):

$$A = \begin{bmatrix} \mathrm{a}\ \mathrm{b} \\ \mathrm{c}\ \mathrm{d} \end{bmatrix} \tag{1}$$

where DWT processes work non-overlappingly using a matrix of size 2 × 2 as follows [14]:

$$\mathrm{LL} = \frac{1}{2}(\mathrm{a} + \mathrm{b} + \mathrm{c} + \mathrm{d})$$

$$\mathrm{LH} = \frac{1}{2}(\mathrm{a} - \mathrm{b} + \mathrm{c} + \mathrm{d})$$

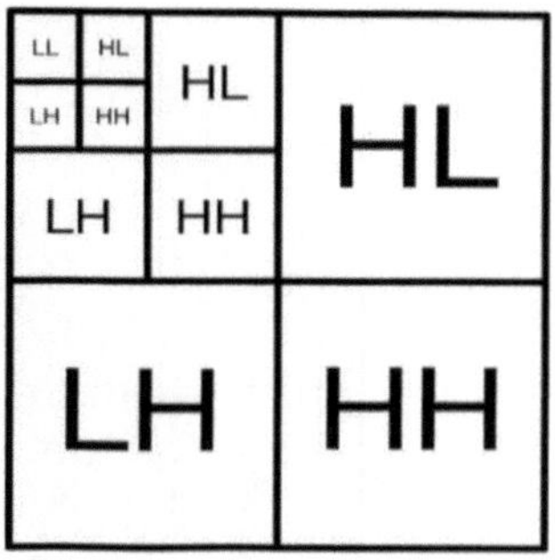

Fig. 1 Three level DWT scheme

$$HL = \frac{1}{2}(a + b - c + d)$$

$$HH = \frac{1}{2}(a + b + c - d) \tag{2}$$

3.2 *Discrete Cosine Transform*

Signals are frequently transformed from the spatial domain (or the 2D domain for images) to the frequency domain using the mathematical process known as DCT. In the frequency domain, the signal can be analyzed and manipulated more easily using various signal processing techniques. DCT is widely used for data compression, pattern recognition, and other image processing [28]. Watermarking with a DCT domain works by dividing the image into several 8 × 8 blocks that do not overlap each other, which then converts those blocks into 64 different DCT coefficients by dividing the image into various sub-frequencies—low (LF), middle (MF), and high (HF) [32]. The advantages of transformation with DCT are better computational efficiency, lower computing costs, and high image compression capabilities [33].

Based on the result of DCT coefficients, the DCT-based coding scheme will go through the zigzag path for zigzag sorting and generate a vector that represents the order of DCT coefficients from LF to HF [28], as shown in Fig. 2. The watermark is embedded by selecting pixels from the vector of zigzag results. [29].

In the DCT transformation, the upper left part is a LF region, while the lower right is a HF region that represents the sharp details of an image that can be easily removed with blurring attacks and Gaussian noise [34]. Hence, to acquire reliable watermarking techniques that are resistant to multiple attacks, the watermark or secret message is encoded in the middle-frequency DCT coefficient (MF). The DCT transformation with a size of 8 × 8 is written in the following calculation [32]:

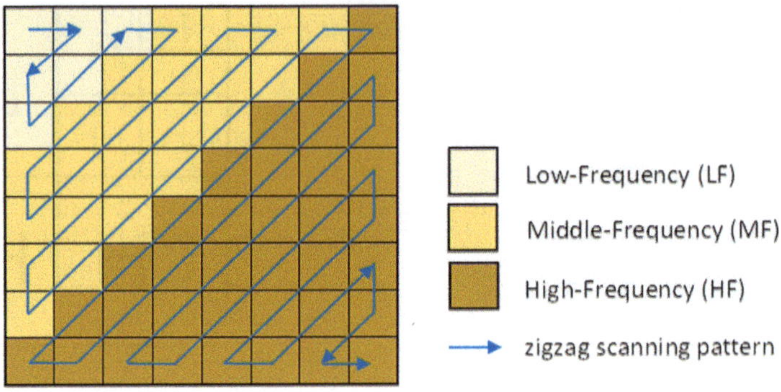

Fig. 2 DCT block region with a zigzag scan pattern

$$F(u, v) = \frac{1}{4}c(u)c(v)\sum_{x=0}^{7}\sum_{y=0}^{7} f(x, y)\cos\left[\frac{(2x+1)u\pi}{16}\right]\cos\left[\frac{(2y+1)u\pi}{16}\right] \quad (3)$$

where, u = 0, 1, 2,…,7, v = 0, 1, 2,…,7 and $c(k) = 1/\sqrt{2}$ for k = 0 and c(k) = 1 for k is otherwise.

3.3 Hessenberg Decomposition

HD is a method for decomposing square matrices with an order of $n \times n$ [5].

The second row and third row in the second column of the HD orthogonal matrix can be modified for the watermark embedding area [35]. There are several advantages of using HD in the watermarking scheme. When HD has functioned as a matrix transformation, the watermark resistance increases. Compared to other decomposition techniques, HD exhibits a reduced level of computational complexity.

HD is defined as an orthogonal HD matrix as follows [4]:

$$HD = PHP^{T} \quad (4)$$

with

$$P = (Q_1Q_2 \ldots Q_{n-2})^{T}/X(Q_1Q_2 \ldots Q_{n-2}) \quad (5)$$

$$Q = (I_n - 2\mu\mu^{T})/\mu^{T}\mu \quad (6)$$

$$H = P^{T}XP \quad (7)$$

$$X = PHP^T \tag{8}$$

where the matrix P is orthogonal, and the matrix H is in upper Hessenberg form. $h_{ij} = 0$ when $i > j + 1$. HD is usually calculated by the householder matrix Q expressed in eq. (6) when μ is a non-zero vector in R^n and I^n is an identity matrix with the order $n \times n$. There is $n - 2$ which is the step in the whole procedure, then HD is calculated on eq. (7).

3.4 Singular Value Decomposition

SVD is an operation in linear algebra that is used to break a symmetric matrix into three submatrices so that singular values can be obtained in the form of a diagonal matrix. The three submatrices are an orthogonal matrix (U) of size $M \times N$, a diagonal matrix (S) of size $M \times N$, and an orthogonal matrix (V) of size $N \times N$ [4]. The basis for decomposition using SVD is finding singular values in the host image, then modifying these values for watermark embedding [11]. The mathematical representation of the matrix A with the order $M \times N$ is as follows [19]:

$$A = USV^T \tag{9}$$

SVD has advantages and disadvantages in the use of decomposition in watermarking schemes. The advantage of SVD decomposition is that the single embedding that SVD uses does not change the visual quality of the original image, SVD has good energy compaction properties [4]. SVD can be applied to non-square matrices because of its non-symmetric decomposition properties [36]. Meanwhile, the disadvantage of SVD decomposition is an error in extracting messages due to an attack that damages the watermark text. As a result, the watermark bit is moved to another detection region. One drawback of the watermark extraction process is the occurrence of a false-negative error, which happens when the watermark is not detected. The other disadvantage is a false-positive error, which takes place when the watermark detector identifies a false watermark signal instead of the genuine one, after detecting an error [26].

3.5 Genetic Algorithm

Genetic algorithms are applied to watermarking to search for optimal parameters to improve watermarking performance. Genetic algorithms are computational models that imitate evolutionary models based on natural selection and natural genetics that carry out various processes such as recombination, mutation, and selection commonly used for function optimization cases. The genetics of this algorithm name

Individu	Fitness
1	0,7
2	0,3
3	0,3
4	0,6
5	0,7
6	0,3
7	0,3
8	0,6

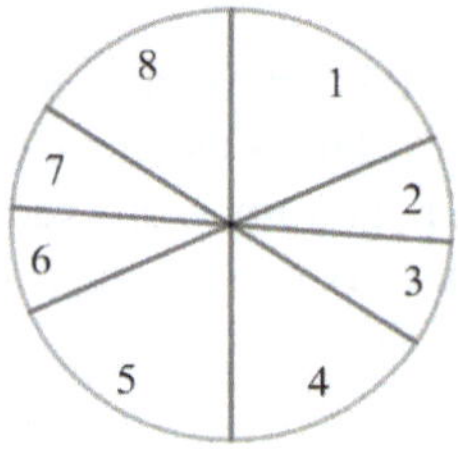

(a)

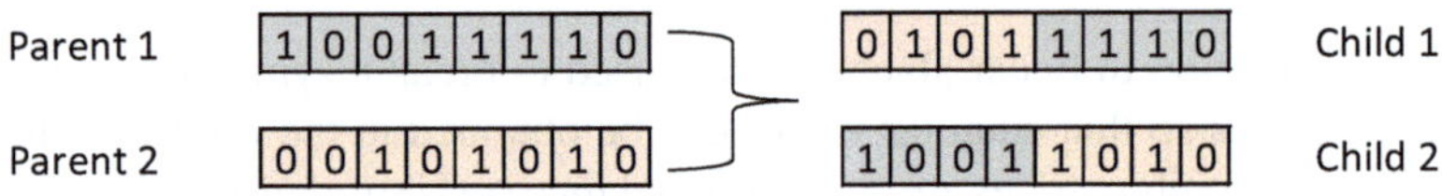

(b)

Fig. 3 **a** Parental selection, **b** Crossover

means getting the best parameter or individual from a population with a collection of individuals. Then the individual becomes the solution to an initial problem [28].

The process of a genetic algorithm starts with a population containing chromosomes or a series of gene values from a potential solution. From the population, two members of the population (parents) will be selected, which are then mated by cross-over to produce a potential solution or child (see Fig. 3). After obtaining a solution candidate, the child undergoes a mutation which is then seen whether it is appropriate for the population. If so, the child replaces the members of the population with a replacement scheme. The process repeats until a condition is reached or has reached a certain total iteration. The following are the components contained in the genetic algorithm [20]:

- Fitness function

  In genetic algorithms, an evaluation function is needed that can assess the advantages of an individual or prospective solution. The feasibility value is displayed in the fitness function. Individuals with high fitness scores survive, and individuals with low fitness scores will die. This condition is under the theory of evolution, where the superior will survive. The same is the case with the fitness function in genetic algorithms to get the best solution from a population.
- Reproduction

  After being assessed through the fitness function, individuals with good fitness scores in their population are chosen to be parent individuals. Reproduction in GA is the process by which individuals are reused into individuals ready to become parents using cross-over and mutation mechanisms. Reproduction with the selection of parents helps provide reproductive opportunities for the best individuals in a population. There are various ways to choose parents, one of which is with the Roulette Wheel.

- Crossover

  Crossover is one way to get an individual with better fitness scores than his parents. Cross-switching occurs by exchanging genes between individual parents by looking for cut points between the genes of cross-transferred parents, after which the genes are exchanged for the position. The number of cut points on the cross move affects the quality of the solution obtained. The more cross-moving operations, the more likely it is those good chromosomes will become damaged.
- Mutation

  The mutation is an operation to restore bit information lost due to the previous cross-switching. Mutations are used with a very small probability because if mutations are carried out too often, the resulting individual becomes weak because the configuration of the bits on the good chromosomes will be damaged. Mutations are performed by generating random values with intervals of 0 to 1 by flipping genes worth 1 to 0 and vice versa.
- Selection

  In the genetic algorithm, there are two types of individual replacement or selection in the population, namely steady-state and generational replacement. A steady state is the process of replacement when two children are produced from the result of cross over.

4 Proposed Method

This section is the design of a watermarking method on digital images that uses decomposition based on wavelet decomposition using the DWT-DCT-HD-SVD method with the optimization using genetic algorithms. The general watermarking system, as depicted in Fig. 4, consists of two primary processes: the embedding and the extraction. This paper employs an embedding process to insert the watermark information or watermark image into the host image, resulting in a watermarked image. The extraction process involves separating the watermarked image to restore the embedded information. The robustness of the watermarking system can be assessed by subjecting the watermarked images to assorted attacks.

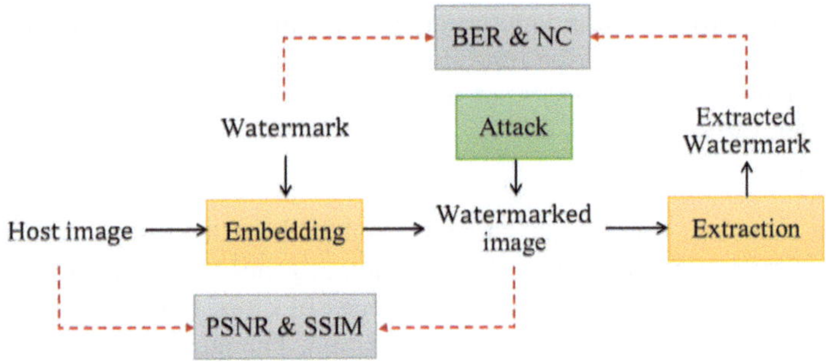

Fig. 4 Watermarking general scheme

4.1 Embedding Process

The embedding method takes the original image or host X and the watermark W as its inputs. The dimensions of X and W are N × N and M × M, respectively. The result obtained from utilizing the embedding technique is X*, which has a dimension of M × M. In the test, three watermark sizes are used that will be adjusted to the decomposition level used at the R-Level DWT on the host image. Figure 5a illustrates the embedding process

The watermark embedding procedure:

Step 1: Perform an R-Level DWT transformation on the host image to divide the image into four different frequency sub-bands: LL, LH, HL, and HH with the

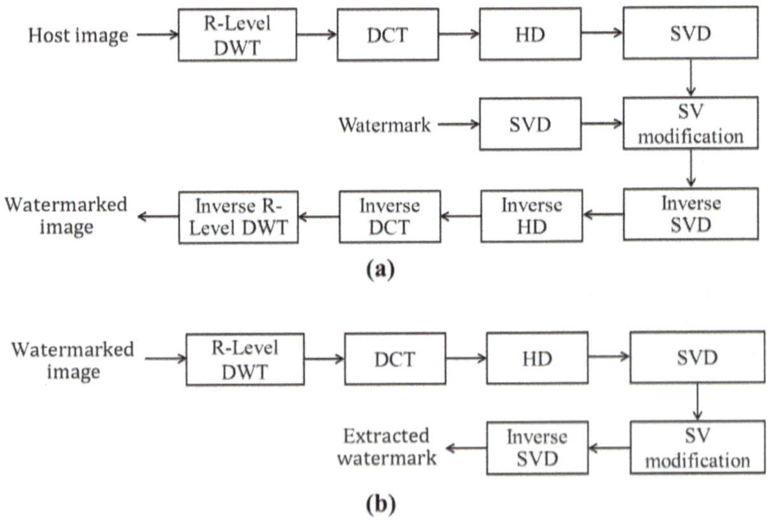

Fig. 5 Proposed scheme **a** Embedding, **b** Extraction

following equation.

$$\mathrm{R} = \log_2 \frac{\mathrm{M}}{\mathrm{N}} \tag{10}$$

$$\mathrm{LL_X LH_X HL_X HH_X} = \mathbf{DWT}(\mathrm{X}) \tag{11}$$

where R is the DWT level.

Step 2: Conduct the DCT transformation to the LL sub-band by scanning the image zigzag to get the DCT coefficient using the following formula:

$$\mathrm{D} = \mathbf{DCT}(\mathrm{LL_X}) \tag{12}$$

where D is the DCT coefficient for the LL sub-band of the host image.

Step 3: Perform HD decomposition to determine the matrix *P*, *H* on D.

$$PHP^{\mathrm{T}} = \mathbf{HD}(\mathrm{D}) \tag{13}$$

Step 4: Perform SVD decomposition to determine the U_H, S_H and V_H matrix values on the *H* matrix obtained from the previous decomposition.

$$U_H S_H V_H^T = \mathbf{SVD}(H) \tag{14}$$

Step 5: Apply SVD decomposition on the watermark image to determined U_W, S_W and V_W matrix.

$$U_W S_W V_W^T = \mathbf{SVD}(\mathrm{W}) \tag{15}$$

Step 7: Modify the singular value (SV modification) with the formula:

$$\mathrm{S_{md}} = S_H + \alpha \times S_W \tag{16}$$

Step 8: Inverse SVD, HD, R-Level DWT and DCT.

Step 9: Watermarked image X* is obtained.

4.2 Extraction Process

The extraction procedure involves taking an X* watermarked image as input and producing a set of N × N extracted watermarks W* as output. The steps for extracting the watermarks are illustrated in Fig. 5b:

Step 1: The watermarked imagery performed R-Level DWT with R using the Eq. (9) to isolate the image into four frequency sub-bands, namely LL_{X*}, LH_{X*}, HL_{X*}, and HH_{X*} with the following equation.

$$\mathrm{LL}_{X^*}\mathrm{LH}_{X^*}\mathrm{HL}_{X^*}\mathrm{HH}_{X^*} = \mathbf{DWT}\left(X^*\right) \tag{17}$$

Step 2: Apply DCT transformation on LL_{X*} with formula:

$$D^* = \mathbf{DCT}(\mathrm{LL}_{X^*}) \tag{18}$$

Step 3: Perform HD decomposition with the following equation on D^*.

$$\mathrm{P}_{D^*}\mathrm{H}_{D^*}\mathrm{P}_{D^*}^{\mathrm{T}} = \mathbf{HD}(\mathrm{H}_{D^*}) \tag{19}$$

Step 4: Perform SVD decomposition on H_{D^*} to determine the matrix values of U_{HD*}, S_{HD*}, V_{HD*}, and V_{HD*} with the following equation.

$$\boldsymbol{U}_{HD^*}\boldsymbol{S}_{HD^*}\boldsymbol{V}_{HD^*}^{T} = \mathbf{SVD}(\mathrm{H}_{D^*}) \tag{20}$$

Step 5: Modify the S value (SV modification) with the formula:

$$\mathbf{S}_{\mathrm{md}}^{*} = \frac{\boldsymbol{S}_{HD^*} - \boldsymbol{S}_{H}}{\alpha} \tag{21}$$

where S_H is the singular value obtained from (14).

Step 6: Perform inverse SVD get watermark extraction.

Step 7: W* watermark image is recovered.

4.3 Optimization Using Genetic Algorithm

This sub-section discusses parameter optimization of the proposed watermarking technique using a genetic algorithm. Some of the parameters that are optimized include: alpha, sub-band, and decomposition level. All parameters become input genetic algorithms to find the best value of each parameter to obtain optimum performance on the image. Testing without and with attacks will be optimized by finding optimal values at alpha, sub-band, and decomposition levels using a population number of 30, mutation rate 0.2, and cross rate 0.4 with a total iteration of 100. The output of the optimization algorithm is imperceptibility with PSNR and SSIM parameters and robustness with NC and BER parameters. PSNR and SSIM are defined as follows [7]:

$$\mathrm{PSNR} = 10 \times \log_{10}\left(\frac{255^2}{\mathrm{MSE}}\right) \tag{22}$$

$$\mathrm{MSE} = \frac{1}{\mathrm{M} \times \mathrm{N}} \sum_{p=0}^{M-1} \sum_{q=0}^{N-1} \left[X(p, q) - X^*(p, q)\right]^2 \tag{23}$$

where the mean square error (MSE) is a calculation of the average squared difference between the host and the watermarked image. It takes into account the pixel values of the host image (represented by X(p, q)) and the watermark image (represented by), both of which have a size of M × N.

PSNR is a parameter used to compare the maximum power that allows power to be distorted. The optimal value of PSNR is above 28 dB, so the image can be said to be good. Meanwhile, SSIM is a technique employed for evaluating the likeness between two images by taking into account human visual system perception, including factors such as lighting, contrast, and image structure. The SSIM value is stretched from 0 to 1 and is said to be good if the value is close to 1. Here's the equation to get an SSIM [13].

$$\mathrm{SSIM} = i(p, q).c(p, q).s(p, q) \tag{24}$$

with the variables i, c, and s represent the lighting, contrast, and comparison of image structure, respectively.

NC is a parameter that measures the correlation or similarity between two images. NC values range from the range of 0 to 1 with an optimal value of 1 or close to 1. Here's the NC formula [9].

$$\mathrm{NC} = \frac{\sum_{p=0}^{m-1} \sum_{q=0}^{n-1} X(p, q) X^*(p, q)}{\sum_{p=0}^{m-1} \sum_{q=0}^{n-1} X^*(p, q)^2} \tag{25}$$

On the receiving side, a watermarked image is extracted from the watermarked image. BER is a parameter that evaluates errors during the watermark image extraction process by ratio error bits (EB) with total bits (TB). The best value for the BER parameter is 0, which implies that there is no loss or alteration between the original and extracted watermark. To determine the BER, the subsequent formula can be used [12]:

$$\mathrm{BER} = \frac{\mathrm{EB}}{\mathrm{TB}} \tag{26}$$

Here, optimization steps using the Genetic Algorithm in the proposed watermarking method (see Fig. 6):

Step 1: Initialize the parameters of the genetic algorithm, namely the number of populations (JP), iterations (JI), cross rate (CR), and mutation rate (MR). Initialization is the initial process by which an individual is built or resurrected randomly as much as the number of the population. The starting values of parameters in the genetic algorithm are determined through testing in order to obtain the most effective values of these parameters during the testing and analysis phase.

Step 2: Set up the genes in the individual as many as three genes, namely the value of α, sub-band and decomposition level.

Step 3: Perform the first iteration for the genetic algorithm process.

Step 4: Carry out the reproduction process, where there are two processes: crossover and mutation. The crossover process requires a cross rate and a population number. On the crossover, two parents with different parent values are used. Furthermore, the crossover process produces two individual children.

Step 5: Mutation process, where mutation rate and population numbers are required. In the mutation process, it only uses one parent which produces one individual child.

Step 6: After getting the population, child of crossover and child of mutation, then enter into the embedding and extraction process on the watermark as in Fig. 5.

Step 7: After calculating PSNR, SSIM, NC, and BER, the parameter values become input for optimization through the fitness function formula. The following is a fitness function formula in the optimization of watermarking scheme algorithm used.

$$\mathrm{FF} = \mathrm{PSNR} + (100 \times \mathrm{SSIM}) + (100 \times \mathrm{NC}) + (100 \times (1 - \mathrm{BER})) \tag{27}$$

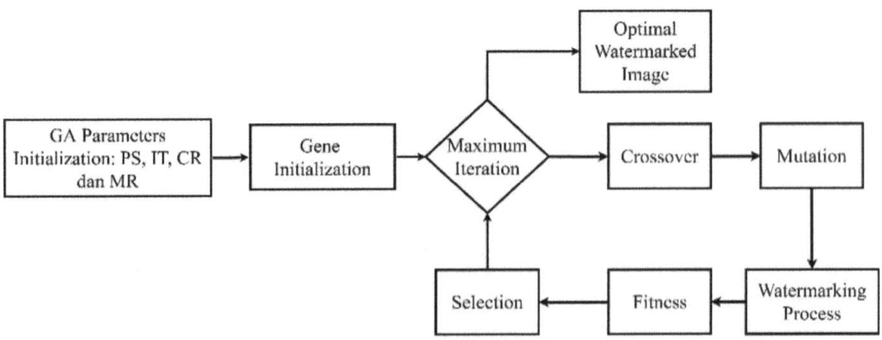

Fig. 6 Optimization of watermarking techniques using genetic algorithms

Step 8: Next, enter the selection process to pass the best individuals as much as the population number.

Step 9: The process repeats to the maximum iteration and gets the optimal value.

5 Experimental Results and Analysis

In this part, the proposed technique for watermarking was examined regarding its ability to invisible and robust. The testing was conducted using grayscale images that were 512 × 512 pixels in size as the host images. One of the host images used is a medical image that aims to be a host image with a logo watermark image that can be seen on Fig. 7. Test the proposed watermarking method, and whether the watermarking method with DWT-DCT-HD-SVD can produce good results also against medical images. Medical image sourced from MedPix TM Medical Image Database.

Three types of grayscale images are used for watermark images with multiple sizes, 64 × 64, 128 × 128, and 256 × 256 pixels. The attacks used for robustness testing are shown in Table 1. The attack will be aimed at the watermarked image, with various attacks, i.e.: filter, noise, compression, geometric, and signal processing. The attack filter is a median, mean, and Low pass Gaussian filter with a 3 × 3 kernel. Variance 0.001 on noise attack with Gaussian, salt pepper, and speckle noises. Testing with compression attack using three quality factors (QF) in JPEG compression, namely QF 50, 70, and 90. The geometry attacks used are six attacks,

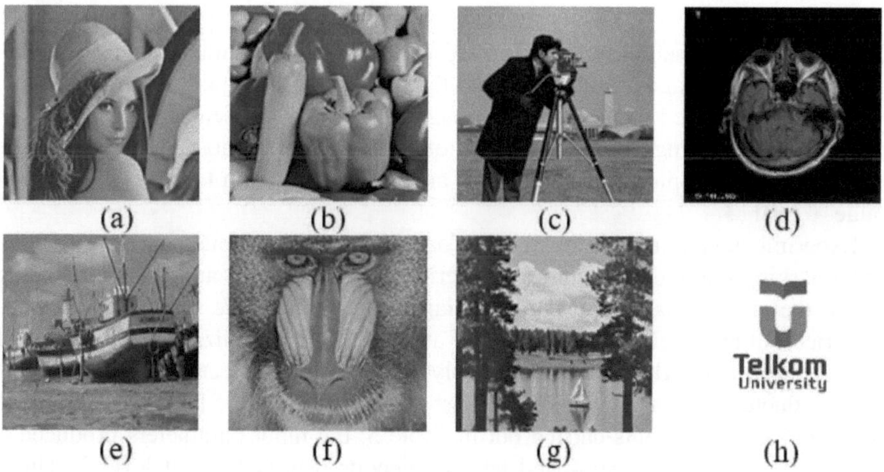

Fig. 7 Host images: **a** Lena, **b** Pepper, **c** Cameraman, **d** MRI, **e** Boat, **f** Baboon, **g** Lake, **h** Watermark image

Table 1 Cardiovascular parameters

Attack	Specification
Filtering	Median Filter/MF (3 × 3)
	Gaussian Low Pass Filter/GF (3 × 3)
Noise	Gaussian Noise/GN (V = 0.001)
	Salt & Pepper Noise/SPN (V = 0.001)
	Speckle Noise/SN (V = 0.001)
Compression	JPEG (QF = 50)
	JPEG (QF = 70)
	JPEG (QF = 90)
Geometric	Cropping/Crop (2%)
	Rescaling (0.25; 1.5; 4)
	Flip Horizontal
	Flip Vertical
Signal processing	Sharpening (0.8)
	Blurring (Len = 7, Theta = 4)
	Histogram Equalization

namely translation, cropping, rescaling, rotation, and flip, which include horizontal flips and vertical flips.

5.1 Optimization Parameter

Optimization parameters were performed with no-attack optimization experiments on all seven host images. Table 2 displays the findings of the experiment, and the best results were obtained using the Lena image. Figure 8(a) presents the optimization outcomes of applying a fitness function of 374.7114 in the host image Lena. The input parameter employs a sub-band LL and a decomposition level of 3, with an α value of 0.0114.

Experimental parameters are also performed to obtain optimal parameters on the watermarking method with attacks. Experiments with attacks are carried out using one of the attacks, namely, using Gaussian noise attacks. The selection of attacks is carried out by looking at the results of attacks before optimization, where one of the attacks with a fairly good and relatively robust result is selected. The experiment was conducted with an image of the host Lena with 512 × 512 pixels. From the optimization experiments carried out in Table 3, the input parameters produced α = 0.3590 using the LL sub-band and the decomposition level at level 3. Those input parameters are used in other attacks to test the watermarking scheme after optimization.

Table 2 Parameter optimization without attack

Host	α	Sub-band	Level	FF	PSNR	SSIM	NC	BER
Lena	0.0114	LL	3	374.7114	74.712	1	1	0
Pepper	0.0082	HH	3	361.4427	61.469	0.9997	1	0
Cameraman	0.0024	HL	2	365.9980	65.999	1	1	0
MRI	0.0074	LH	3	362.7407	62.751	0.999	1	0
Boat	0.0015	HH	2	370.2387	70.232	1	1	0
Baboon	0.0119	LL	3	373.0975	73.115	1	1	0
Lake	0.0016	HL	1	363.1525	63.155	1	1	0

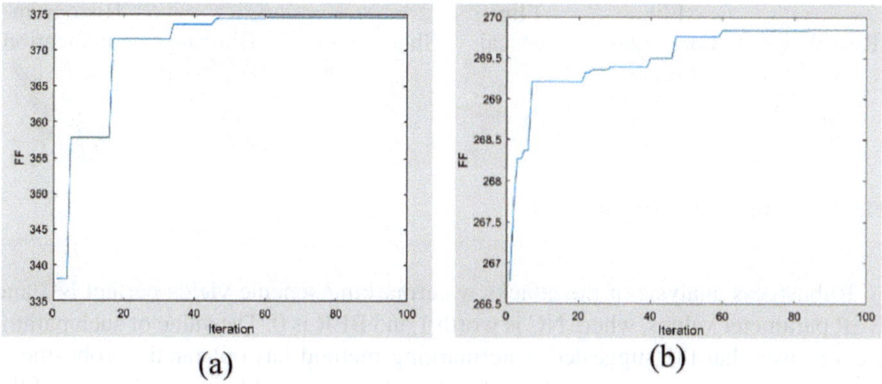

Fig. 8 GA plot for "Lena" as host image **a** without attack **b** Gaussian attack

Table 3 Parameter optimization with Gaussian attack

Host	α	Sub-band	Level	FF	NC	BER
Lena	0.3590	LL	3	286.1358	1	0.1135

5.2 Imperceptibility and Robustness Analysis

A watermarking technique must be undetectable by the human visual system in order for it to be regarded effective. This requirement ensures that the information is kept secure by ensuring that the embedded watermark image is invisible. From the no-attack experiment with optimizations seen in Table 2, the best results were obtained on the Lena host image with a PSNR of 74.7120 dB, and the SSIM is 1. Based on this value, it can be concluded that the suggested watermarking approach has outstanding imperceptibility. Figure 9 illustrates the watermarked images under attacks, providing a clear visual representation concerning invisibility.

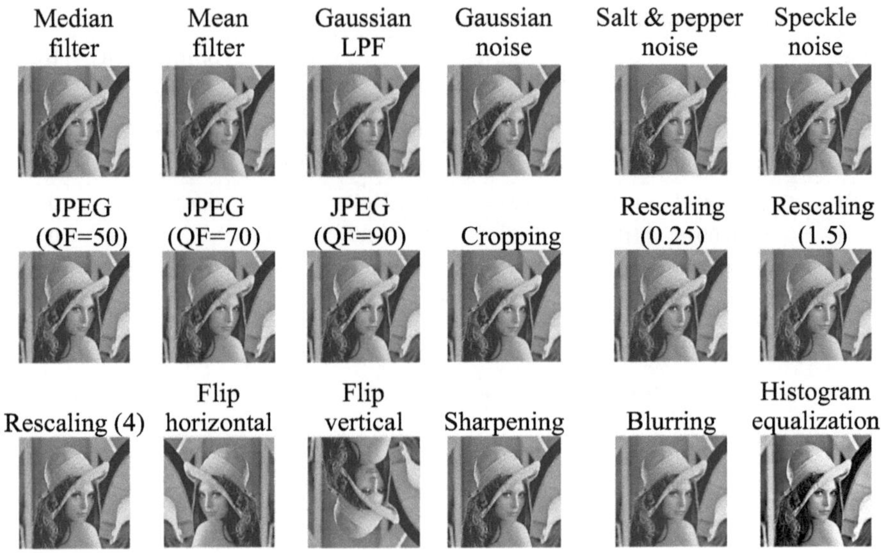

Fig. 9 Watermarked image with attack

Robustness analysis of the attacks watermarking scheme yields perfect NC and BER parameter values, where NC is worth 1 and BER is 0. The value of such parameters proves that the suggested watermarking method has outstanding robustness. Robustness experiments were also carried out by testing with attacks. Testing of the watermarking system with attacks with optimization parameters in Table 3 was done using 512 × 512 pixels of Lena images under attacks.

In watermarking schemes, robustness is the ability to maintain watermark images in the extraction process in various attacks. The results of experiments on robustness with attacks yielded good robustness values. Robustness experiments were also carried out by testing with attacks. Testing the watermarking system with attacks with optimization parameters in Fig. 9 was achieved using Lena host images measuring 512 × 512 pixels with filter attacks, noise attacks, geometric attacks, signal processing attacks, and combined attacks. The results of the experiment with the attack can be seen in Fig. 10 for the extraction watermark image with the test results of the NC and BER parameters.

In watermarking schemes, robustness is the ability to maintain watermark images in the extraction process in various attacks. The results of the experiment with the attack can be seen in Fig. 10 for the extraction watermark image with the test results of the NC and BER parameters. The results of experiments on robustness with attacks produced good robustness values with optimal values in Gaussian noise, salt & pepper, Speckle noise, JPEG QF 50, QF 70, QF 90, Rescaling 1.5, Rescaling 4, Flip horizontal, and Flip vertical attacks with a value of NC 1. From Fig. 10, it can be seen that the proposed watermarking scheme has robust properties against almost all

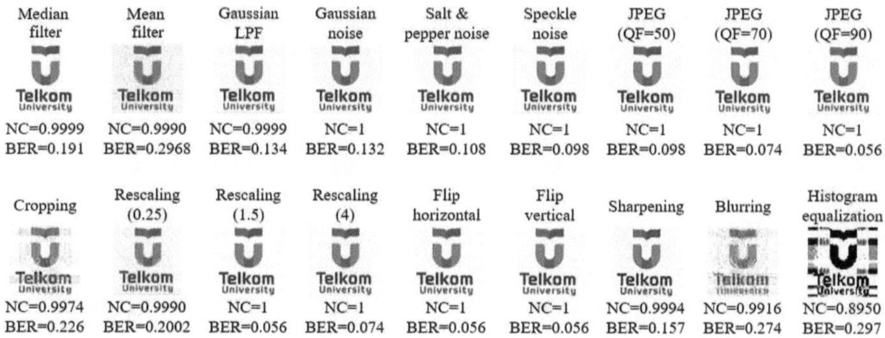

Fig. 10 Robustness results with attack

attacks, but the results are not good concerning robustness in translation, rotation, and combined attacks.

5.3 *Comparative Analysis*

In this part, the outcomes of the suggested method for watermarking are compared to those of previous research. The research on [4] used the DWT-HD-SVD method with fly fruit optimization. In this study, we also tried a watermarking technique using the DWT-HD-SVD method with optimization of the genetic algorithm used in this study. Table 4 displays the findings of the initial comparison for imperceptibility on all three schemes. The test results prove that the method in this work has a much good imperceptibility than before where the highest PSNR value is 74.7120 dB with an SSIM of 1 on the Lena host image with a watermark size of 64 × 64.

Performance comparison for robustness was carried out by comparing [4, 19], DWT-HD-SVD with optimization of genetic algorithms in this study. The comparison of the resilience of the watermarking approach employing tests with a Lena host image size 512 × 512 pixels and a watermark image 64 × 64 pixels is shown in Table 5. The testing findings showed that the watermarking scheme in this study was more robust against median filter attacks, Gaussian LPF, Gaussian noise, salt & pepper noise, speckle noise, JPEG QF 50, cropping, rescaling, and blurring attacks.

Table 4 Imperceptibility comparison of the proposed method with [4] and DWT-HD-SVD-GA

Watermark size	DWT-HD-SVD [4]		DWT-HD-SVD-GA		Proposed	
	PSNR	SSIM	PSNR	SSIM	PSNR	SSIM
64 × 64	38.2477	0.9991	56.7493	0.9997	74.7120	1
128 × 128	38.1625	0.9992	50.6961	0.9987	49.7542	0.9999
256 × 256	38.1621	0.9992	44.6793	0.9941	45.4847	0.9998

Table 5 Robustness (NC) comparison of the proposed method with [4] and DWT-HD-SVD-GA

Attack	[4]	DWT-HD-SVD-GA	Proposed
MF	0.9993	0.9997	**0.9999**
GF	0.9992	0.9993	**0.9999**
GN	0.9997	0.9999	**1**
SPN	0.9997	0.9999	**1**
SN	0.9998	0.9999	**1**
JPEG (QF = 50)	0.9997	0.9999	**1**
Crop	0.9760	0.9954	**0.9974**
Rescaling 0.25	0.9995	0.9983	0.9990
Rotation	0.9306	0.4328	0.8941
Sharpening	1	0.9989	0.9994
Blurring	0. 9147	0. 9854	**0. 9916**
Histogram Equalization	0.9878	0.9073	0.9548

Table 6 Robustness (NC) comparison of the proposed method with [19] and DWT-HD-SVD-GA

Attack	[19]	DWT-HD-SVD-GA	Proposed
JPEG (QF = 90)	0.9265	1	**1**
SPN (v = 0.1)	0.9473	0.9369	**0.9488**
GN (v = 0.01)	0.9770	0.9850	**0.9909**
MF (2 × 2)	0.9570	0.9989	**0.9992**
GF	0.9992	0.9993	**0.9999**
Sharpening	0.9890	0.9989	**0.9994**
Histogram Equalization	0.8909	0.9073	**0.9548**

Table 6 compares robustness (NC) comparison of the FDCuT-DCT-SVD scheme [19] and DWT-HD-SVD-GA using an MRI medical host image measuring 512 × 512 pixels and a watermark image measuring 128 × 128 pixels. The test results showed that the scheme in this study had better robustness than the other two comparison schemes with tests using JPEG compression (QF 90), salt & pepper noise, Gaussian noise, median filter, Gaussian LPF, sharpening, and histogram equalization.

6 Conclusions

In this work, we improved watermarking method in image using DCT in the DWT-HD-SVD scheme optimized with a genetic algorithm. The utilization of Genetic Algorithms is applied to explore the most favorable values for alpha, sub-band, and decomposition levels. Based on experiments that have been carried out with watermarking schemes without attacks resulted in a mean PSNR of 66.23295 dB, SSIM of 0.9999, NC 1 and BER 0 which proves that this watermarking scheme has good imperceptibility and robustness. Good robustness and imperceptibility are also obtained using various watermark sizes. Based on comparisons with related studies, the watermarking scheme in this study resulted in better robustness in filter attacks, noise attacks, compression attacks, geometric attacks, and signal processing attacks. In future research, alternative optimization techniques [37] could be employed to enhance the proposed watermarking method's resilience to hybrid attacks, as well as translation and rotation.

Acknowledgements The authors express their gratitude to the Directorate of Research and Community Service at Telkom University for supporting and funding this work.

References

1. Dixit A, Dixit R (2017) A review on digital image watermarking techniques. Int J Image, Graph Signal Process 9:56
2. Borra S, Thanki R, Dey N (2018) Digital image watermarking: theoretical and computational advances. CRC Press
3. Anand A, Singh AK (2020) An improved DWT-SVD domain watermarking for medical information security. Comput Commun 152:72–80
4. Liu J, Huang J, Luo Y, Cao L, Yang S, Wei D, Zhou R (2019) An optimized image watermarking method based on HD and SVD in DWT domain. IEEE Access 7:80849–80860
5. Su Q, Wang G, Lv G, Zhang X, Deng G, Chen B (2017) A novel blind color image watermarking based on Contourlet transform and Hessenberg decomposition. Multimed Tools Appl 76:8781–8801
6. Jain P, Ghanekar U (2018) Robust watermarking technique for textured images. Procedia Comput Sci 125:179–186
7. Mousavi SM, Naghsh A, Abu-Bakar SA (2014) Watermarking techniques used in medical images: A survey. J Digit Imaging 27:714–729
8. Parashar P, Singh RK (2014) A survey: Digital image watermarking techniques. Int J Signal Process, Image Process Pattern Recognit 7:111–124
9. Agarwal N, Singh AK, Singh PK (2019) Survey of robust and imperceptible watermarking. Multimed Tools Appl 78:8603–8633
10. Sultan K, Aldhafferi N, Alqahtani A, Mahmud M (2018) Reversible and fragile watermarking for medical images. Comput Math Methods Med
11. Singh G, Goel N (2016) Entropy based image watermarking using discrete wavelet transform and singular value decomposition. In: 2016 3rd International Conference on Computing for Sustainable Global Development (INDIACom), pp. 2700–2704, IEEE.
12. Selvam P, Balachandran S, Iyer SP, Jayabal R (2017) Hybrid transform based reversible watermarking technique for medical images in telemedicine applications. Optik 145:655–671
13. Begum M, Uddin MS (2020) Analysis of digital image watermarking techniques through hybrid methods. Adv Multimed 2020:1–12
14. Abdulrahman AK, Ozturk S (2019) A novel hybrid DCT and DWT based robust watermarking algorithm for color images. Multimed Tools Appl 78:17027–17049
15. Fazli S, Moeini M (2016) A robust image watermarking method based on DWT, DCT, and SVD using a new technique for correction of main geometric attacks. Optik 127:964–972

16. Takore TT, Kumar PR, Devi GL (2016) Efficient gray image watermarking algorithm based on DWT-SVD using genetic algorithm. In: 2016 International Conference on Information Communication and Embedded Systems (ICICES) 2016:pp. 1–6, IEEE
17. Begum M, Uddin MS (2020) Digital image watermarking techniques: A review. Information 11:110
18. Song J, Song J, Bao Y (2012) A blind digital watermark method based on SVD and chaos. Procedia Eng 29:285–289
19. Novamizanti L, Wahidah I, Wardana NP (2020) A robust medical images watermarking using FDCuT-DCT-SVD. Int J Intell Eng Syst 13:266–278
20. Budiman G, Novamizanti L, Iwut I (2017) Genetics algorithm optimization of DWT-DCT based image Watermarking. J Phys: Conf Ser 795:012039. IOP Publishing
21. Variengien A, Nichele S, Glover T, Pontes-Filho S (2021) Towards self-organized control: Using neural cellular automata to robustly control a cart-pole agent. Innov Mach Intell (IMI) 1:1–14
22. Gupta N, Khosravy M, Patel N, Senjyu T (2018) A bi-level evolutionary optimization for coordinated transmission expansion planning. IEEE Access 6:48455–48477
23. Khosravy M, Nakamura K, Hirose Y, Nitta N, Babaguchi N (2022) Model inversion attack by integration of deep generative models: Privacy-sensitive face generation from a face recognition system. IEEE Trans Inf Forensics Secur 17:357–372
24. Khosravy M, Nakamura K, Nitta N, Dey N, Crespo, RG, Herrera-Viedma E, Babaguchi N (2022). Social IoT approach to cyber defence of a deep-learning-based recognition system in front of media clones generated by model inversion attack. IEEE Trans Syst, Man, Cybern: Syst
25. Joshi A, Khosravy M, Gupta N (2020). Machine learning for predictive analysis. Proc ICTIS
26. Maheshwari DB (2018) An analysis of wavelet based dual digital image watermarking using SVD. In: 2018 International Conference On Advances in Communication and Computing Technology (ICACCT), pp. 69–73, IEEE
27. Wu JY, Huang WL, Xia-Hou WM, Zou WP, Gong LH (2020) Imperceptible digital watermarking scheme combining 4-level discrete wavelet transform with singular value decomposition. Multimed Tools Appl 79:22727–22747
28. Agarwal N, Singh PK (2022) Discrete cosine transforms and genetic algorithm based watermarking method for robustness and imperceptibility of color images for intelligent multimedia applications. Multimed Tools Appl 81:19751–19777
29. Al-Haj A, Mohammad A, Amer A (2017) Crypto-watermarking of transmitted medical images. J Digit Imaging 30:26–38
30. Abduldaim AM, Waleed J, Mazher AN (2020) An efficient scheme of digital image watermarking based on Hessenberg factorization and DWT. In: 2020 International Conference on Computer Science and Software Engineering (CSASE), pp. 180–185, IEEE
31. Sejpal S, Borse D (2019) Comparative performance analysis of color image watermarking scheme using Hessenberg decomposition and Schur decomposition. In: 2019 5th International Conference On Computing, Communication, Control and Automation (ICCUBEA), pp. 1–6, IEEE
32. Novamizanti L, Suksmono AB, Danudirdjo D, Budiman G (2023) Robust and imperceptible watermarking on medical images using coefficient pair modification. IIUM Eng J 24(1):88–105
33. Rabie T, Kamel I (2016) On the embedding limits of the discrete cosine transform. Multimed Tools Appl 75:5939–5957
34. Iwut I, Budiman G, Novamizanti L (2016) Optimization of discrete cosine transform-based image watermarking by genetics algorithm. Indones J Electr Eng Comput Sci 4:91–103
35. Mohammed NS, Abduldaim AM (2021) Algebraic Hessenberg decomposition method optimized by genetic algorithm for zero watermarking technique. Int J Math Comput Sci 16:1497–1514
36. Ahmadi SB, Zhang G, Wei S (2020) Robust and hybrid SVD-based image watermarking schemes. Multimed Tools Appl 79:1075–1117

Chapter 14
Genetic Algorithm Application in Continuum Robot Optimization

Atoosa Davarpanah, Alireza Takolpour Saleh, and Amir Lotfavar

1 Introduction

In today's world, robots have become an integral part of our daily lives, enabling us to accomplish tasks that would otherwise be challenging or impossible. While there are various types of robots, certain categories are more widely recognized than others. In a broad classification, robots can be divided into two groups: rigid-link robots and continuum robots. Each group possesses distinct characteristics, and numerous robots can belong to either category.

Rigid-link robots are well-known for their precise and rigid movements, but these characteristics can limit their workspace. On the other hand, continuum robots exhibit a flexible and adaptable nature, allowing them to operate effectively in diverse environments, including geometrically complex spaces. They can conform and adjust to their surroundings, making them particularly well-suited for tasks in narrow or constrained areas. Continuum robots have found extensive applications in fields such as medical surgeries, inspections, and more. Their unique qualities and capabilities make them an exciting and active area of research.

The primary focus of this paper revolves around a specific type of continuum robot known as the Concentric Tube Continuum Robot (CTR), also referred to as telescoping pre-curved tubes. This particular robot design features a backbone composed of multiple concentric pre-curved tubes, capable of translating and rotating

A. Davarpanah (✉) · A. T. Saleh · A. Lotfavar
Department of Mechanical Engineering and Aerospace, Shiraz University of Technology, Shiraz, Iran
e-mail: falseatoosa.dvrpnh@gmail.com

A. T. Saleh
e-mail: tavakolpour@sutech.ac.ir

A. Lotfavar
e-mail: lotfavar@sutech.ac.ir

M. Khosravy et al. (eds.), *Frontiers in Genetics Algorithm Theory and Applications*, Springer Tracts in Nature-Inspired Computing,
https://doi.org/10.1007/978-981-99-8107-6_14

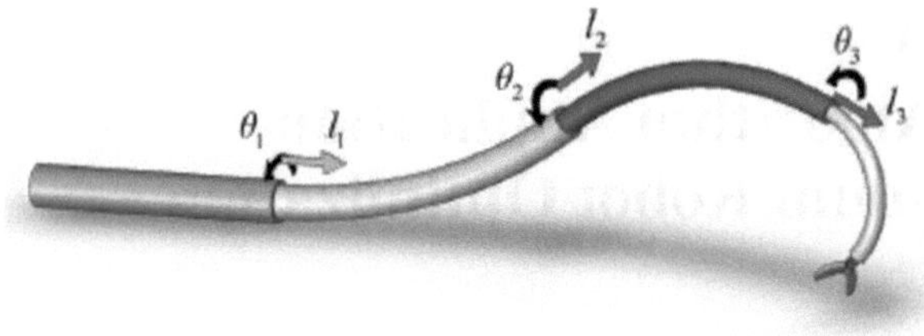

Fig. 1 The diagram depicting the CTR configuration with four tubes illustrates the key components of this robot. In the schematic, the angle of rotation is represented by the variable θ, while the translation length is denoted by l [1]

relative to one another. The overall mechanism resembles that of a telescope, with the robot's structure extending through longitudinal translational sliding of the tubes. The utilization of pre-curved tubes enables the robot to achieve more intricate and complex motions. To aid in comprehending the CTR, a schematic representation of this robot type is presented in Fig. 1. The depicted robot in Fig. 1 consists of four main tubes that can transfer and rotate in relation to one another.

The application of CTR in the medical field is particularly advantageous due to its ability to operate effectively within confined spaces, such as within organs. Medical-grade CTRs are typically constructed using materials that exhibit a compliance level similar to biological matter. This compliance allows the robot to interact with delicate tissues and structures without causing damage. By leveraging the flexible and adaptable nature of CTR, medical professionals can perform minimally invasive procedures with enhanced precision and reduced trauma to the patient. The unique properties of CTR make it a promising technology for various medical applications.

Concentric Tube Continuum robots, also known as Active Cannulas in medical applications, find extensive application in complex surgical procedures [1]. In certain research papers, the curved tubes of CTR are simplified as segments of circles to facilitate kinematic problem-solving [2–5]. In 2008, Webster et al. proposed a novel approach for obtaining the kinematics of telescoping tubes [6]. Walker et al., in the same year, presented a geometry-based method for the inverse kinematics of CTR [7]. Webster et al., in 2009, employed the Minimum Energy method to determine the CTR configuration [8]. The design of Concentric Tube Continuum Robots was further explored by Webster and Rucker in 2010 [9]. Dupont et al., in 2011, defined an algorithm for CTR design [10]. In 2013, Walker categorized continuum robots in research [11]. Su et al., in 2016, investigated the modeling and design of MRI using CTR [12].

In 2019, A. Davarpanah et al. utilized the Genetic Algorithm to optimize the structure of a specific type of Concentric Tube Continuum Robot, enabling it to reach different target points [13]. In the same year, they also optimized the robot parameters to navigate through complex paths, employing various geometrical equations [14]. A. Davarpanah et al., in 2020, focused on optimizing CTR structure parameters with the aim of enhancing accuracy and stability using the Genetic Algorithm [15]. M.

Khazaee et al. in 2021 investigated behavior-based navigation of an autonomous hexapod robot using a hybrid automation [16].

Due to the redundancy present in the inverse kinematics of CTR, determining an optimized solution becomes a challenging task. However, it is of utmost importance to overcome these challenges and achieve an optimal solution.

In this research paper, we propose the utilization of the Genetic Algorithm (GA), which is a type of AI method, to address the optimization of both the robot parameters and the inverse kinematic solution. By employing GA, we aim to leverage its powerful search and optimization capabilities to navigate the complex solution space and identify the most optimal configurations.

The Genetic Algorithm operates based on principles inspired by natural selection and genetics. It involves the creation and evolution of a population of potential solutions, represented as individuals, through selection, crossover, and mutation operations. These operations mimic the process of genetic evolution, allowing for the generation of improved solutions over successive iterations.

By harnessing the power of Genetic Algorithm, we aim to overcome the challenges posed by the redundancy in the inverse kinematics of CTR. Through the optimization of robot parameters and the inverse kinematic solution, we strive to achieve an optimized configuration that enhances the overall performance and control of CTR.

By incorporating the Genetic Algorithm into our research, we aim to demonstrate its effectiveness as an AI method for optimizing the complex kinematic properties of CTR and paving the way for improved control and operation of these robotic systems.

It seems necessary to introduce Genetic Algorithms and imply to some papers using Genetic Algorithm for different applications.

The realm of Artificial Intelligence (AI) encompasses a diverse array of study areas, encompassing an assortment of methodologies. Currently, developers are actively refining and employing various Artificial Intelligence Methods across an extensive spectrum of applications. Several prevalent AI techniques comprise Machine Learning, Deep Learning, Natural Language Processing (NLP), Computer Vision, Reinforcement Learning (RL), Evolutionary Algorithms, and Expert Systems.

Genetic Algorithm (GA) serves as an Evolutionary algorithm designed to address optimization problems. Taking inspiration from Charles Darwin's theory of natural evolution, GA operates on the principle of natural selection. It entails modifying the population to tackle specific optimization challenges. GA finds utility in resolving intricate optimization problems that prove challenging for conventional methodologies, spanning domains like engineering, computer science, and economics. Its efficacy is particularly notable when confronted with scenarios featuring multiple optimal solutions. The GA process commences by generating a population of potential solutions in a randomized manner. Each solution is subsequently assessed based on its fitness, with the fittest solutions chosen for reproduction. The new solutions undergo crossover and mutation, which introduce fresh variations into the population. This iterative process continues until the algorithm converges upon the optimal solution.

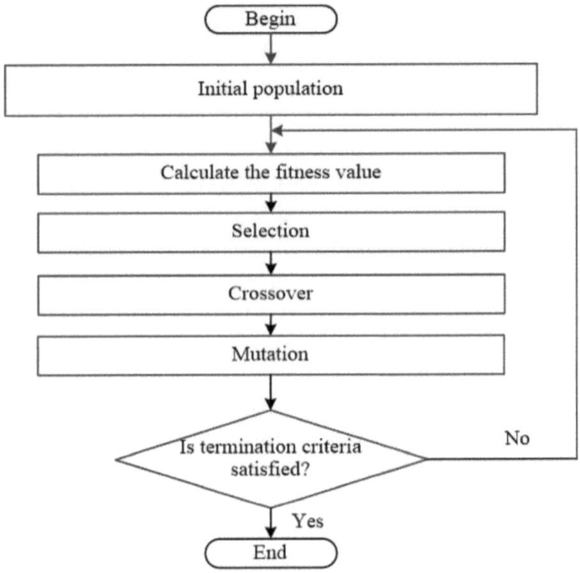

Fig. 2 Standard Genetic Algorithm flowchart [17]

Figure 2 illustrates a straightforward flowchart outlining the Genetic Algorithm procedure.

This paper explores the application of Genetic Algorithm, an AI method, to optimize the structure of a concentric tube continuum robot. The external structure of the continuum robot consists of curved tubes, allowing for variable length and rotation angles of each tube. To determine the inverse kinematics of the robot accurately, three variables are considered for each tube: curvature, rotation angle, and length. This assumption applies to three-dimensional movements, while 2D movements exclude rotation. Due to the backbone structure and variable parameters of continuum robots, redundancy is inherent, resulting in multiple solutions for each problem. Consequently, obtaining the inverse kinematics of continuum robots poses a significant challenge for researchers. To solve this challenge, this paper proposes the use of Genetic Algorithm to optimize the robot's structure, mitigating redundancy issues. By employing Genetic Algorithm, the paper aims to derive an optimized solution for the robot's inverse kinematics. Acquiring knowledge of the inverse kinematics is crucial for effectively controlling the robot, thereby addressing subsequent challenges encountered by researchers.

First, we introduce the concentric tube continuum robots and discuss the kinematics and theory that is used for these types of robots.

The paper then examines the robot's capability to reach different target points, providing illustrative examples. We discuss the robot's movement in 3D spaces, where the rotation of each tube is taken into account. Introducing tube rotation adds complexity to the parameters governing the robot's movement, thereby posing a more challenging inverse kinematics problem.

To solve the challenge, the authors of the paper propose a MATLAB code as the solution for solving the inverse kinematics of the concentric tube continuum robot. The code utilizes Genetic Algorithm (GA) for optimization purposes.

Various target points have been incorporated into the optimization of the robot to assess the effectiveness of the solution. Moreover, by repeatedly addressing the problem solely from the perspective of precision, diverse optimal solutions have been discovered. Moving forward, the problem introduces the length parameter, enabling the identification of the most compact robot that not only exhibits superior accuracy but also demonstrates exceptional stability.

The subsequent phase of this research paper involves the fine-tuning of parameters to enable the concentric tube continuum robot to reach different target points. This a type of problem that deals with the real condition of using continuum robots.

In Sect. 3 robot structure parameters are optimized specifically for the goal of reaching the target point a. in addition to the accuracy of the robot, the stability of robot is also considered. In Sects. 4 and 5 the challenging problems get more complicated. We consider optimization of a concentric tube continuum robot for reaching three different target points with good accuracy. In Sect. 4 the number of parameters that needs to be optimized reaches number 21 and in Sect. 5, it reaches number 23. These two differences relate to this fact that we assumed that the three tube curvatures are constant in Sect. 4 while for the optimization of the robot in Sect. 5 we consider the middle tube's curvature variant. This assumption makes the problem more complicated but it helps the accuracy and flexibility of the robots. Optimization of 21 or 23 parameters at the same time seems impossible without using AI methods such as Genetic Algorithm.

In summary, our approach introduces a novel methodology for solving the inverse kinematics of concentric tube continuum robots (CTR), enabling the identification of the most suitable solution tailored to specific situations, target points, and efficient navigation through tight workspaces.

The method employed in this study relies on the utilization of minimum energy principal equations. It is assumed that the robot operates within a free space environment, thereby neglecting any external forces acting on the robot. Additionally, the curvature of each tube is assumed to remain constant along its length. The results obtained from this approach are simulated using MATLAB software. The kinematics and equations involved in the process are summarized concisely. Furthermore, the optimization of robot parameters is accomplished through the implementation of a Genetic Algorithm.

2 Kinematics

The application of Cosserat rod theory is relevant in this research as it provides a suitable framework for analyzing the kinematics of elongated and slender objects, such as rods. Consequently, this theory is also adopted in the present study. The assumption is made that each tube within the robot can be modeled as an individual rod. To describe the robot's configuration, a homogeneous transfer function denoted as (g(s)) is utilized.

$$g(s) = \begin{pmatrix} R(s) & P(s) \\ 0 & 1 \end{pmatrix} \quad (1)$$

Equation (1) introduces the relationship between the direction of the robot at a given point, denoted as R(s), and the position of that point along the centerline of the tube, denoted as P(s). Both R(s) and P(s) are functions dependent on the parameter s, which represents the absolute length of the robot from its base to the desired point. This parameter s serves as a reference to specify different locations along the robot's structure.

Figure 2 depicts a segment of a concentric tube continuum robot, illustrating the concept of the Cosserat Rod Theory. The schematic diagram highlights the line of centroids of the curved tube, represented by vector r, with the coordination center denoted as O. In certain papers, the assumption is made that each curved section corresponds to a portion of a circle, allowing the coordination center for that section to coincide with the center of the circle. The arrows depicted in Fig. 3 provide a visual aid in comprehending how g(s) describes the position of each point along the curved tube.

The diagram shows various vectors that play a significant role in this theory. The vector r corresponds to the line of centroids of the curved tube, while the vector u denotes the tangent direction along the centerline of the tube. Additionally, the vector v represents the cross-sectional rotation, and the vector n signifies the normal direction to the cross-section. These vectors are fundamental in characterizing the behavior and kinematics of the curved tube within the Cosserat Rod Theory.

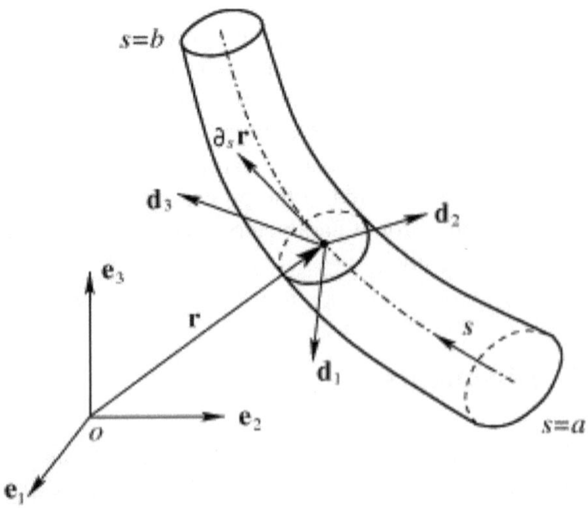

Fig. 3 Presents a cross-sectional view of a curved tube segment within the framework of the Cosserat Rod Theory

Figure 4 illustrates a basic curve situated in the plane. The depicted curve incorporates specific parameters that are employed in determining the transfer matrix associated with the curve. By considering these parameters, the transfer matrix can be calculated, enabling a comprehensive understanding of the geometric properties and transformations of the curve.

The paper adopts the assumption that the outer tubes possess greater stiffness compared to the inner tubes. As a result, the shape of the robot within each section mirrors that of the outer tube present in that specific section. This concept is visually depicted in Fig. 5, providing a clear representation of the alignment between the robot's shape and the outer tube within each section.

Equation (2) represents the general transfer matrix obtained by multiplying three matrices together to obtain the transfer matrix.

$$T = \begin{bmatrix} \cos(\theta^O + \theta^S) & -\sin(\theta^O + \theta^S) & 0 & -d\sin(\theta^O) \\ \sin(\theta^O + \theta^S) & \cos(\theta^O + \theta^S) & 0 & d\cos(\theta^O) \\ 0 & 0 & 1 & 0 \\ 0 & 0 & 0 & 1 \end{bmatrix} \tag{2}$$

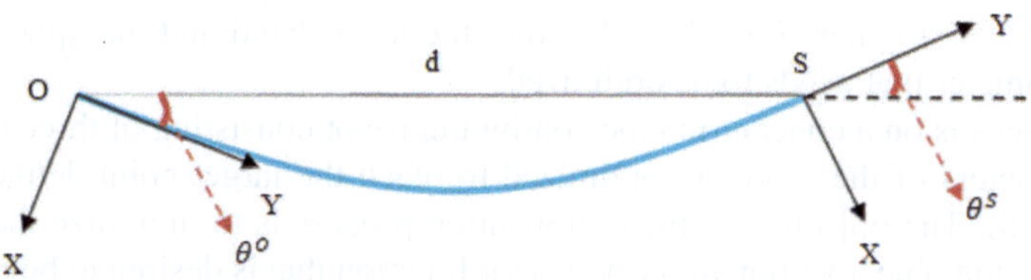

Fig. 4 A curved tube in 2D

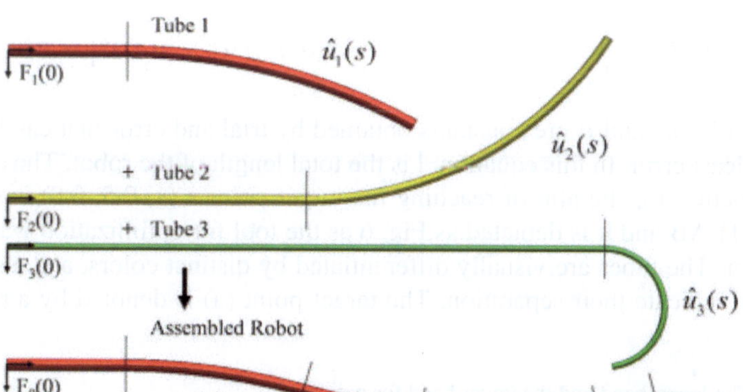

Fig. 5 The configuration of a dominating-stiffness tube pair, **a** where the retracted robot conforms to the shape of the stiff outer tube and **b** a portion of the extended inner tube relaxes to its initial curvature [1]

The determination of vector (r) and the direction of ∂_s r for each point on the path is achieved by solving the general transfer matrix. Understanding the inverse kinematics of the robot provides insights into the appropriate robot structure for desired movements. However, due to redundancy, inverse kinematics of continuum robots pose a challenge for researchers. For solving this problem, optimization using Genetic Algorithm in our MATLAB code is employed to obtain the optimal solution for inverse kinematics in various scenarios. This code gives us the optimal solution for some defined problems and it helps the researchers to control the robot as a result.

3 Optimization of Robot Parameters by Genetic Algorithm for a Target Point

To determine the inverse kinematics of the robot accurately, three variables are considered for each tube: curvature, rotation angle, and length. This assumption applies to three-dimensional movements. In this section robot structure the parameters are optimized specifically for the goal of reaching the target point a (1, 0.5, 0.8). The Genetic Algorithm employed in this study utilizes a population size of 100 and reaches the 51st generation. Table 1 shows the lower band and the upper band for each parameter that needs to be optimized.

The focus is on a concentric tube continuum robot consisting of three tubes, and the parameters of the robot are optimized to reach the target point defined as a = (1, 0.5, 0.8). The objective of the optimization process is to minimize the error. In this paper, the objective function is the error function that is desired to be minimum. The error function, as stated in Eq. (3), is utilized to quantify the deviation from the desired target.

$$OF = \frac{1}{m_a}\left[(x - x(a))^2 + (y - y(a))^2 + (z - z(a))^2\right] + \mu l \tag{3}$$

In Eq. (3), m_a and μ are constants obtained by trial and error that can help us to have the least error. In this equation, l is the total length of the robot. The optimized robot structure for the aim of reaching the target point a (1, 0.5, 0.8) is simulated with MATLAB and it is depicted as Fig. 6 as the tool for optimization was Genetic Algorithm. The tubes are visually differentiated by distinct colors, and a small line is used to indicate their separation. The target point (a) is denoted by a prominent

Table 1 The lower band and the upper band for parameters

Tube number	Rotation angle limit (degree)	Curvature limit (1/dm)	Optimal length (dm)
1	[−180, 180]	[−2, 2]	[0, 2]
2	[−180, 180]	[−5, 5]	[0, 2]
3	[−180, 180]	[−20, 20]	[0, 0.6]

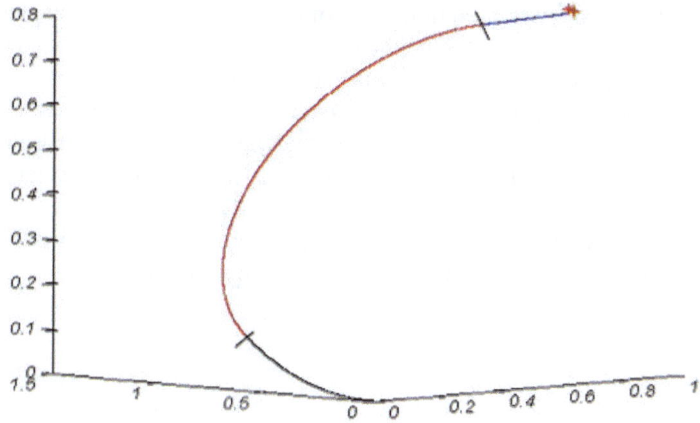

Fig. 6 The simulation of optimal robot for reaching target point a

red star. Notably, the first tube, characterized by the highest stiffness, is depicted in black, followed by the red-colored second tube, and finally the blue-colored third tube.

Equation (4) is the Geometric matrix related to the position and the orientation of the tip of the robot. The last column of the matrix is showing the position of the robot's tip. As it is can be seen the tip position is near the point (a) with a great accuracy.

$$g = \begin{bmatrix} 0.7936 & 0.4628 & -0.3950 & 0.9936 \\ 0.3608 & -0.8807 & -0.3071 & 0.4924 \\ -0.4900 & 0.1012 & -0.8658 & 0.7898 \\ 0 & 0 & 0 & 1 \end{bmatrix} \tag{4}$$

9 parameters have been optimized using GA for the problem and the final structure of the robot is obtained. These 9 parameters are the length of each tube, the curvature of each section and the rotation angle for each tube. The structure is made of three curved tubes.

4 Efficient Multi-Point Navigation with Genetic Algorithm-Optimized Robot Structure: Exploring the Three-Tube Design with Consistent Curvature

Real robots usually have constant curvatures and we can only change the robot length and the rotation angle for each tube. So, in this section we consider the curvature of each tube of the robot as a constant parameter has been optimized using Genetic Algorithm. By assuming that the curvature remains constant for different target points, we also optimized the length and rotation angle of the tubes simultaneously. During this optimization stage, the primary objective is to discover a concentric tube continuum robot capable of precisely reaching multiple predefined target points. The robot design incorporates three tubes' curvature and other structural parameters. For this optimization process, a population size of 30 and a generation of 100 are taken into account. The ranges considered for the rotation angle, curvature, and length of the tubes are outlined in Table 2.

Equation (5) is showing the error in this problem which is also our objective function. We are supposed to minimize the objective function.

$$e = \sum \left[(x - x_t)^2 + (y - y_t)^2 + (z - z_t)^2 \right] \tag{5}$$

In Eq. (5), the symbol $\sum$ is used to indicate that the equation is applicable for robot navigation considering multiple target points. In this case, there are three target points denoted as (x_t, y_t, z_t), representing their respective coordinates. The goal is to achieve precise accuracy in reaching each target point.

The error in the equation represents the sum of the squared distances between each target point and the robot's tip, denoted as (x, y, z).

The optimal parameters for constructing the robot's structure, such as the length and rotation angle of the tubes, are determined using Genetic Algorithm, taking into account the position of each target point. This means that the optimal values for these parameters may vary for different target points.

In this optimization problem 21 parameters have been optimized simultaneously via Genetic Algorithm which is approximately impossible without using AI. When the number of parameters is increased the time that Genetic Algorithm spend on the solution is also increased. So, the problems with less parameters take less time.

Figure 7 is illustrating the Genetic Algorithm process for determining the best fitness of parameters to achieve the defined goal. The main goal in this problem is finding a robot with a specific structure that can reach different target points with good accuracy. In this section three different target points are defined, and the robot construction is obtained according to those three target points.

Figure 7 is showing the step-by-step process of solving the problem through multiple generations. The diagram depicts how the population evolves over time, with each generation undergoing selection, crossover, and mutation to produce a new population. The fitness values, representing the quality of each solution, are

Table 2 The lower band and the upper band for parameters in multi-point navigation optimization of CTR with three constant curvature tubes

Tube number	Rotation angle limit (degree)	Curvature limit (1/dm)	Optimal length (dm)
1	[−180, 180]	[−3, 3]	[0, 3]
2	[−180, 180]	[−5, 5]	[0, 1.5]
3	[−180, 180]	[−20, 20]	[0, 0.8]

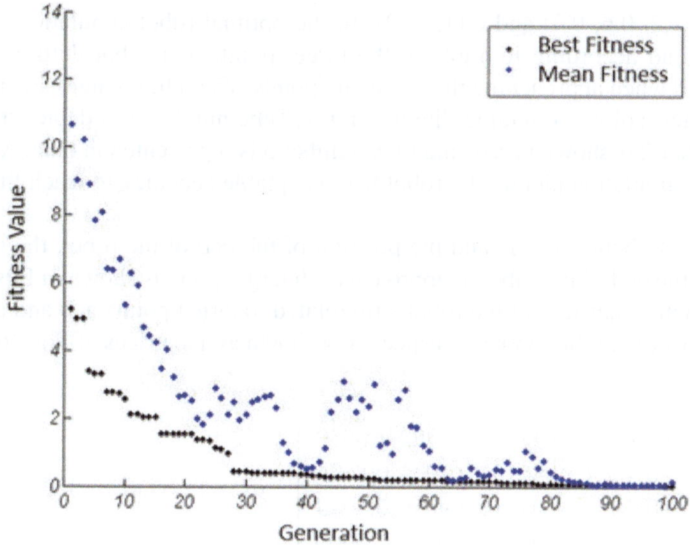

Fig. 7 Genetic algorithm diagram of best fitness objective finction versus generation (for the case with three target points and constant curvature tubes)

also shown for each generation. The diagram provides a visual representation of how the genetic algorithm optimizes the solution through iterative improvements.

21 Optimized parameters of this concentric tube continuum robot is gathered in Table 3 as you can see the optimized curvature of each tube is determined in first raw of the table and the other parameters are optimized related to these curvatures.

The optimal parameters for this problem are shown in Table 3 where these structural parameters are curvature, length and rotation angle of each tube.

Figure 8 is showing the simulation of the optimal robot's structure and its navigation for three different target points. These three target points are also depicted in Fig. 8 by red stars. The position of the three pre-defined target points are a (1, 0.8,

Table 3 Optimal parameters of the robot's structure obtained using Genetic Algorithm. The parameters are optimized based on the goal of reaching three different target points with the same curvature for each tube

Target point	Optimal parameters	Tube number 1	Tube number 2	Tube number 3
Point a, b, c	Optimal curvature (1/dm)	1.626	2.4050	3.04
Point a	Optimal rotation angle (deg)	107.28	79.56	55.98
Point a	Optimal tube length (dm)	1.206	0.6945	0.4088
Point b	Optimal rotation angle (deg)	143.1	133.74	93.24
Point b	Optimal tube length (dm)	2.619	0.3735	0.4312
Point c	Optimal rotation angle (deg)	86.22	121.68	159.66
Point c	Optimal tube length (dm)	1.38	1.056	0.7976

0.5), b (−1.2, 0.6, 0.3) and c (1, 1, 2). In the optimal robot simulation, as shown in Fig. 8 and according to Table 3, the target points and robot deformations are determined when approaching three specific points. The tube changes are marked by altering their color and adding dividing lines. Tube number 1 is depicted in black, tube number 2 is shown in red, and tube number 3 is represented in blue. As you can see in the simulation picture the robot has acceptable accuracy to reach three target points.

In order to better understand the position of the end of the robot, the geometric matrix of the end of the robot to approach each target point is shown in Eqs. (6)–(8). The geometric matrices of the robot's tip related to target points a, b and c and also the end position of the robot are respectively shown as a matrices of Eqs. (6), (7) and (8).

$$g_a = \begin{bmatrix} 0.8430 & 0.3722 & 0.3882 & 1.0530 \\ -0.4694 & 0.8615 & 0.1933 & 0.8626 \\ -0.2625 & -0.3452 & 0.9010 & 0.5044 \\ 0 & 0 & 0 & 1.0000 \end{bmatrix} \quad p_a = [1.0530,\ 0.8626,\ 0.5044] \tag{6}$$

$$g_b = \begin{bmatrix} 0.1341 & -0.1890 & -0.9728 & -1.1919 \\ -0.5677 & -0.8192 & 0.0809 & 0.5524 \\ -0.8122 & 0.5414 & -0.2172 & 0.2673 \\ 0 & 0 & 0 & 1.0000 \end{bmatrix} \quad p_b = [-1.1919,\ 0.5524,\ 0.2673] \tag{7}$$

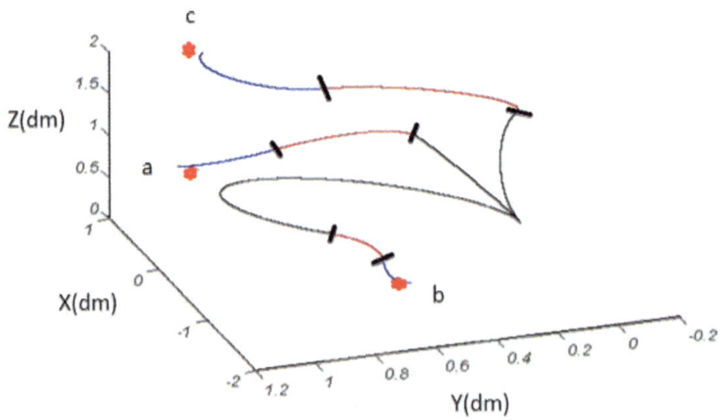

Fig. 8 The simulation of optimal robot for reaching three target points (point a, point b and point c). Target points are determined by red stars and for each state the tubes are separated by small black lines. In this simulation the curvature of each tube remains constant for different target points

$$g_c = \begin{bmatrix} 0.6508 & -0.1107 & 0.7511 & 0.9013 \\ 0.6545 & -0.4196 & -0.6290 & 0.9325 \\ 0.3848 & 0.9009 & -0.2007 & 1.8870 \\ 0 & 0 & 0 & 1.0000 \end{bmatrix} \quad p_c = [0.9013,\ 0.9325,\ 1.8870] \tag{8}$$

In this problem the error which is our objective function according to Eq. (5) is 0.036 which shows good accuracy for the robot in reaching three different target points. This problem is solved using Genetic Algorithm and as you can see the error is minimized by solving the inverse kinematics of the robot. The problem of the inverse kinematics for this case is very complicated because we were looking for set 21 parameters together in order to have the best optimal result. Solving the inverse kinematics of robots with redundancy is a challenging problem for researchers even when the degree of redundancy is one. So, it would be very difficult if the degree of redundancy increases to this high level. The difference between this work and previous works is that we solved a very complicated optimization problem of a continuum robot via Genetic Algorithm. It seems that the only useful problem solvers in these types of complicated problems are AI methods. In next section, the problem remains constant but it is assumed that the middle tube has variant curvature.

5 Efficient Multi-Point Navigation with Genetic Algorithm-Optimized Robot Structure: Exploring the Three-Tube Design with Two Constant Curvature Tube and One Variant Curvature Tube

In this section another problem is considered which is mostly similar to the previous problem in Sect. 4. In the optimization problem of reaching three specific target points, there is a modification in the middle tube, which now has a variable curvature, while the other two tubes maintain a constant curvature as before. The lower and upper bounds, population size, generation process, and error calculation methodology remain the same as in the previous scenario. The ranges considered for the rotation angle, curvature, and length of the tubes are outlined in Table 4.

Table 4 The lower band and the upper band for parameters in multi-point navigation optimization of CTR with two constant curvature tubes and one variant curvature tube

Tube number	Rotation angle limit (degree)	Curvature limit (1/dm)	Optimal length (dm)
1	[−180, 180]	[−3, 3]	[0, 3]
2	[−180, 180]	[−5, 5]	[0, 1.5]
3	[−180, 180]	[−20, 20]	[0, 0.8]

For this optimization process, a population size of 30 and a generation of 100 are taken into account. The error function is defined as Eq. (5). The error in the equation represents the sum of the squared distances between each target point and the robot's tip, denoted as (x, y, z). There are three target points denoted as (x_t, y_t, z_t), representing their respective coordinates. The goal is to achieve precise accuracy in reaching each target point.

The optimization problem to reach three specific target points has been successfully addressed once again. However, this time, there is a difference in the configuration. The middle tube now exhibits a variable curvature, while the other two tubes have a constant curvature, similar to the previous case.

In this optimal state, the error is calculated as 0.006, which is reduced compared to the previous case. The reduction in error can be attributed to the altered properties of the tubes, specifically the variable curvature implemented in the middle section. The variant curvature can improve the flexibility of the robot for reaching different target points and as a result, it increases the accuracy of solving a problem of navigation.

Figure 9 is illustrating the Genetic Algorithm process for determining the best fitness of parameters to achieve the defined goal. The main goal in this problem is finding a robot with a specific structure that can reach different target points with good accuracy. In this section three different target points are defined, and the robot construction is obtained according to those three target points. The difference between this optimization and the previous one is that the middle tube's curvature is variable and it is optimized for each specific target point separately. So, 23 parameters are optimized simultaneously instead of the 21 parameters.

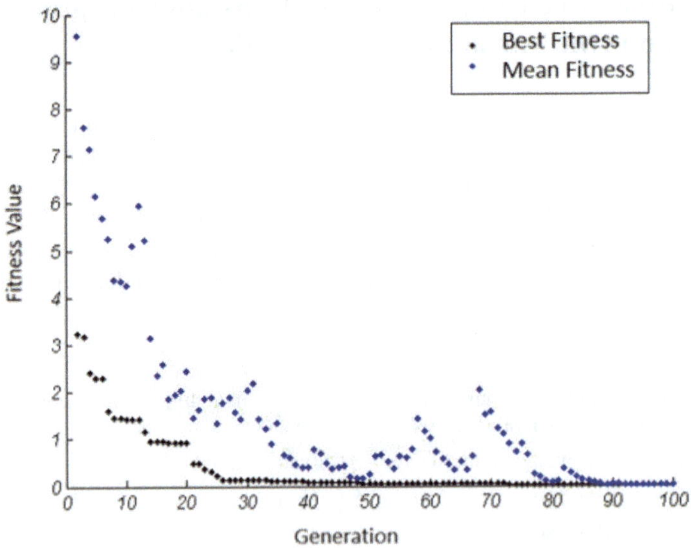

Fig. 9 Genetic algorithm diagram of best fitness objective function versus generation (for the case with three target points, two constant curvature tubes and one variant curvature tube)

In Fig. 9, the quality of each solution is depicted through fitness values, which are presented for every generation. The diagram visually illustrates the iterative improvements made by the genetic algorithm in optimizing the solution.

Optimum values of curvature, length and angle of rotation for all three tubes to reach each of the three target points are shown in Table 5. In this case, the number of parameters is 23 that are optimized at the same time.

Figure 10 illustrates the optimal robot simulation based on Table 5. The figure depicts the target points with red stars and the robot's deformation as it approaches these points. Notably, the changes in the tubes are highlighted through color variations and dividing lines. Specifically, tube number 1 is represented in black, tube number 2 in red, and tube number 3 in blue. The improved accuracy of the robot's shape compared to the previous state is evident. Thus, one effective approach to enhance accuracy in this robot, particularly in scenarios involving multiple distinct target points, is the utilization of both tubes with variable curvature and tubes with constant curvature.

This assumption makes the problem more complicated, but it helps the accuracy and flexibility of the robots. Optimization of 21 or 23 parameters at the same time seems impossible without using AI methods such as Genetic Algorithm.

The results of optimal parameters show the complexity of these types of inverse kinematic problems with high degree of redundancy. While one degree of redundancy can be a challenging problem, this high degree of redundancy provides a lot of complexity. Using Genetic Algorithm can help solving these types of problems which are very complicated for being solved by traditional methods. AI methods are very

Table 5 Optimal parameters of the robot's structure obtained using Genetic Algorithm. The parameters are optimized based on the goal of reaching three different target points with the two constant curvature tubes and one variant curvature tube

Target point	Optimal parameters	Tube number 1	Tube number 2	Tube number 3
Point a, b, c	Optimal curvature (1/dm)	1.890	Point a 1.73 Point b 1.95 Point c 0.555	4.14
Point a	Optimal rotation angle (deg)	124.02	106.92	22.5
Point a	Optimal tube length (dm)	1.347	0.075	0.7848
Point b	Optimal rotation angle (deg)	179.28	−168.84	71.64
Point b	Optimal tube length (dm)	2.223	0.6795	0.3208
Point c	Optimal rotation angle (deg)	102.24	144.36	−121.68
Point c	Optimal tube length (dm)	0.8220	1.494	0.3928

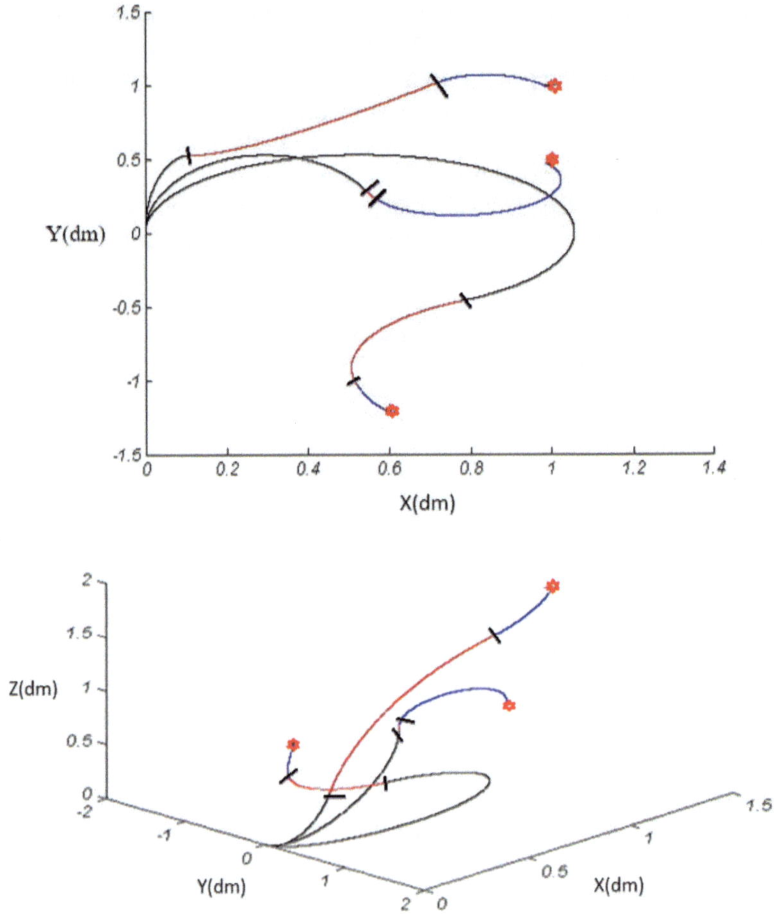

Fig. 10 The simulation of optimal robot for reaching three target points (point a, point b and point c). Target points are determined by red stars and for each state the tubes are separated by small black lines. In this simulation the curvature of two tubes remain constant for different target points and the curvature of the middle tube is optimized according to the target points and other structure parameters

applicable in different fields of research and can change the technology world in many aspects.

In pipes with variable curvature, the curvature can change from the minimum value to the maximum value. It is no longer necessary that the curvature of the middle tube, which has a variable curvature, is greater than the curvature of the first tube. Comparing the results of this section to the results in the previous section, it can be concluded that a tube with variant curvature can help the accuracy of the robot and it can improve the flexibility of the continuum robot which is the main desired property in these types of robots.

6 Conclusions

In this research paper, a comprehensive optimization process was conducted to enhance the performance of the robot across various target points. The inclusion of the length parameter was crucial, as it enabled the acquisition of an optimal robot design that not only exhibited remarkable accuracy but also ensured superior stability.

In this study, the robot structure parameters are optimized by Genetic Algorithm that can enhance the efficiency of robot optimization. Using Genetic Algorithm as a solver of these optimization problems for continuum robots is a great novelty that can improve the solutions in solving inverse kinematic problems of these types of robots. It can make complicated problems much easier. But the time for solving these types of problems with a lot of parameters need not be very small to be used in online control of this robot type.

At first a robot structure is optimized for a specific target point while 9 parameters were optimized using Genetic Algorithm. Then, in the next sections the continuum robot is optimized according to three different target points. In first problem, it is assumed that the curvature of each tube remains constant for different target points which is a real assumption for designing CTR. IN this section, 21 parameters are optimized at the same time in the last section, the problem is similar to the previous one. The only difference is that the middle tube's curvature is no longer constant and as a result, two other parameters are added to the previous problem. The inverse kinematic of this robot is very challenging because of high order of redundancy in these types of problems and optimization and even solving the inverse kinematics of the robot seems to be impossible without using AI methods such as GA.

The accuracy attained by the robot was exceptional, with minimal errors on the order of 0.01 mm in each part. This novel methodology enables the customization and optimization of the concentric tube continuum robot to suit various workspaces, thereby expanding its applicability. The performance of the optimized robot in each section was simulated using the MATLAB program.

Our method offers a notable advantage by incorporating stability considerations into the optimization process, distinguishing it from other existing approaches. As a result, the optimized robot not only achieves exceptional accuracy but also exhibits remarkable stability. This distinctive feature enhances the overall performance and reliability of the robot.

Notably, the achieved accuracy of the optimal robot stands as a significant advantage of our method. The precision attained through the optimization process ensures that the robot's movements and actions align closely with the desired targets, contributing to successful task completion and reliable performance.

In summary, our method stands out by offering the combined benefits of stability, practicality, and exceptional accuracy. It empowers researchers and practitioners to optimize the concentric tube continuum robot to suit diverse workspaces while attaining the desired level of precision and stability.

Nomenclatures

k	Curvature
$P(s)$	Centerline of the curve
$R(s)$	Direction of the curve
s	The length from the base to the intended point of the robot

References

1. Dupont PE, Lock J, Itkowitz B, Butlar E (2010) Design and control of Concentric-Tube Robots. IEEE Trans Robot 26(2)
2. Xu K, Simaan N (2008) An investigation of the intrinsic force sensing capabilities of continuum robots. IEEE Trans Rob 24(3):576–587
3. Webster RJ, Romano JM, Cowan NJ (2009) Mechanics of Precurved-Tube Continuum Robots. IEEE Trans Robot 25(1)
4. Nepalli S, Jones BA (2007) Design, construction and analysis of a continuum Robot: IEEE/RSJ International Conference on Intelligent Robots and Systems, pp 1503–1507
5. Tan N, Gu X, Ren H (2018) Design, characterization and applications of a novel soft actuator driven by flexible shafts. Mech Mach Theory 122:197–218
6. Webster RJ, Romano JM, Cowan NJ (2008) Kinematics and calibration of active cannulas. In: IEEE international conference on robotics and automation, Pasadena, CA, USA, 19–23 May
7. Neppalli S, Csecsits MA, Jones BA, Walker I (2008) A geometrical approach to inverse kinematics for continuum manipulators. In: IEEE/RSJ international conference on intelligent robots and systems, 22 February
8. Jones B, Gray R, Tulapati K (2009) Three dimensional statics for Continuum Robotics. In: IEEE/RSJ international conference on intelligent robots and systems, pp 2659–2664
9. Webster RJ, Jones BA (2010) Design and kinematic modeling of constant curvature continuum robots: a review. In: The international journal of robotics research online, 10 June
10. Anor T, Madsen JR, Dupont P (2011) Algorithms for design of Continuum Robots using the Concentric Tubes approach: a neurosurgical example. In: IEEE international conference on robotics and automation, Shanghai international conference center Shanghai, China, 9–13 May
11. Walker ID (2013) Continuous vackbone "Continuum" robot manipulator: review article. Hindawi Publishing Corporation, ISRN Robotics, ISRN Robotics, Article ID, p 726506
12. Su H, Li G, Rucker DC, Webster RJ, Fischer GS (2016) A Concentric Tube Continuum Robot with piezoelectric actuation for MRI-guided closed-loop targeting. Ann Biomed Eng 44(10):2863–2873
13. Davarpanah A, Lotfavar A, Takolpour Saleh A (2019) The optimization of Concentric Tube Continuum Robot based on Target Point via Genetic Algorithm. In: ISME the 27th annual international conference of Iranian society of mechanical engineers, Tehran, Iran, 30 April–2 May
14. Davarpanah A, Lotfavar A, Takolpour Saleh A (2019) The optimization of CTR for transferring in complicated and tight spaces via Genetic Algorithm. In: ISME the 27th annual international conference of Iranian society of mechanical engineers, Tehran, Iran, 30 April–2 May
15. Davarpanah A, Khazaee M, Moslemi A, Khadem SE (2020) Optimization of concentric tube continuum robot based on accuracy and overall length of the robot via genetic algorithm. Ann Robot Autom 4(1):007–012
16

16. Khazaee M, Sadedel M, Davarpanah A (2021) Behavior-based navigation of an autonomous hexapod robot using a hybrid automation. J Intell Rob Syst 102(2):29
17. Höschel K, Lakshminarayanan V (2018) Genetic algorithms for lens design: a review. J Opt 48:134–144

MIX
Papier aus verantwortungsvollen Quellen
Paper from responsible sources
FSC® C105338
FSC
www.fsc.org

If you have any concerns about our products,
you can contact us on
ProductSafety@springernature.com

In case Publisher is established outside the EU,
the EU authorized representative is:
Springer Nature Customer Service Center GmbH
Europaplatz 3, 69115 Heidelberg, Germany

Printed by Libri Plureos GmbH
in Hamburg, Germany